PHYSICS OF RADIO-FREQUENCY PLASMAS

Low-temperature radio-frequency (RF) plasmas are essential in various sectors of advanced technology, from micro-engineering to spacecraft propulsion systems and efficient sources of light. The subject lies at the complex interfaces between physics, chemistry and engineering. Focusing mostly on physics, this book will interest graduate students and researchers in applied physics and electrical engineering.

The book incorporates a cutting-edge perspective on RF plasmas. It also covers basic plasma physics, including transport in bounded plasmas and electrical diagnostics. Its pedagogic style engages readers, helping them to develop physical arguments and mathematical analyses. Worked examples apply the theories covered to realistic scenarios, and over 100 in-text questions let readers put their newly acquired knowledge to use and gain confidence in applying physics to real laboratory situations.

PASCAL CHABERT is Research Director within CNRS. He currently leads the Low-Temperature Plasmas group of the 'Laboratoire de Physique des Plasmas' at Ecole Polytechnique. His expertise is in plasma physics and plasma processing.

NICHOLAS BRAITHWAITE is Professor of Engineering Physics at The Open University, where his research group works on the physics of 'technological' plasmas. He has been on the editorial board of the journal of *Plasma Sources Science & Technology* since 1998.

PHYSICS OF RADIO-FREQUENCY PLASMAS

PASCAL CHABERT

CNRS

NICHOLAS BRAITHWAITE

The Open University

CAMBRIDGE
UNIVERSITY PRESS

University Printing House, Cambridge CB2 8BS, United Kingdom

Cambridge University Press is part of the University of Cambridge.

It furthers the University's mission by disseminating knowledge in the pursuit of
education, learning and research at the highest international levels of excellence.

www.cambridge.org
Information on this title: www.cambridge.org/9780521763004

© P. Chabert and N. St. J. Braithwaite 2011

This publication is in copyright. Subject to statutory exception
and to the provisions of relevant collective licensing agreements,
no reproduction of any part may take place without the written
permission of Cambridge University Press.

First published 2011
3rd printing 2014

A catalogue record for this publication is available from the British Library

Library of Congress Cataloguing in Publication data
Chabert, Pascal, 1969–
Physics of Radio-Frequency Plasmas / Pascal Chabert, Nicholas Braithwaite.
    p.   cm.
Includes bibliographical references and index.
ISBN 978-0-521-76300-4 (hardback)
1. Low temperature plasmas.   2. Radio frequency.   I. Braithwaite, Nicholas (Nicholas St. J.)   II. Title.
TA2020.C43   2011
621.044 – dc22   2010042728

ISBN 978-0-521-76300-4 Hardback

Cambridge University Press has no responsibility for the persistence or accuracy
of URLs for external or third-party internet websites referred to in this publication,
and does not guarantee that any content on such websites is, or will remain,
accurate or appropriate.

Contents

	Acknowledgements	*page* vii
1	Introduction	1
	1.1 Plasmas	1
	1.2 Plasma processing for microelectronics	3
	1.3 Plasma propulsion	9
	1.4 Radio-frequency plasmas: E, H and W-modes	14
	1.5 What lies ahead	17
2	Plasma dynamics and equilibrium	18
	2.1 The microscopic perspective	19
	2.2 The macroscopic perspective	37
	2.3 Global particle and energy balance	41
	2.4 The electrodynamic perspective	45
	2.5 Review of Chapter 2	55
3	Bounded plasma	59
	3.1 The space charge sheath region	61
	3.2 The plasma/sheath transition	72
	3.3 The plasma region: transport models	78
	3.4 Review of Chapter 3	90
4	Radio-frequency sheaths	96
	4.1 Response times	97
	4.2 Ion dynamics	102
	4.3 Electron dynamics	110
	4.4 Analytical models of (high-frequency) RF sheaths	116
	4.5 Summary of important results	130
5	Single-frequency capacitively coupled plasmas	131
	5.1 A constant ion density, current-driven symmetrical model	133
	5.2 A non-uniform ion density, current-driven model	146

	5.3 Global model	154
	5.4 Other regimes and configurations	165
	5.5 Summary of important results	174
6	Multi-frequency capacitively coupled plasmas	176
	6.1 Dual-frequency CCP in the electrostatic approximation	177
	6.2 Electromagnetic regime at high frequency	187
	6.3 Summary of important results	218
7	Inductively coupled plasmas	219
	7.1 Electromagnetic model	222
	7.2 Impedance of the plasma alone	233
	7.3 The transformer model	236
	7.4 Power transfer efficiency in pure inductive discharges	241
	7.5 Capacitive coupling	243
	7.6 Global model	246
	7.7 Summary of important results	252
	7.8 Further considerations	253
8	Helicon plasmas	260
	8.1 Parallel propagation in an infinite plasma	264
	8.2 Helicon wave propagation in a cylinder	268
	8.3 Conditions for existence of the helicon modes	276
	8.4 Wave power absorption: heating	277
	8.5 E–H–W transitions	283
	8.6 Summary of important results	286
9	Real plasmas	287
	9.1 High-density plasmas	288
	9.2 Magnetized plasmas	293
	9.3 Electronegative plasmas	298
	9.4 Expanding plasmas	313
10	Electrical measurements	318
	10.1 Electrostatic probes	319
	10.2 Electrostatic probes for RF plasmas	340
	10.3 A retarding field analyser (RFA)	348
	10.4 Probing with resonances and waves	354
	10.5 Summary of important results	365
	Appendix: Solutions to exercises	368
	References	375
	Index	383

Acknowledgements

The authors are grateful to many colleagues who have been interested in the progress of this book from concept to reality. In particular, we acknowledge detailed advice and guidance from Jean Paul Booth, Valery Godyak, Mike Lieberman and Jean Luc Raimbault. Additional feedback and encouragement has come from Rod Boswell, Mark Bowden, Christine Charles, Bill Graham and Alex Paterson.

The perspectives we have of plasma physics also owe much to our various post-docs and students, many of whom have played a key part in defining the content and style of our text. At Ecole Polytechnique, PC acknowledges his past and present PhD students Jaime Arancibia, Emilie Despiau-Pujo, Claudia Lazzaroni, Gary Leray, Pierre Levif, Laurent Liard, Amélie Perret and Nicolas Plihon and former post-docs Ane Aanesland, Cormac Corr and Albert Meige. At the Open University, NB acknowledges his past and present (low-pressure plasma) PhD students Gareth Ingram, Sasha Goruppa, Suidong Yang, Pierre Barroy, Paulo Lima, Eva Vasekova and Vladimir Samara and former post-docs Charlie Mahony, Alec Goodyear, Jafar Alkuzee and Tashaki Matsuura.

We wish also to record that we have drawn much inspiration over many years of professional acquaintance with John Allen, Raoul Franklin, Al Lichtenberg, Leanne Pitchford and Miles Turner. All the above are joined by other international colleagues too numerous to mention, with whom we have exchanged ideas at conferences and workshops over the last 10–15 years; many of those whose work we cite have been kind enough to provide original data.

We also gratefully acknowledge the support of various organizations. Many hours of discussion were facilitated by the hospitality of the members' room at the IoP in London and our respective laboratories, The Laboratoire de Physique des Plasmas at the Ecole Polytechnique in Paris and the Atomic, Molecular and Plasma Physics group at The Open University. Equally important has been the financial support of our national research councils, CNRS and EPSRC, and various other funders without whom there would have been less to write about.

In spite of all the wisdom that surrounds us, there will inevitably be misunderstandings and errors in our work. We take full responsibility for these and will try harder next time.

1
Introduction

1.1 Plasmas

A plasma is an ionized gas containing freely and randomly moving electrons and ions. It is usually very nearly electrically neutral, i.e., the negatively charged particle density equals the positively charged particle density to within a fraction of a per cent. The freedom of the electric charges to move in response to electric fields couples the charged particles so that they respond collectively to external fields; at low frequencies a plasma acts as a conductor but at sufficiently high frequencies its response is more characteristic of a dielectric medium. When only weakly ionized (the most common situation for industrial applications) a plasma also contains neutral species such as atoms, molecules and free radicals. Most of this book is about weakly ionized plasmas that have been generated at low pressure using radio-frequency (RF) power sources.

Plasma is by far the most common condition of visible matter in the universe, both by mass and by volume. The stars are made of plasma and much of the space between the stars is occupied by plasma. There are big differences between these plasmas: the cores of stars are very hot and very dense whereas plasmas in the interstellar medium are cold and tenuous. Similar contrasts also apply to artificially produced plasmas on Earth: there are hot dense plasmas and colder less dense plasmas. In the former class are the fully ionized media encountered in research into controlled thermonuclear fusion for power generation, where the challenge is to confine a plasma that is hot enough and dense enough, for long enough so that light nuclei will fuse, liberating huge amounts of energy. The other class – the colder, weakly ionized plasmas also called low-temperature plasmas – includes those used in various industrial applications from lighting to semiconductor processing. Low-temperature plasmas are readily produced by electrical discharges through gases using sources ranging from DC to microwave frequencies (GHz). The gas pressure is typically between a fraction of one pascal and a few times atmospheric pressure (10^5 Pa).

When working in a DC mode and with emitting electrodes, atmospheric pressure discharges tend to operate in a high current regime. The current is carried in a narrow channel in which there is a plasma with its charged and neutral constituents in near-thermal equilibrium (all species comprising the medium having roughly the same temperature of about 10 000 K). Familiar examples can be seen in the giant sparks of lightning and in the arcs of electrical torches for welding and cutting. Arcs are not suitable for the treatment of soft surfaces because the neutral gas is too hot. However, severe gas heating can be avoided in atmospheric discharges if the conditions that allow thermal equilibrium are inhibited, leading to a general class of so-called 'non-thermal plasmas', in which electrons are markedly hotter than the ions and the gas atoms. One way to do this is with an RF-excited dielectric barrier discharge (DBD), in which electrodes are covered by a dielectric material so that charge build-up on the surface automatically extinguishes the discharge before the formation of an arc. These discharges operate with short repetitive pulses, often in a filamentary mode. Each filament carries a very weak current, but the local electron density and temperature are sufficient to dissociate and ionize a small but significant fraction of the gas. The neutral gas remains cold and the medium does not have time to reach thermal equilibrium during a current pulse. DBDs are growing in importance for low-cost industrial applications such as the sterilization of clinical materials and the removal of volatile organic compounds from air. In some circumstances some gases exhibit a more diffuse mode of DBD. A related class of discharge confines the plasma in a space that is too small for a thermal equilibrium to be established. At atmospheric pressure these are termed micro-discharges as the characteristic dimensions are sub-millimetre.

Low-temperature, non-thermal equilibrium plasmas are more easily generated on larger scales at lower pressure. The system is then composed of a vacuum chamber, typically several centimetres across, a throughput of feedstock gas and electrodes (or antennas) to inject electrical power. At low pressure, the discharge operates in the so-called glow regime, in which the plasma occupies the chamber volume as opposed to the filamentary modes generally observed at atmospheric pressure. Most of the volume is occupied by quasi-neutral plasma that is separated from the chamber walls and other surfaces by a narrow region of positive space charge. These boundary layers, or 'sheaths', typically extend over a distance of less than a centimetre. They form as a consequence of the difference between the mobility of electrons and positive ions. The potential structure in the plasma tends to confine electrons and to expel positive ions into the sheaths.

The absence of thermal equilibrium in low-pressure plasmas is important for their commercial applications since the electrical energy is preferentially transferred to electrons that are heated to tens of thousands of kelvin, while heavy particles remain nearly at room temperature. A distribution of electrons that has a temperature of 10 000 K contains a significant fraction that have enough energy to dissociate molecules of the feedstock gas into reactive species (atoms, free radicals and ions). The plasma thus converts electrical energy into chemical and internal energy that can be directed, for instance, into surface modification. Sheaths are also of major importance since they in turn locally convert electric field energy derived from the power supply into the directed kinetic energy of ions reaching the surfaces. Electric fields in the sheaths tend to accelerate ions perpendicular to the surfaces. The energy of ions bombarding any particular surface is a major parameter of process control that can readily be raised to thousands of times the energy that binds atoms together in small molecules and extended solids. These non-thermal phenomena account for the rich variety of plasma processing technology, from the surface activation of polymers to the implantation of ions in semiconductors.

Plasma processing technology is used in many manufacturing industries, especially in the surface treatment of components for the automotive, aerospace and biomedical sectors. Plasma technologies offer advantages in terms of environmental impact, through reduced use of toxic liquids, and in terms of engineering scale, through their compatibility with nanoscale fabrication. The biggest impact has certainly been in microelectronics, for which very large-scale integrated (VLSI) circuits could not be fabricated without plasma-based technologies. In the following sections some industrial applications of low-pressure radio frequency-plasmas will be described, setting the context for the more detailed analyses of later chapters.

1.2 Plasma processing for microelectronics

Integrated circuits (ICs) consist of several layers of carefully engineered thin films of semiconductors, dielectrics and conductors, fashioned *in situ* and interconnected by a very complex architecture of conducting tracks (see Figure 1.1). The thin films are deposited by means of plasma processes and are etched by reactive plasmas in order to form patterns of the order of a few tens of nanometres, i.e., a hundred times smaller than a human hair.

The basic element in the design of a large-scale integrated circuit is the metal-oxide-semiconductor field effect transistor (MOSFET) – see Figure 1.2. Most commonly the transistor is made in a layer of high-quality silicon grown onto a substrate of single-crystal, silicon semiconductor. The device regulates the flow of current from a 'source' region to a 'drain' region via a channel that is controlled by a gate electrode. The gate electrode is isolated from the channel by means of a few nanometres of dielectric, usually silicon dioxide. The MOSFET can operate as a very efficient switch for current flowing between the source and drain. The

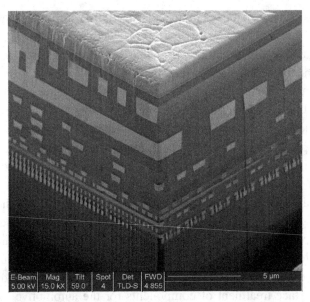

Figure 1.1 Multilevel metal dielectric interconnects in VLSI circuits.

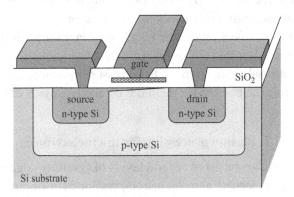

Figure 1.2 The structure of a silicon IC in the region of a MOSFET in which a gate electrode controls the formation of a channnel of n-type silicon between source and drain regions.

switch is activated by biasing the gate. The gate size is the critical dimension in determining the level of integration and the speed of the device. In the technology known as CMOS (complementary metal-oxide semiconductor), the building blocks of memory and logic circuitry are based on devices that incorporate a MOSFET that has an n-type channel with one that has a p-type channel. CMOS technology is a dominant semiconductor technology for microprocessors, memories and application-specific integrated circuits. The main advantage of CMOS is its relatively low power dissipation.

1.2 Plasma processing for microelectronics

A common description of the evolution of microelectronics technology is based on what is known as Moore's law. The prediction made by Gordon Moore in 1965 was that the number of transistors on the most complex integrated circuit chip would roughly double every two years. It has proved to be a remarkably good guide to the development of the IC market – a development that has been enabled by plasma-based processes.

1.2.1 Plasma etching

The principles of plasma etching are the following. In a first step, the substrate that bears the material which is to be etched is covered by a thin (<1 μm) layer of 'photoresist'. The photoresist is then patterned by exposure to UV through a contact mask, which creates a high contrast in the solubility of photoresist between areas that are exposed and areas that are shaded. The shadow pattern of the mask is next developed into the photoresist by means of wet chemistry to obtain open areas in the resist layer. The patterned wafer is then transferred to a plasma reactor. In the case of silicon-based materials the process gas is usually composed of one or more types of halogen-containing molecules (e.g. CF_4, SF_6, Cl_2 or HBr). The gas is introduced into the reactor where, on the formation of a plasma, it is dissociated by electron impact to form reactive species. In the case of SF_6, for instance, there are reactions with electrons (e^-) such as

$$e^- + SF_6 \rightarrow SF_5 + F + e^-,$$
$$e^- + SF_6 \rightarrow SF_4 + 2F + e^-,$$
$$e^- + SF_6 \rightarrow SF_2 + F_2 + 2F + e^-,$$
etc.

The F atoms in the gas phase (g) are an effective silicon etchant, reacting with a surface (solid phase, s) to form a volatile etch product that may be pumped away:

$$4F(g) + Si(s) \rightarrow SiF_4(g).$$

In the absence of bombarding ions and crystallographic effects, the etching will proceed equally in all directions, i.e., isotropically, since the etchant atoms arrive without any specific directional influence, as shown in Figure 1.3(a). Isotropic profiles are also obtained in wet etching and are not suitable for the high aspect ratio etching required for dense integration (the aspect ratio is the ratio of the width of a feature to its depth).

In 1979, using a combination of atom and ion beams, Coburn and Winters [1] demonstrated that energetic ions arriving at a surface increase the effectiveness of etching with neutral atoms by more than an order of magnitude. This synergy is easily exploited in plasma reactors because plasmas naturally provide active

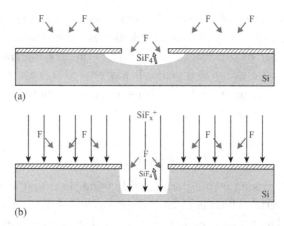

Figure 1.3 (a) Isotropic chemical etching; (b) anisotropic reactive ion etching.

neutral radicals as well as energetic ions that have been accelerated in the sheaths. Furthermore, owing to the sheaths, the ions have trajectories perpendicular to the surface, and they are found to increase considerably the etch rate perpendicular to the surface but to have little influence on the side-wall etching. Therefore, etching in the presence of ion bombardment tends to be anisotropic. The deliberate use of energetic ions to enhance the rate of etch reactions is known as reactive ion etching.

> **Q** Suggest two factors that would link increased plasma density to higher (anisotropic) etch rates.
> **A** The increased electron density will tend to increase the supply of radicals and therefore is likely to lead to a higher etch rate; the increased flux of ions to the surface will also tend to enhance the anisotropic etching.

Although ions can usefully introduce anisotropy to the etching, it is usually not sufficient to achieve the very high level of profile control required in CMOS technology. Polymerizing chemistries have therefore been introduced to add an etch-inhibiting coating to certain surfaces. When CF_4 is used as the feedstock gas the free radicals formed in the plasma, for instance CF and CF_2, tend to contribute to the growth of a polymeric film on the side walls of a feature, providing a so-called passivation layer. This layer does not form on the plasma-facing areas because it is continuously disrupted by the arrival of energetic ions, from a perpendicular direction. Fluorocarbon gases like CHF_3, CF_4, C_2F_6, C_4F_8 are known to be polymerizing and are routinely used to etch dielectric materials in microelectronics. In order to control the degree of polymerization, oxygen is often added to the gas to promote the formation of CO_2 on the surface, thereby competing with film growth. Polymerization is also a very efficient way to control the selectivity of a process, i.e., the ability to etch a specific material without affecting the underlying layer made of another material. A classic example is the capability of CF_4/O_2 plasmas to change the relative etch rates of Si and SiO_2 and hence the selectivity. An oxygen-rich mixture will etch pure Si faster than SiO_2 while an oxygen-lean mixture will etch SiO_2 faster than pure Si. Fluorocarbon plasmas have received a great deal of attention because a large number of the process steps in the manufacture of silicon ICs involve the differential etching of silicon and silicon dioxide [2–4].

Other halogen-based etching schemes are also important: one of the critical steps in CMOS is the etching of the gate stack, which is usually achieved in $Cl_2/HBr/O_2$ plasmas. Passivation this time involves the formation of silicon-based films of SiO_xCl_y [5]. Unwelcome process drifts have been linked to the deposition of these passivation layers on the reactor walls [6].

Plasma etching is also a key enabling technology in optoelectronics and photonics. For instance, the high aspect ratio, deep ridge InP-based heterostructure that is a vital building block in photonic device fabrication is readily manufactured by plasma processing. For this purpose, an etching process is required that can produce narrow, single-mode, ridge waveguides with smooth side-walls, free from undercuts or notches to minimize the optical scattering losses [7].

For removing large quantities of material in so-called 'deep etching' (on the order of tens of micrometres deep), plasma etching is again useful [8]. This is used in the fabrication of micro-electro-mechanical systems (MEMS), based on miniature gears, pivots, linkages, cantilevers, fluid channels and other components etched into silicon substrates; for harsh environments silicon carbide is preferred. Deep etching of these materials calls for high-density plasma sources to keep processing times within bounds. An example of the deep etching of silicon carbide is shown in Figure 1.4, where a helicon plasma was used (see Section 1.4) to form a dense plasma in a mixture of SF_6 and O_2 [9–11].

1.2.2 Plasma deposition

Plasma-enhanced chemical vapour deposition (PECVD) allows the deposition of a variety of thin films at lower temperatures than those utilized in classical CVD reactors. For example, whereas ordinary CVD of high-quality silicon dioxide films requires temperatures in the range of 650–850°C, similar quality films can be deposited at 300–350°C via a plasma-enhanced process. Again, the advantage of using plasmas comes from the fact that they are able to fragment molecules into reactive radicals, even at room temperature. For deposition, radicals that condense on the substrate are required to contribute to film growth (in contrast to etching where the chemistry is chosen so that radicals react with the surface to form volatile products).

Figure 1.4 Deep micrometre-scale structure etched in SiC using a SF_6/O_2 helicon plasma.

Apart from microelectronics, one of the most important applications of PECVD is the fabrication of flat-panel displays [12]. Liquid crystal displays (LCDs) in particular have emerged as favourites for laptops and flat-panel monitors. When combined with a transistor switch at each pixel, the so-called active matrix display (AMLCD) readily achieves high resolution (several million pixels), large size, full colour and TV-compatible response times. An AMLCD is made of two glass sheets between which is a thin layer of liquid crystal. On one glass sheet is an array of thin film transistors (TFTs). Each TFT switches the voltage on a small indium tin oxide (ITO) transparent electrode that defines a pixel. The other sheet is covered by colour filters and a common electrode. The TFT array is manufactured by a series of plasma-based process steps that alternate thin-film deposition with patterning. A major challenge for the design of plasma systems that will form the TFTs is to do with maintaining control of the uniformity of the plasma over the entire area of the display – the larger the better so far as the market is concerned. Related issues of scalability will be discussed in more detail in later chapters.

Plasmas are also used in the physical deposition process known as sputtering that is commonly used for depositing metal layers on semiconductor circuits. In sputter deposition systems a low-pressure plasma provides ions (typically Ar^+) that are accelerated onto a metallic target that is negatively biased. The ions are given a kilovolt or so of energy by the acceleration so that when they collide with the target, atoms of the target are dislodged, or 'sputtered', forming a plume of ejected material. Sputtering is a purely physical, unpatterned, etching process. A substrate placed near a sputtering target is effectively sprayed with atoms from the

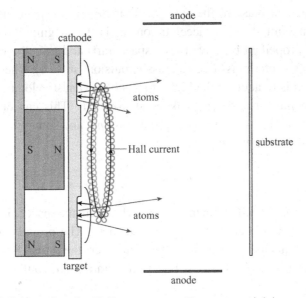

Figure 1.5 Schematic of a DC magnetron. Target material is sputtered most intensely from adjacent plasma created by the spiralling ring of electrons whose drift around the ring constitutes the Hall current.

target, building up a few tens of nanometres in a matter of minutes. A common configuration for sputtering uses a so-called magnetron arrangement, where a magnetic field is arranged parallel to the target surface so that electrons spiral round the field creating a locally intensified and efficient ionization of the gas. This is illustrated in Figure 1.5. A high-density plasma forms in a ring adjacent to the target, which is in turn more aggressively eroded in this region than elsewhere, owing to the intense ion bombardment. The forces on the electrons from the combination of the magnetic field and the electric field in the plasma push the spiralling electrons around the ring of intensified plasma, driving what is known as a 'Hall current'. The geometry of electric and magnetic fields also occurs in one of the plasma thrusters discussed in the next section, for which the Hall current is a key feature.

1.3 Plasma propulsion

A rocket-propelled spacecraft in free flight receives its acceleration from expelling mass (the propellant). Its equation of motion is derived from a force equation that balances the rates of change of momentum of the spacecraft and the ejected matter:

$$m\dot{v} = -\dot{m}v_g, \qquad (1.1)$$

where m is the total mass of the spacecraft (including unspent fuel) at a given time, $\dot{v}$ is the magnitude of its acceleration, v_g is the magnitude of the exhaust velocity of the propellant (relative to the spacecraft) and $\dot{m}$ is the rate of change of the spacecraft's total mass due to mass expulsion ($\dot{m} < 0$). The challenge for space propulsion is to achieve the highest possible exhaust velocities and to fully ionize the propellant in order to make best use of it. This can clearly be seen after integration of (1.1) between any initial mass m_0 and some final mass m_f, for constant exhaust speed, which gives

$$\Delta v = v_g \ln \frac{m_0}{m_f}, \quad (1.2)$$

showing that for a given reduction in mass, the change in the spacecraft speed during a given period of acceleration is proportional to v_g. The propulsion community usually uses two quantities to characterize a thruster: the thrust $T = \dot{m} v_g$ and the specific impulse $I_s = v_g/g$, where g is the acceleration due to the Earth's gravity at sea level.

1.3.1 Conventional plasma thrusters

Electric propulsion techniques [13] may be separated into three categories: (i) electrothermal propulsion, in which the propellant is electrically heated and then expanded thermodynamically through a nozzle; (ii) electrostatic propulsion, in which ionized propellant particles are accelerated by an electric field; (iii) electromagnetic propulsion, in which current driven through a propellant plasma interacts with an internal or external magnetic field to provide a stream-wide body force. Here is a short description of the most common systems.

Resistojets and arcjets

These belong to the first category. In a resistojet the gas is heated via the chamber wall or a heater coil whereas in an arcjet the gas is heated by an electric arc. The propellant gas is then accelerated downstream through a nozzle. These thrusters have limited specific impulse (less than 1000 s) and face technological challenges owing to the high temperatures required.

Electrostatic ion thrusters

Positive ions are created in a plasma (DC, RF or microwave, usually magnetized) and accelerated out through a DC-biased grid. In order to maintain an overall charge balance, the ion beam must be neutralized downstream, via a thermionic filament or some other source of electrons. Electrostatic thrusters have been successfully demonstrated and they can provide very high specific

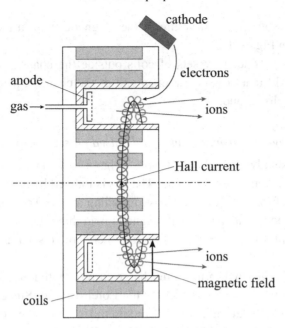

Figure 1.6 Schematic of a Hall effect thruster.

impulse. However, erosion of the accelerating grids by energetic ions limits the lifetime.

Hall effect thrusters

In a Hall effect thruster [14], electrons emitted from an external cathode enter an annular channel where they are partially confined in the opening by a radial component of magnetic field as they slowly diffuse in a region of nearly axial electric field, towards the anode at the closed end of the channel; see Figure 1.6. The electrons spiral around the magnetic field because of the Lorentz force that acts on charged particles crossing a magnetic field. Under the combined influence of the perpendicular electric and magnetic fields there is a net drift of electrons around the annular channel forming the so-called Hall current which is perpendicular to the electric and magnetic fields. These confined electrons ionize the gas (typically xenon) to create positive ions that are in turn accelerated out of the open end of the channel by the axial field that sustains the discharge between anode and cathode. The acceleration region is filled with magnetized, quasi-neutral plasma and some of the electrons emitted by the external cathode serve to neutralize the ion beam as it escapes downstream. Improvements in performance are primarily aimed at reducing erosion rates, and lowering beam divergence.

> **Q** Identify the source and structure of the magnetic field in the Hall thruster illustrated in Figure 1.6.
>
> **A** The field is produced by sets of coils outside the channel which create a magnetic field that crosses the open end of the channel, passing round the back of the closed end.

Magnetoplasmadynamic and pulsed plasma thrusters

The magnetoplasmadynamic thruster (MPDT) has a coaxial geometry with a cylindrical anode and a central cathode rod. The gas is ionized by an electric discharge between the tip of the cathode and the surrounding anode. The current flowing in the electrodes produces an azimuthal magnetic field that interacts with the current in the discharge plasma. Note that in this case there are no separate magnetic field coils.

The moving charges that form the discharge current, both ions and electrons, are forced out of the mouth of the device by the Lorentz force. Since neutral plasma is expelled, a separate neutralizer is not required. However, the power required to reach high efficiency is very high (100 kW), so MPDTs are only considered as an option for high-power propulsion.

The magnetic field of the discharge current in the electrodes is also used in a pulsed plasma thruster. Here a spring-loaded slab of PTFE is pushed into part of the space formed by planar electrodes, between which a high-voltage pulsed discharge is triggered. The propellant may include material ablated from the PTFE; it is then ionized by the discharge itself and accelerated, as in the MPDT, by the combined action of the electric and magnetic fields.

1.3.2 Newer concepts

New plasma thruster concepts are being studied, such as the variable specific impulse plasma rocket developed by NASA [15], the double-layer thruster developed by the Australian National University [16], and an electronegative plasma thruster developed by the Ecole Polytechnique in France [17].

All these concepts use a high-density plasma source for the ionization stage based on the so-called helicon source. Helicon waves are electromagnetic disturbances that propagate in a magnetized plasma. It turns out that energy can be transferred efficiently from these waves into the plasma electrons, creating a strong source of ionization. The Plasma Research Laboratory at the Australian National University has been one of the prime developers of this kind of source. Chapter 8 will focus on the details of the helicon source operation.

Variable-impulse thruster

The variable specific impulse magnetoplasma rocket motor takes the plasma generated by a helicon source and then energizes the ions by exciting more electromagnetic waves that resonate with the gyration of ions around the magnetic field lines. Further downstream the magnetic field lines diverge, effectively forming a magnetic nozzle in which ion gyromotion is transferred into axial motion, creating an escaping plume of plasma and directional thrust. A characteristic of this arrangement is that the gas inflow and the resonant heating together determine the behaviour of the motor, allowing a high degree of control of the specific impulse.

Double-layer thruster

Even without additional energy input, a magnetized plasma can be configured to produce thrust. Solenoids create an expanding magnetic field that is roughly uniform at a few tens of millitesla in the source tube, decreasing to a fraction of a millitesla a few centimetres away from the source. The high-density plasma formed in this way is prevented from leaving the source by a non-linear structure known as a current-free electric double layer that arises spontaneously in the plasma near the exit of the source tube. This structure is in fact two adjacent layers of space charge, one positive, the other negative, that can be thought of as a thin standing shock wave across which there exists an electric potential jump. The resulting electric field accelerates ions from the source plasma to high exhaust velocities, creating thrust. The double layer is purely the result of plasma expansion so no accelerating grids are required. Also, it turns out that there is an equal flux of electrons and positive ions from the thruster, so there is no need for a neutralizer.

Dual-ion thruster

In a classical electrostatic (ion) thruster, the thrust is provided by the positively charged particles, while the negatively charged particles, namely the electrons, are not used for thrust. The electrons are used for ionization in the plasma production region, and neutralization downstream. The LPP at Ecole Polytechnique (France) have proposed a new concept named PEGASES, in which an electronegative gas is used as the propellant. Positive and negative ion beams are produced simultaneously, obviating the need for a neutralizer. The main plasma is produced in a long cylinder in which a longitudinal magnetic field confines the electrons but not the ions (typically tens of milliteslas). A helicon source provides a very high-density plasma core, of several centimetres in radius, creating a column of fully ionized plasma. In an electronegative gas, it has been shown that this configuration produces a stratified plasma with an electropositive core (a plasma with a significant

Table 1.1 *Frequency ranges for plasma sources*

Type	Range
DC or low-frequency	$f < 1\,\text{MHz}$
radio-frequency	$1 < f < 500\,\text{MHz}$, commonly 13.56 MHz
microwave	$0.5 < f < 10\,\text{GHz}$, commonly 2.45 GHz

fraction of electrons), and an ion–ion (electron-free) plasma at the periphery. Ions are extracted radially from the ion–ion plasma, via sets of accelerating grids, biased positively, to extract the negative ions, and negatively to extract the positive ions.

1.4 Radio-frequency plasmas: E, H and W-modes

Plasma reactors used for etching and/or PECVD are often driven at frequencies lying between 1 MHz and 200 MHz, that is within the radio-frequency (RF) domain. In particular, 13.56 MHz and its harmonics are popular choices, having been set aside for industrial and medical applications – most other parts of the spectrum are allocated to telecommunications. The helicon sources used for space plasma propulsion are also driven in the RF domain, usually at 13.56 MHz. Processing plasmas can also be generated by DC and low-frequency discharges, or by microwaves. Table 1.1 gives a frequency classification for the various types of common reactor.

The RF range is of particular interest as at the lower end all except the most massive ions in gas discharge plasmas are able to follow the instantaneous RF fields and at the higher end, all ions are inertially constrained, responding only to the time-averaged fields. Throughout the RF range, electrons are able to respond instantaneously to fields.

In the microwave region there is a convenient 'cyclotron' resonance for electrons moving in relatively modest magnetic fields. The natural frequency with which electrons gyrate in a field of 86.6 mT is 2.45 GHz, so this makes an attractive combination of an easily achieved flux density with an inexpensive power source (thanks to domestic microwave ovens).

RF electromagnetic fields can be generated in many ways, for instance by applying an RF voltage across two parallel electrodes or by circulating RF currents in coils or antennas, either immersed in the plasma or separated from it by a dielectric window. The electromagnetic fields will couple to the electrons in the plasma and transfer energy to them to sustain the plasma. The efficiency with which power is coupled from the power supply into the charged particles and the plasma uniformity both strongly depend on the design of the RF excitation. The

1.4 Radio-frequency plasmas: E, H and W-modes

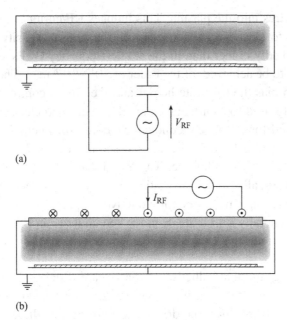

Figure 1.7 Schematic of reactors (a) capacitively coupled and (b) inductively coupled.

two classical RF reactors used in industry are the capacitively coupled plasmas (CCP) reactor, a schematic of which is shown in Figure 1.7(a), and the inductively (or transformer) coupled plasmas (ICP, TCP) reactor shown in Figure 1.7(b).

The CCP reactor developed in the 1970s comprises two parallel electrodes, separated by a gap of a few centimetres, immersed in a vacuum chamber. The electrodes are driven by an RF power source of typically ~ 1 kW and at a typical frequency of 13.56 MHz; the plasma density in a CCP is typically around 10^{15}–10^{16} m^{-3}.

The substrate to be etched is usually placed on the powered electrode. Gentler conditions tend to prevail at the ground electrode. In such reactors, the RF power level applied to the electrodes controls simultaneously the ion flux to the substrate and the energy with which the ions arrive there. The lack of independent control of ion flux and ion energy is a severe limitation of single-frequency CCPs. To overcome this limitation, dual-frequency systems have been introduced, for which the applied RF waveform is the sum of two independently controllable components. This arrangement has proved useful for certain etching steps in microelectronic fabrication. CCP reactors will be discussed in Chapters 5 and 6.

ICP systems also often use two RF power supplies. The first drives a coil, usually external to the plasma and separated from it by a dielectric window. The RF current flowing in the coil launches an evanescent disturbance that decays over a distance of a few centimetres into the plasma. This induces RF current in the plasma and transfers energy to electrons; i.e., it controls the plasma density. The coupling efficiency is markedly higher than in the single-frequency CCPs, enabling higher plasma densities to be achieved, of the order of 10^{16}–10^{18} m^{-3}. The second power supply is used to bias the substrate holder and thereby to control the ion energy. ICPs are routinely used to etch metals and silicon in microelectronic fabrication. They are also used for III–V semiconductor etching for photonics, and for deep etching in MEMS technology. ICPs are discussed in Chapter 7.

These two different reactor families, CCPs and ICPs, are usually associated with two regimes, the so-called E (electrostatic) mode for capacitive coupling and the so-called H (electromagnetic) mode for inductive coupling. Inductive reactors with an external coil generally start in the E-mode and undergo an E–H transition when the plasma density reaches a critical level as power to the coil is increased [18].

> Q Figure 1.7 suggests that inductive discharges are sustained by RF current in a coil, so how is it that the plasma initially forms in the electrostatic (E) mode?
> A In order for RF current to be driven through the coil there must be an RF voltage across it and this voltage produces an additional electrostatic electric field – indeed this field is associated with the RF voltage on the coil and would exist even if the coil windings were broken so that the current was zero. The E-mode forms a low-density plasma – the H-mode does not take over until the plasma density achieves sufficient conductivity for the electromagnetic mechanism to predominate.

The E–H transition is unstable when electronegative gases (attaching gases which lead to negative ion production in the plasma) are used [19, 20]. Studies have also shown that CCPs may also experience mode transitions (from E to H) if they are driven at high frequency because of an induced field parallel to the electrode [21, 22]. Both of these phenomena will be discussed in later chapters.

Finally, there is a third regime that couples energy from the RF fields to the plasma, labelled W for wave. For electron densities higher than those found in typical CCP and ICP discharges, disturbances at radio frequencies do not propagate in the absence of a static magnetic field. For RF-wave-sustained plasmas there has to be a background (steady) magnetic field. An antenna is used to launch a type of propagating disturbance called a helicon wave which propagates away from the antenna into the plasma bulk where the wave energy is absorbed by electrons. The coupling of energy in the W-mode achieves densities above 10^{19} m^{-3} and in larger volumes than can be achieved by an H-mode ICP. Helicon reactors have therefore been used for plasma processing applications demanding high ion fluxes, for example in the context of deep etching of hard materials [11]. They also seem

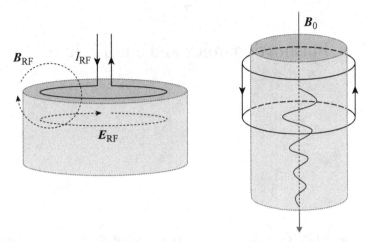

Figure 1.8 Cylindrical columns of plasma excited by external currents. Left: the plasma is sustained with currents driven by the RF electric field induced in the plasma in response to external RF current – the field strength diminishes with distance from the external coil. Right: the external antenna sets the current path such that helicon waves are launched, propagating axially – the plasma is sustained by energy absorbed from these waves.

very promising as the ionization stage of plasma thrusters because of their ability to produce highly ionized plasmas. Figure 1.8 illustrates the difference between inductive and helicon wave excitation.

1.5 What lies ahead

This opening chapter has introduced various situations where RF power sources are used to create low-pressure, non-equilibrium plasmas for specific technological applications. There are many more examples than those given here, even within the low-pressure domain; at higher pressure yet more can be found. However, there is already more than enough background to provide the context for the detailed physics-based analysis that is the primary concern of this book.

The next chapter will rehearse some basic plasma physics, starting with a particle description and culminating in a view of plasmas that treats them as interacting fluids of electrons, ions and neutral gas. In the chapters that follow, the fluid equations are analysed in growing levels of sophistication to describe plasmas that are created and sustained within boundaries and at low pressure. This regime characterizes the technological plasmas described above. Electrical models will be developed to allow comparative analyses of RF plasmas into which power is coupled via electrostatic, induction and wave mechanisms.

2
Plasma dynamics and equilibrium

One way to model the dynamics of the plasma contained in a reactor would be to calculate rigorously the trajectory of each of the charged particles using Newton's laws. This is not feasible for many reasons: (i) the number of charged particles is too large given the typical densities (10^{16}–10^{18} m^{-3}) and the reactor volume (a few litres); (ii) charged particles move in response to the electromagnetic (Lorentz) force associated with electromagnetic fields, which in this case are generated by the presence and motion of all the other charged particles – that is, by local space charge and currents; the problem is non-linear and should be solved self-consistently; (iii) particles experience collisions that modify their velocities and energies on very short time scales.

Q (i) How many ions are there in a cubic millimetre ($V = 10^{-9}$ m^3) of plasma of charged particle density $n = 10^{16}$ m^{-3}?
(ii) How far will an electron travel in $t = 0.1$ μs when accelerated in vacuum from rest by an electric field of $E = 10^2$ V m^{-1}?
(iii) In a typical low-pressure, electrical discharge plasma a large fraction of electrons have speeds around $v = 10^6$ m s^{-1} and collide with gas atoms typically every $\lambda \sim 10^{-1}$ m, depending on the pressure; what is the average time between successive collisions?

A (i) $N = n \times V = 10^7$.
(ii) $s = \frac{1}{2}(eE/m)t^2 \approx 10^{-1}$ m.
(iii) $\tau = \lambda/v \sim 10^{-7}$ s.

The first level of simplification of the above problem is achieved in particle-in-cell (PIC) computer simulations. The basic idea behind the PIC method is indeed to solve Newton's law and the electromagnetic fields simultaneously, including collisions between particles. However, the difference between a simulated plasma and a real plasma lies in the representation of the charges, the fields and the space-time in which the phenomena occur. In a PIC simulation a large number of neighbouring charged particles are represented by a 'super-particle'; it is always multiply charged and has the same charge-to-mass ratio as that of the actual particles. The large number of charges in a plasma are thus replaced by a much smaller number of these super-particles. Time and space are discretized and the calculations of electromagnetic fields and super-particle motions are done iteratively until a steady state is reached. PIC simulations are useful to understand subtle kinetic phenomena, but the computational time required is often too long to model the general macroscopic behaviour using purely numerical schemes.

From an analytical point of view, there are two approaches to the modelling of the plasma dynamics: one based on kinetic theory and the other based on fluid theory. The first is a microscopic approach and relies on statistical physics. Velocity (or energy) distribution functions are introduced, $f(\mathbf{r}, \mathbf{v}, t)$, and the evolution of these distributions is solved using conservation laws. Kinetic theory is useful to model non-linear wave–particle interactions and collisionless phenomena such as stochastic heating. Knowledge of the velocity distribution function is also important in detailed calculations of transport and reaction coefficients. However, kinetic calculations are too complicated to describe the macroscopic behaviour of a plasma reactor. Most of the fundamental properties described in this text do not require a kinetic treatment and will be addressed by a macroscopic fluid theory (hydrodynamics). For this, macroscopic quantities such as the fluid density n, the fluid velocity, $\mathbf{u}$, etc. are obtained from integrations over velocity of the distribution function $f(\mathbf{r}, \mathbf{v}, t)$.

In the following the basic ideas of kinetic theory will be introduced along with definitions of distribution functions, thermal equilibrium distributions, and various averages over these distributions. Some basic concepts of collisions and reactions will also be presented. The fluid equations will then be introduced – the exact derivation of these equations, starting from kinetic equations, is beyond the scope of this text (details can be found in many plasma physics textbooks such as [23]). The fluid equations will then be combined to obtain particle and energy balance equations that are the building blocks of the physics described in this book. Finally, the fluid equations will be linearized to examine the propagation of electromagnetic and electrostatic perturbations.

2.1 The microscopic perspective

2.1.1 Distribution functions and Boltzmann equation

The kinetic theory of gases is a useful starting point from which to appreciate the microscopic view of plasmas. Consider N particles with a random distribution of positions (**r**) and velocities (**v**). The velocity distribution function $f(\mathbf{r}, \mathbf{v}, t)$ defines the number of particles being at a given time t inside the six-dimensional elementary volume of phase space $dx\,dy\,dz \times dv_x\,dv_y\,dv_z$. It is sometimes convenient to express this elementary volume in a more compact notation, namely $d^3\mathbf{r}\,d^3\mathbf{v}$. The number of particles dN in the volume $d^3\mathbf{r}\,d^3\mathbf{v}$ in the neighbourhood of the position **r**, with velocity around **v**, is thus

$$dN = f(\mathbf{r}, \mathbf{v}, t)\,d^3\mathbf{r}\,d^3\mathbf{v}. \tag{2.1}$$

Having defined the velocity distribution function, one can then calculate macroscopic quantities by averaging over the velocity coordinates. These macroscopic quantities are determined by taking the velocity moments of the distribution function. They are the basic variables of the fluid theory presented in Section 2.2. The first of these is the particle density defined as

$$n(\mathbf{r}, t) = \int_{-\infty}^{\infty}\int_{-\infty}^{\infty}\int_{-\infty}^{\infty} f(\mathbf{r}, \mathbf{v}, t)\,d^3\mathbf{v}. \tag{2.2}$$

The average value of any quantity in a distribution of particles is found in statistical mechanics by integrating over the distribution weighted by that quantity, divided by the total number of particles in the distribution. It is usual to denote this process by angled brackets so for example the mean velocity, also called the drift velocity, is

$$<\mathbf{v}(\mathbf{r},t))> = \frac{\int_{-\infty}^{\infty}\int_{-\infty}^{\infty}\int_{-\infty}^{\infty} \mathbf{v} f(\mathbf{r}, \mathbf{v}, t)\,d^3\mathbf{v}}{\int_{-\infty}^{\infty}\int_{-\infty}^{\infty}\int_{-\infty}^{\infty} f(\mathbf{r}, \mathbf{v}, t)\,d^3\mathbf{v}};$$

the drift velocity is often given the more concise notation $\mathbf{u}(\mathbf{r}, t)$. The total particle flux can therefore be defined as

$$\Gamma(\mathbf{r}, t) = n(\mathbf{r}, t)\mathbf{u}(\mathbf{r}, t) = \int_{-\infty}^{\infty}\int_{-\infty}^{\infty}\int_{-\infty}^{\infty} \mathbf{v} f(\mathbf{r}, \mathbf{v}, t)\,d^3\mathbf{v}. \tag{2.3}$$

Similarly, the total kinetic energy density in the distribution is given by

$$w = n(\mathbf{r}, t) <\frac{1}{2}mv^2> = \frac{1}{2}m \int_{-\infty}^{\infty}\int_{-\infty}^{\infty}\int_{-\infty}^{\infty} v^2 f(\mathbf{r}, \mathbf{v}, t)\,d^3\mathbf{v}, \tag{2.4}$$

where m is the particle mass. It turns out that the kinetic energy density can be divided into two components, one associated with the random motion of the particles and the other associated with the net drift:

$$w = \frac{3}{2}p(\mathbf{r}, t) + n(\mathbf{r}, t)\frac{1}{2}m\,\mathbf{u}(\mathbf{r}, t)^2; \tag{2.5}$$

the first term is identified with the internal energy density, so $p(\mathbf{r}, t)$ is the isotropic pressure, and the second term is due to the net flow of momentum. When the drift velocity is zero, that is for symmetrical distribution functions, the net momentum flow is zero and the kinetic energy density is just proportional to the pressure.

Distribution functions obey a conservation equation that has the form of a continuity equation. Particles enter and leave an elementary volume and can be produced by ionizing collisions, or destroyed by recombination, within this volume. The equation governing the evolution of the distribution is called the Boltzmann equation, and is given by (see for example [2])

$$\frac{\partial f}{\partial t} + \mathbf{v} \cdot \nabla_r f + \frac{\mathbf{F}}{m} \cdot \nabla_v f = \left.\frac{\partial f}{\partial t}\right|_c, \qquad (2.6)$$

where the force acting on charged particles is $\mathbf{F} = q\,[\mathbf{E} + \mathbf{v} \times \mathbf{B}]$, with q the particle charge, and $\mathbf{E}$ and $\mathbf{B}$ the local electric and magnetic fields, respectively. The right-hand side of Eq. (2.6) is a symbolic representation of collision processes and in practice it can be difficult to set up a model for what this symbol represents (e.g., see [2]). The velocity moments of this equation allow one to construct the fluid equations, described in Section 2.2.

2.1.2 Thermal equilibrium distributions

Equation (2.6) effectively follows the continuous evolution of the distribution function in response to the electromagnetic forces acting on the charged particles and to the various relaxation processes including many types of collisions. Nevertheless, within a plasma, the distribution function of electrons in particular is often near a thermal equilibrium distribution called the Maxwellian distribution (also known as a Maxwell–Boltzmann distribution). The Maxwellian distribution conveniently relates a characteristic electron temperature to the average energy of electrons and to the mean speed of electrons. However, in the calculation of ionization or excitation coefficients, it is sometimes important to take account of the deviation of the actual distribution of electron energies from a Maxwellian.

In the remainder of this section the spatial and temporal dependence of the distribution function will not be written explicitly, so $f(\mathbf{r}, \mathbf{v}, t) \to f(\mathbf{v})$.

Q Distinguish between $\mathbf{v}$, v and v_x.
A $\mathbf{v}$ is the velocity vector, $v = (v_x^2 + v_y^2 + v_z^2)^{1/2}$ is the magnitude of the velocity vector (also called the speed) and v_x is the x-component of the velocity vector (effectively the speed in the x-direction).

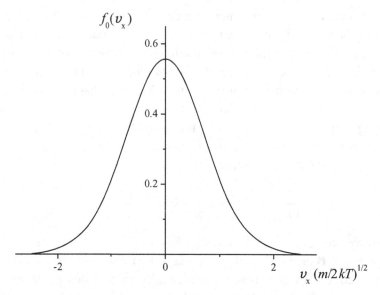

Figure 2.1 A one-dimensional Maxwellian velocity distribution normalized so that the area under the curve is unity: $f_0(v_x) = (m/2\pi kT)^{1/2} \exp\left(-mv_x^2/2kT\right)$.

The Maxwellian three-dimensional velocity distribution is given by

$$f(\mathbf{v}) = n\left(\frac{m}{2\pi kT}\right)^{3/2} \exp\left(-\frac{m\left(v_x^2 + v_y^2 + v_z^2\right)}{2kT}\right), \tag{2.7}$$

where n is the particle number density defined in Eq. (2.2). The distribution function $f(\mathbf{v})$ is proportional to the number of particles with velocities between $\mathbf{v}$ and $\mathbf{v} + d\mathbf{v}$. Figure 2.1 shows the one-dimensional version, the component velocity distribution, that is obtained by integrating over v_y and v_z:

$$f(v_x) = n\left(\frac{m}{2\pi kT}\right)^{1/2} \exp\left(-\frac{mv_x^2}{2kT}\right).$$

Using Eqs (2.3) and (2.4), one can evaluate important averaged (mean) quantities. First note that the net particle flux, Eq. (2.3), in any particular direction must be zero, because the distribution is symmetrical and thus the drift velocity is zero. One can still evaluate a characteristic speed by averaging $|\mathbf{v}| = v$ over the distribution:

$$<v> = \left(\frac{m}{2\pi kT}\right)^{3/2} \int_{-\infty}^{\infty}\int_{-\infty}^{\infty}\int_{-\infty}^{\infty} \left(v_x^2 + v_y^2 + v_z^2\right)^{1/2}$$
$$\times \exp\left(-\frac{m\left(v_x^2 + v_y^2 + v_z^2\right)}{2kT}\right) dv_x dv_y dv_z. \tag{2.8}$$

2.1 The microscopic perspective

> **Q** What does the condition of thermal equilibrium require of the mean of a distribution of particle velocities?
>
> **A** The mean velocity must be zero otherwise there would be a net flow and therefore internal processes would not be in equilibrium.

Since the Maxwellian velocity distribution is isotropic (the same in all directions), the distribution can also be expressed entirely in terms of the scalar speed rather than the velocity vector, **v**, and its components, v_x, v_y, v_z. This simplifies the integral in Eq. (2.8).

The speed distribution $f_s(v)$ gives the proportion of particles with speeds between v and $v + dv$:

$$f_s(v) = n \left(\frac{m}{2\pi kT}\right)^{3/2} 4\pi v^2 \exp\left(-\frac{mv^2}{2kT}\right), \tag{2.9}$$

where the factor of 4π represents an integration over all the angles in which particle trajectories may point. The density is now recovered by integrating over all possible speeds,

$$n = \int_0^\infty f_s(v)\, dv.$$

The mean speed of a particle is then defined by

$$<v> = \left(\frac{m}{2\pi kT}\right)^{3/2} 4\pi \int_0^\infty v^3 \exp\left(-\frac{mv^2}{2kT}\right) dv. \tag{2.10}$$

This average (or mean) speed, $<v>$, is also often given the symbols $\bar{v}$ or $\bar{c}$; the former will be used here. Evaluating the integral in Eq. (2.10) gives

$$\bar{v} = \left(\frac{8kT}{\pi m}\right)^{1/2}. \tag{2.11}$$

> **Q** According to Figure 2.2, what is the most probable speed for a particle in a Maxwellian distribution?
>
> **A** The figure has a peak that corresponds with the most probable speed at $v\,(m/2kT)^{1/2} = 1$. This corresponds with $v = (2kT/m)^{1/2}$, which is clearly not the same as the mean speed $\bar{v}$ which is about 13% larger.

Electrons have a small mass and, in gas discharge plasmas, a high temperature. Using the typical value of $T \approx 30\,000$ K leads to $\bar{v}_e \approx 10^6\,\mathrm{m\,s^{-1}}$. This is much larger than the typical drift speeds observed in the plasma. By contrast, ions are heavy particles and are close to room temperature, typically $T \approx 500$ K, so that for

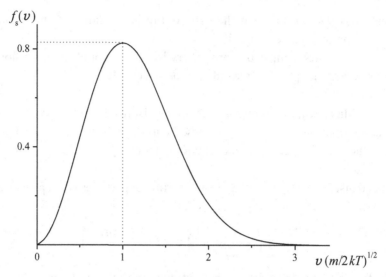

Figure 2.2 A Maxwellian speed distribution normalized so that the area under the curve is unity: $f_s(v) = (4/\sqrt{\pi})\,(mv^2/2kT)\exp(-mv^2/2kT)$.

argon ions $\bar{v}_i \approx 500 \text{ m s}^{-1}$. In Chapter 3 it will be shown that ions leave the plasma with drift speeds significantly larger than $\bar{v}_i$. Therefore, except in the very central region of the plasma, ions are far from thermal equilibrium.

In a similar way, the isotropic distribution of particle speeds can be recast as a distribution in energy space with $f_e(\varepsilon)$ being the number of particles with kinetic energy between ε and $\varepsilon + d\varepsilon$:

$$f_e(\varepsilon) = \frac{2n}{\sqrt{\pi}} \left(\frac{1}{kT}\right)^{3/2} \varepsilon^{1/2} \exp\left(-\frac{\varepsilon}{kT}\right). \tag{2.12}$$

Q What is the most probable energy for a particle in a Maxwellian distribution (Figure 2.3)?
A The most probable energy corresponds with the peak at $\varepsilon = kT/2$.

The kinetic energy density can be found from the velocity distribution by multiplying the energy distribution by $\varepsilon = mv^2/2$ and integrating over all energies:

$$w = \frac{2n}{\sqrt{\pi}} \left(\frac{1}{kT}\right)^{3/2} \int_0^\infty \varepsilon^{3/2} \exp\left(-\frac{\varepsilon}{kT}\right) d\varepsilon = \frac{3}{2}nkT. \tag{2.13}$$

Since $w \equiv n <\varepsilon>$, the average kinetic energy of a particle is $3kT/2$. The distribution is isotropic and any particle is free to move in three independent directions,

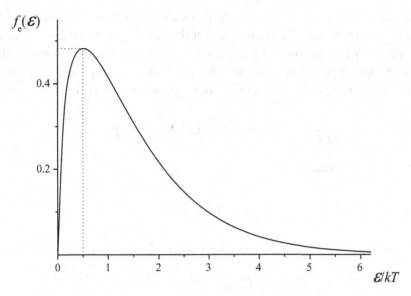

Figure 2.3 A Maxwellian energy distribution normalized so that the area under the curve is unity: $f_e(\varepsilon) = (2/\sqrt{\pi})(\varepsilon/kT)^{1/2} \exp(-\varepsilon/kT)$.

suggesting that the average energy corresponds with $kT/2$ in each of the three translational degrees of freedom.

> **Q** How can the mean kinetic energy per particle of a Maxwellian distribution be obtained, considering only energy associated with its motion in the x-direction?
>
> **A** Multiply the velocity distribution by $mv_x^2/2$ and integrate over all velocities to get the total kinetic energy associated with the x components of motion and then divide by n to get the average energy per particle:
>
> $$\left< \frac{mv_x^2}{2} \right> = \left(\frac{m}{2\pi kT}\right)^{3/2} \int_{-\infty}^{\infty}\int_{-\infty}^{\infty}\int_{-\infty}^{\infty} \frac{mv_x^2}{2}$$
>
> $$\times \exp\left(-\frac{m(v_x^2 + v_y^2 + v_z^2)}{2kT}\right) dv_x dv_y dv_z.$$
>
> The integrals are standard ones and the result confirms the suggestion that each degree of freedom has a mean thermal energy of $kT/2$ associated with it. Note that the characteristic temperature T of a Maxwellian distribution gives a measure of thermal energy.

Although the random thermal flux of particles is zero for a Maxwellian distribution, it is useful to have a local measure of the flux crossing any particular plane at any time as a consequence of the thermal motion of particles. For the particles crossing the $x-y$ plane in the positive z-direction, this is determined by an integral over all x and y components of velocity, but only positive z components:

$$\Gamma_{\text{random}} = n \left(\frac{m}{2\pi kT}\right)^{3/2} \int_{-\infty}^{\infty} dv_x \int_{-\infty}^{\infty} dv_y \int_{0}^{\infty} v_z \exp\left(-\frac{mv^2}{2kT}\right) dv_z. \quad (2.14)$$

Evaluating this integral yields

$$\Gamma_{\text{random}} = n \left(\frac{kT}{2\pi m}\right)^{1/2}.$$

Using the expression for the random speed in Eq. (2.11), this can also be written

$$\Gamma_{\text{random}} = \frac{n\bar{v}}{4}. \quad (2.15)$$

Given the very large difference between the electron average speed and the ion average speed, the thermal flux of electrons heading towards the plasma boundaries is very large compared to the thermal flux of ions leaving the plasma. Ions and electrons are created at the same rate within the plasma volume and the main loss mechanism is often recombination at the walls. So, to maintain the flux balance at the wall in the steady state, as will be seen later, the potential in the plasma must be higher than the potential at the wall. In effect, close to the wall the potential falls by $\Delta\phi$ with respect to the plasma. In that case only electrons with sufficient perpendicular velocity, $v_z > \sqrt{2e\Delta\phi/m}$, can reach the wall. The *particle* flux leaving the plasma is the same as that reaching the wall; that is,

$$\Gamma_{\text{wall}} = n \left(\frac{m}{2\pi kT}\right)^{3/2} \int_{-\infty}^{\infty} dv_x \int_{-\infty}^{\infty} dv_y \int_{\sqrt{2e\Delta\phi/m}}^{\infty} v_z \exp\left(-\frac{mv^2}{2kT}\right) dv_z. \quad (2.16)$$

Evaluating the integral Eq. (2.16) yields

$$\Gamma_{\text{wall}} = \frac{n\bar{v}}{4} \exp\left(-\frac{e\Delta\phi}{kT}\right). \quad (2.17)$$

The *energy* flux leaving the plasma can also be calculated in a similar manner:

$$Q = n \left(\frac{m}{2\pi kT}\right)^{3/2} \frac{m}{2} \int_{-\infty}^{\infty} dv_x \int_{-\infty}^{\infty} dv_y \int_{\sqrt{2e\Delta\phi/m}}^{\infty} v^2 v_z \exp\left(-\frac{mv^2}{2kT}\right) dv_z. \quad (2.18)$$

2.1 The microscopic perspective

This can be shown to give

$$Q = \left[\frac{n\bar{v}}{4} \exp\left(-\frac{e\Delta\phi}{kT}\right)\right](2kT + e\Delta\phi). \tag{2.19}$$

The energy flux leaving the plasma is not equal to the energy flux reaching the wall because some of the energy is deposited in the electrostatic field at the plasma boundary. The amount of energy flux reaching the wall is only

$$Q_w = \left[\frac{n\bar{v}}{4} \exp\left(-\frac{e\Delta\phi}{kT}\right)\right] 2kT. \tag{2.20}$$

The term in square brackets is just the number of particles lost to the wall per square metre per second. The average kinetic energy carried out by each particle that escapes therefore is $2kT$.

> **Q** The SI unit for energy is the joule (J); in atomic, molecular and plasma physics an alternative energy unit, the electron volt (eV), is formed by dividing the quantity in joules by the magnitude of the electronic charge, e, so that $1\,\text{eV} \equiv 1.602 \times 10^{-19}\,\text{J}$. What is the equivalent temperature in eV of a distribution with $kT = 3.2 \times 10^{-19}\,\text{J}$?
>
> **A** The temperature is said to be "2 eV" because $kT/e = (3.2 \times 10^{-19}/e)\,\text{V} \approx 2\,\text{V}$.

Exercise 2.1: Electron energy flux to a wall For a Maxwellian electron population of $10^{16}\,\text{m}^{-3}$ with mean energy $2\,\text{eV}$, calculate the rate of energy transfer to a wall that is at $-10\,\text{V}$ with respect to the plasma.

2.1.3 Collisions and reactions

The different types of particle in a plasma (electrons, ions, atoms, free radicals, molecules) interact in the volume via collision processes that occur on very short time scales. These collisions can be *elastic* (without loss of total kinetic energy) or *inelastic* (with transfer between the kinetic energy and the internal energy of the colliding particles).

In the simple situation of weakly ionized plasmas in noble gases, the most frequent collisions involving charged particles are elastic encounters with neutral atoms.

Collisions between charged particles (electron–electron, electron–ion and ion–ion) are not frequent and direct electron–ion recombination is usually negligible in the volume of low and medium-density plasmas at low pressure. Consequently, the charged particles tend to be generated in the plasma volume by ionization and lost at the vessel walls (and other surfaces). In most of this book the process of ionization by electron impact will be supposed to be a single-step process. In practice in some instances multi-step ionization occurs, but this is a rare process unless there are intermediate longer-lived (metastable) states that can act as energy reservoirs pending the arrival of subsequent electrons.

> **Q** It is usually the case that in low-pressure, electrical discharges through noble gases, *binary*, *elastic* encounters between charged and neutral particles far outnumber other kinds of charged particle interaction. Explain why this should be so.
>
> **A** First, plasmas in these circumstances are weakly ionized, so the neutral particles provide the most likely target for any charged particle. Second, the commonest interactions are binary collisions (i.e., involving two particles) because the chances of three or more particles simultaneously encountering each other are much lower. Finally, all particles in a distribution can participate in elastic collisions, whereas the inelastic processes like ionization and excitation require the total initial kinetic energy to exceed some minimum (threshold) value. This usually excludes the majority of those present.

Etching or deposition systems use plasmas formed in molecular gases. The situation is now even more complicated, with chemical reactions between atoms, free radicals and molecules also playing a large role. These additional species interact in the volume and at surfaces. As a result, the charged-particle dynamics are somewhat modified. Electrons can be attached to molecules in the gas phase, which is another path of electron loss, and positive ions may recombine in the plasma volume with negative ions – both of these frustrate the simplifying assumption of volume production and wall loss.

The complete mathematical description of collisions is complex [2, 24] and is certainly beyond the scope of this book. The aim here is to develop the simplest models through which general insight can be gained, rather than to cover every eventuality. To this end, in the next section fundamental quantities such as cross-sections, mean free paths and collision frequencies will be defined. As an example, the different collisions processes that may be important in plasma etching with molecular gases will be reviewed.

Cross-section, mean free path and collision frequency

The simplest descriptions of binary collisions, whether electron–atom, atom–atom or ion–atom, suppose the situation to be that of a hard-sphere projectile interacting with a hard-sphere target. Consider a slab of gas, dx thick and of area A, containing

n_g identical gas atoms per unit volume. Suppose that the slab is bombarded by a uniform beam of small projectiles, with particle flux Γ. Each of the $n_g A dx$ gas atoms in the slab presents a cross-sectional target area which will be given the symbol σ. The cross-section is proportional to the probability of a collision – collisions are twice as likely if the cross-section of the targets is doubled. In collision physics those particles that hit the targets are said to have been scattered out of the beam. On passing through the slab, the uniform flux of projectiles will be diminished in proportion to the total area of the targets.

Considering the proportion of the beam scattered by collisions with targets in the slab, the loss of flux after passing through the slab is

$$\frac{d\Gamma}{\Gamma} = -n_g A dx \frac{\sigma}{A} = -n_g \sigma dx. \tag{2.21}$$

Integrating Eq. (2.21) shows that the beam flux decays exponentially:

$$\Gamma = \Gamma_0 \exp\left(-\frac{x}{\lambda}\right), \tag{2.22}$$

with a characteristic decay length, $\lambda = 1/(n_g \sigma)$, which is called the mean free path between collisions. If the particles in the beam all travel at a speed v, the characteristic time between collisions is $\tau = \lambda/v$, and the collision frequency is

$$\begin{aligned} \nu = \tau^{-1} &= n_g \sigma v \\ &= n_g K, \end{aligned} \tag{2.23}$$

(be careful to distinguish between frequency ν and speed v) where $K = \sigma v$ is the rate of interaction per atom of gas (more generally termed the rate coefficient for collisions). Unfortunately, this very simple situation is not quite sufficiently realistic to describe collisions even in a weakly ionized plasma.

Firstly, the effective cross-section for binary collisions is a function of the magnitude of the impact velocity, that is the relative speed between the target and the projectile (or equivalently, the total kinetic energy in the interaction). The cross-sections for elastic and inelastic collisions between electron and argon atoms are shown in Figure 2.4. Since the mass ratio between electrons and atoms is very large, the elastic collisions transfer very little kinetic energy (in much the same way that a football bounces off a wall). There is no threshold for this process, though there is a strong energy dependence with a pronounced minimum (the 'Ramsauer' minimum) that is characteristic of the low-energy impacts in the noble gases. At higher impact energy the elastic cross-section tends to decrease as the relative speed (or collision energy) increases. Naïve models equate the electron-impact cross-section to the physical size of an atom, on which basis argon would have a

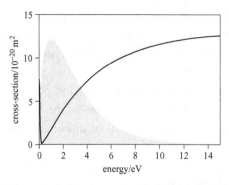 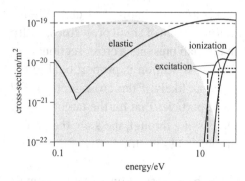

Figure 2.4 Elastic and inelastic cross-sections in argon – schematic. Broken lines indicate useful approximations. The grey tone indicates the shape of a 1 eV Maxwellian.

cross-section of about 3×10^{-20} m^2, but as Figure 2.4 shows, the size perceived by an electron appears to depend on its energy.

> **Q** According to Figure 2.4, for a 14 eV electron, which is the more likely type of collision: elastic or inelastic (excitation)?
> **A** The cross-section of elastic collisions at 14 eV is around 20 times greater than that for excitation (but below the threshold for ionization, which is at 15.6 eV), and so at this energy only 1 in 20 collisions is expected to be inelastic.

Inelastic collisions between an electron and an atom involve the interconversion of kinetic energy and the internal potential energy of the atom, causing either excitation into a higher quantum state, or, if there is sufficient kinetic energy, ionization. These collisions have a threshold which is roughly the quantum energy required for the process, typically 10 to 20 eV for noble gases in the ground state. The inelastic cross-sections sharply increase just after the threshold, pass through a maximum (usually located around twice to three times the threshold energy) and then smoothly fall at higher energy. Since both elastic and inelastic cross-sections are strong functions of energy, and hence the magnitude of the impact velocity, they should therefore be generally written as $\sigma(v_{\text{impact}})$ – in the case of electron collisions this is usually $\sigma(v_e)$.

Secondly, the projectiles are not mono-energetic but are randomly distributed in speed according to a distribution function, that to a first approximation can be taken to be Maxwellian. Neither are the atomic targets stationary (although their thermal motion is usually considerably less energetic than that of the electrons), so neglecting their motion is an acceptable approximation for electron–atom collisions. The calculation of the collision frequency should then be an average based upon an integration over the presumed Maxwellian distributions, taking into account also

2.1 The microscopic perspective

the energy dependence of the cross-section. Therefore, the average frequency of collisions of electrons can be estimated to be

$$\bar{\nu} = n_g \int_0^\infty \sigma(v_e) v_e f_s(v_e) dv_e. \tag{2.24}$$

When the cross-section depends on the impact speed, as it certainly does for real gases (see Figure 2.4), this integral must be evaluated numerically. In the energy range of most interest for most plasmas ($0.5\,\text{eV} < \varepsilon < 10\,\text{eV}$) the argon elastic cross-section varies almost linearly with energy but not much detail is lost if it is assumed to remain at a steady intermediate value $\bar{\sigma}_{el}$ across this range for the sake of evaluating the integral in Eq. (2.24). Then for elastic collisions

$$\nu_m = n_g \bar{\sigma}_{el} \bar{v}_e, \tag{2.25}$$

where $\bar{v}_e$ is the mean speed (see Eq. (2.10)) and the subscript m signifies elastic, 'momentum transfer' collisions (between electrons and neutrals) that are responsible for randomizing any directed momentum. A rate coefficient for elastic momentum transfer collision can be defined such that $K_{el} = \nu_m/n_g \equiv \bar{\sigma}_{el} \bar{v}_e$.

Warning: Strictly under RF conditions the momentum transfer collision frequency is different from Eq. (2.25) and includes a frequency-dependent correction [25]. Nevertheless, in this book Eq. (2.25) will be used.

Q The ion momentum is redistributed through ion–atom collisions at a frequency ν_i – how will this compare with that for electrons ν_m?
A Ions are larger projectiles than electrons so while the target atoms are the same, the probability of collisions (the cross-section) with ions is greater. However, because $m \ll M$, $\bar{v}_e \gg \bar{v}_i$ so in fact $\nu_m \gg \nu_i$.
Comment: In argon, ion–neutral charge exchange collisions (to be discussed below) are as, or even more, frequent than ion–neutral elastic scattering collisions.

When the collision is inelastic the cross-section has a threshold, e.g., ionization or excitation. The following idealized form of the cross-section can be used to estimate the dependence of the ionization frequency on electron energy:

$$\sigma = 0 \text{ when } \varepsilon < \varepsilon_{iz},$$

$$\sigma = \sigma_{iz} \text{ when } \varepsilon > \varepsilon_{iz};$$

that is a step function with threshold ε_{iz}, the ionization energy. In this case, the integral Eq. (2.24) must be calculated with a minimum speed $v = (2e\varepsilon_{iz}/m)^{1/2}$, and gives the following expression:

$$\nu_{iz} = n_g \sigma_{iz} \bar{v}_e \left(1 + \frac{e\varepsilon_{iz}}{kT_e}\right) \exp\left(-\frac{e\varepsilon_{iz}}{kT_e}\right) \tag{2.26}$$

for the average frequency of ionizing collisions.

> **Q** Distinguish in Eq. (2.26) factors relating to electrons from factors relating to the gas.
> **A** ν_{iz} is the number of ionizations produced per second *per electron* of a population with characteristic temperature T_e, in a gas of particle density n_g and ionization energy ε_{iz}.

A rate coefficient for ionization, K_{iz}, can be defined as $K_{iz} = \nu_{iz}/n_g$. Using Eq. (2.26), this rate coefficient has the form

$$K_{iz}(T_e) = K_{iz0} \exp\left(-\frac{e\varepsilon_{iz}}{kT_e}\right); \tag{2.27}$$

the pre-exponential factor

$$K_{iz0} = \sigma_{iz} \bar{v}_e \left(1 + \frac{e\varepsilon_{iz}}{kT_e}\right) \tag{2.28}$$

depends on the cross-section and therefore on the gas used, but its dependence on T_e is weak compared with the exponential part of the rate coefficient. This so-called Arrhenius form of Eq. (2.27) for $K_{iz}(T_e)$ can also be used for the electronic excited states, with the generic form

$$K_{exc}(T_e) = K_{exc0} \exp\left(-\frac{e\varepsilon_{exc}}{kT_e}\right), \tag{2.29}$$

where ε_{exc} is the energy of the excited quantum state under consideration.

In this book argon is used as a typical, electropositive, atomic gas, with the simplified set of properties given in Table 2.1. This set has been chosen as the best fit of the numerical integration of the argon cross-sections over a Maxwellian distribution (note that the real ionization threshold for argon is 15.6 eV, not 17.44 eV) [2]. The ion–neutral mean free path can conveniently be expressed as

$$\lambda_i/\text{mm} = \frac{4.2}{P/\text{Pa}}. \tag{2.30}$$

Exercise 2.2: Comparative frequencies and mean free paths For argon at 10 Pa, 300 K, and Maxwellian populations with temperatures $T_e = 2$ eV and $T_i = 0.05$ eV, calculate: (i) the electron–neutral collision frequencies of

2.1 The microscopic perspective

Table 2.1 *Simplified data set used for argon gas in the global models presented in this book*

Ionization	$K_{iz0}/m^{-3}\,s^{-1}$	5.0×10^{-14}	ε_{iz}/eV	17.44
Excitation	$K_{exc0}/m^{-3}\,s^{-1}$	0.16×10^{-18}	ε_{exc}/eV	12.38
Elastic (electrons)	$\overline{\sigma}_{el}/m^2$	1.0×10^{-19}		
Elastic (ions at 0.05 eV) - including charge exchange	$\overline{\sigma}_i/m^2$	1.0×10^{-18}		

ionization, excitation and momentum transfer and the ion–neutral momentum transfer collision frequency; (ii) electron–neutral and ion–neutral mean free paths for elastic scattering.

Energy transfer in collisions

The cross-section embodies the concept of the probability of a collision of a particular type. It does not, however, specify exactly how much energy is redistributed between the participating particles. The mechanics of collisions, based on the simultaneous conservation of momentum and energy for a pair of colliding particles of masses m and M, shows that:

(i) elastic collisions between dissimilar mass particles, $m \ll M$ (electron–atom), can transfer only a fraction, δ, of the impact energy where

$$\delta \leq 2m/M, \tag{2.31}$$

whereas for equal mass particles (ion–atom) in a head-on collision all the kinetic energy is transferred from one to the other;

(ii) inelastic collisions between dissimilar masses, $m \ll M$ (electron–atom), can transfer all the kinetic energy into internal energy, provided the initial energy is above the threshold for the process, whereas equal masses can only transfer half of the initial energy into internal energy and therefore an ion needs twice the ionization threshold before it can cause impact ionization of an atom.

Inelastic collisions and chemical reactions in molecular gases

In molecular gases it is easily found, for instance by absorption measurements, that infrared and microwave radiation can probe the energy levels associated respectively with modes of molecular vibrations and rotations. The more commonly populated electronic excitation levels of an atom or molecule are typically separated by a few eV of energy, and these states can be probed by photons of equivalent energy, corresponding with visible and ultraviolet radiation.

Inelastic collision interactions between electrons and molecules or atoms lead to energy dissipation in the electron population and should be taken into account when considering the electron power balance (see Section 2.3). These collisions often lead to dissociation, and the fragments may be neutrals or charged particles (positive or negative ions). Furthermore, the fragments may react in the gas phase or at the surface (reactor walls or substrate), sometimes leading to etching or deposition. The complex plasma chemistry resulting from all these interactions is beyond the scope of this book (for details on collisions and reactions, see [2, 24]).

In the following, a reduced set of collisions processes is presented to describe the basic phenomena encountered in the RF plasmas discussed later in this text. In particular, CF_4, Cl_2 and Ar are used as examples. In the interests of clarity, later chapters of this book will not include the detailed effects of plasma chemistry, whether in the volume or at a surface; it is therefore important to use this discussion to reflect from time to time on what effect the inclusion of plasma chemistry would have on any given situation described later on.

Dissociation into neutral fragments

The feedstock gas is dissociated by electron impact to produce neutral reactive fragments known as radicals. These play a fundamental rôle in plasma processing because of their chemical reactivity. They may react in the plasma volume or at the surfaces. This is an abbreviated list of possible dissociation reactions:

$$e^- + CF_4 \rightarrow CF_3 + F + e^-,$$
$$e^- + CF_4 \rightarrow CF_2 + 2F + e^-,$$
$$e^- + CF_4 \rightarrow CF + F_2 + F + e^-,$$
etc.

The radicals can be further dissociated to produce smaller fragments, for instance

$$e^- + CF_3 \rightarrow CF_2 + F + e^-,$$
$$e^- + CF_2 \rightarrow CF + F + e^-,$$
etc.

Note that the fragments may also be in excited states – this is often indicated by '*' after the atom/molecule that is electronically excited or else giving in brackets the molecular excited state, e.g., vibrational ($v = 2$). The energy changes between levels of electronic excitation generally correspond with UV and visible frequencies, whereas vibrational levels map onto the IR part of the electromagnetic spectrum.

Dissociative ionization and attachment

When one of the fragments is a charged particle, the process is called dissociative ionization if the fragment is positively charged (positive ion), or **dissociative**

2.1 The microscopic perspective

attachment if the fragment is negatively charged (negative ion). Typical examples for CF_4 are:

$$e^- + CF_4 \rightarrow CF_3^+ + F + 2e^-,$$
$$e^- + CF_4 \rightarrow CF_3^- + F,$$
$$e^- + CF_4 \rightarrow CF_3 + F^-,$$
etc.

These are quite often the important processes for charged particle production and loss. Direct attachment of an electron to a molecule, without dissociation, is unlikely, except for large and highly electronegative molecules (e.g., SF_6). In chlorine discharges the following processes:

$$e^- + Cl_2 \rightarrow Cl_2^+ + 2e^-,$$
$$e^- + Cl_2 \rightarrow Cl^+ + Cl + 2e^-$$

are in competition.

Vibrational excitation

Molecules have discrete vibrational and rotational energy levels. The higher energy levels can be populated by electron impact excitation. Taking the example of vibrations, we have

$$e^- + Cl_2(v = 0) \rightarrow Cl_2(v = 1) + e^-,$$
$$e^- + Cl_2(v = 0) \rightarrow Cl_2(v = 2) + e^-,$$
$$e^- + Cl_2(v = 1) \rightarrow Cl_2(v = 2) + e^-,$$
etc.

The energy difference between two vibrational levels is significantly less than 1 eV, so that a very large fraction of the electron population can experience such inelastic collisions. The reaction rate for vibrational excitation is therefore high and vibrational excited states may play a rôle in the discharge equilibrium. In particular, the vibrational energy can be coupled to translational motion, leading to significant neutral gas heating [26]. Molecules with more than two atoms have more degrees of freedom and consequently a richer vibrational spectrum.

Chemical reactions between neutrals

The neutral fragments generated by electron impact dissociation may recombine in the gas phase. To facilitate the simultaneous conservation of momentum and energy the recombination process needs a third body, usually denoted M (in weakly dissociated plasmas the third body is the feedstock gas molecule):

$$CF_3 + F + M \rightarrow CF_4 + M,$$
$$CF_2 + F + M \rightarrow CF_3 + M,$$
$$CF + F + M \rightarrow CF_2 + M,$$
etc.

The reaction rate is proportional to the gas pressure and such recombinations are often negligible in the low-pressure regime of etching plasmas where surface reactions (recombination, etching, deposition) are particularly important. Note that other processes, such as exchange reactions ($CF + O \rightarrow CO + F$), may also need to be considered.

Surface reactions

In the typical pressure regime of etching plasmas the chemical reactions in the gas phase are slow and the transit time of radicals to the chamber walls is shorter than the typical reaction time. The interaction of radicals with the surfaces is therefore of primary importance. Taking the example of chlorine, one gets the following reaction:

$$Cl(g) + Cl(ads) \rightarrow Cl_2(g)$$

where (g) denotes an atom or a molecule in the gas phase, and (ads) an atom adsorbed at the surface. Other types of reactions include *etching*, in which a volatile product is formed by chemical reaction at the surface:

$$Cl(g) + SiCl_3(s) \rightarrow SiCl_4(g) \uparrow;$$

or *deposition*, in which the incoming atoms, radicals or ions become bonded into the surface, contributing to the growth of a thin film:

$$SiH(g) \rightarrow Si(s) \downarrow + H(g) \uparrow .$$

Charge exchange and positive–negative ion recombination

Heavy charged particles (positive ions and negative ions) also experience collisions with neutrals or between themselves. A very important process is resonant charge transfer, in which an ionized atom interacts with a neutral of the same species (the resonance being one of having identical quantum structure). For example, in argon plasmas:

$$Ar^+_{fast} + Ar_{slow} \rightarrow Ar_{fast} + Ar^+_{slow}.$$

This process has a large cross-section and the corresponding mean free path is shorter than the mean free path for elastic scattering. The charge transfer may also be non-resonant when species are different (e.g., $O + N^+ \rightarrow O^+ + N$). Another type of charge transfer is ion–ion recombination, for instance

$$\mathrm{CF_3^+ + F^- \to CF_3^* + F.}$$

This process is often the main loss mechanism for negative ions. As will be seen in Chapter 9, unlike positive ions, negative ions are trapped in the plasma volume where they are ultimately destroyed in gas-phase reactions.

2.2 The macroscopic perspective

In many instances, the charged-particle motion can be adequately described by macroscopic equations, called the fluid (or hydrodynamic) equations. By integrating the Boltzmann equation Eq. (2.6) over the velocity coordinates of the distribution functions, one obtains the fluid equations which will only depend on position and time. The plasma is then described as a fluid defined by macroscopic quantities, such as its density $n(\mathbf{r}, t)$, its velocity $\mathbf{u}(\mathbf{r}, t)$, and its pressure $p(\mathbf{r}, t)$.

Warning: Fluid equations are valid for all forms of distribution functions $f(\mathbf{r}, \mathbf{v}, t)$ for which the integration of the Boltzmann equation Eq. (2.6) over velocity space is valid. For simplicity, Maxwellian distributions are often used in this book to evaluate the collision frequencies, and consequently the transport coefficients and reaction rates that appear in fluid equations. Significant departures from Maxwellian distributions will then render such values less appropriate.

2.2.1 Fluid equations

A set of fluid equations for the constituents of a plasma is obtained by taking the velocity moments of the Boltzmann equation Eq. (2.6). There is a hierarchy of such equations, the first of which is a species conservation equation, obtained by integrating Eq. (2.6) over velocity space:

$$\frac{\partial n}{\partial t} + \nabla \cdot (n\mathbf{u}) = S - L; \qquad (2.32)$$

an equation of this form is also called a continuity equation. The first term on the LHS can easily be associated with the first one in Eq. (2.6); it describes the changing density at a particular point in space. The second term corresponds with the next one in Eq. (2.6), the spatial differential operating independently of the velocity integral. This term accounts for density changes associated with flow in or out of the local space. The force term in the Boltzmann equation does not survive in this first integration even for charged species because it concerns values of the distribution function at $v = \pm\infty$, where it is zero. On the RHS, S and L represent the contribution of collisions that cause increases and decreases respectively in the local fluid density, being volume source and volume loss terms. In an atomic, electropositive plasma at low pressure, the source of electrons is produced by ionization in the volume and they are lost at the reactor walls since electron–ion recombination is negligible. Therefore in the electron conservation equation $L = 0$ and, as seen in Section 2.1.3, $S = n_e n_g K_{iz}(T_e)$ where $K_{iz}(T_e)$ is given by Eq. (2.27); note that in this particular instance, a *Maxwellian distribution function* has been assumed. It will be shown in Chapter 9 that S and L take different forms in the case of electronegative plasmas.

Momentum conservation is next. An equation for this is obtained by taking the first moment of the Boltzmann equation (which consists of multiplying the Boltzmann equation by the momentum of a particle, $m\mathbf{v}$, and then integrating over velocity). This leads to an equation involving drift velocity $\mathbf{u}$. With $\mathbf{B} = 0$,

$$nm\left[\frac{\partial \mathbf{u}}{\partial t} + (\mathbf{u} \cdot \nabla)\mathbf{u}\right] = nq\mathbf{E} - \nabla p - m\mathbf{u}\left[n\nu_m + S - L\right], \qquad (2.33)$$

in which p stands for the particle pressure. This equation is equivalent to the Navier–Stokes equation in neutral fluids and represents the equilibrium of forces acting on the fluids, so it is also sometimes referred to as the force balance equation. On the LHS of the equation are the acceleration and the inertial terms. On the RHS there are three types of force, the electric driving force, the pressure gradient force and a friction force. Note that in the last term of the RHS, the particles are assumed to be generated and lost while moving at the drift velocity; again for electrons in a classical electropositive plasma, $S = n_e n_g K_{iz}(T_e)$ and $L = 0$. A detailed discussion of each term of Eq. (2.33) will be given in Chapter 3.

In the case of isotropic (unmagnetized) plasmas, the pressure is a scalar and is related to the density and temperature by a thermodynamical equation of state:

$$p = nkT. \qquad (2.34)$$

The relationships between the fluid variables $(n, \mathbf{u}, p, T)$ and the electric field $(\mathbf{E})$, Eqs (2.32)–(2.34), do not define a closed set.

Q What other equation(s) could be introduced to specify the electric field?
A It is usual to turn to Maxwell's equations in this circumstance; when $\mathbf{B} = 0$, Gauss's law should be sufficient as it provides a relationship between $\mathbf{E}$, n_e and n_i. This serves to highlight that in fact it is necessary to consider at least two fluids, one for electrons and one for ions.
Comment: When there are RF (or higher frequency) components of the electric and magnetic fields it is necessary to include all four of Maxwell's equations.

2.2 The macroscopic perspective

Even with Maxwell's equations the description is incomplete. There are three different ways to close the equation set. The first is to assume that the electron and ion temperatures are constant in space and time, with a value determined by two final equations: these could be as simple as giving a value for T_e and T_i; in the next section, for instance, T_i is set to zero and it is shown that the electron temperature can be effectively set by the system size and the gas pressure, based on a global balance. Many problems, such as the transport theories developed in Chapter 3, use this *isothermal* approximation for which changes in pressure are ascribed only to changes in density:

$$\nabla p = kT \nabla n. \tag{2.35}$$

The second way is to consider that the variations described by the fluid equations are so fast that the fluids are not able to exchange energy with their surroundings within the time frame of interest. In that case, the situation is *adiabatic*, for which thermodynamics supplies a relationship between pressures and densities (one for each fluid):

$$\frac{\nabla p}{p} = \gamma \frac{\nabla n}{n}, \tag{2.36}$$

where γ is the ratio of the specific heat at constant pressure to that at constant volume. For one-dimensional motion, $\gamma = 3$. The adiabatic approximation typically holds for high-frequency waves.

The third and most thorough approach is to consider the flow of heat and internal energy in the plasma, based on another moment of the Boltzmann equation, moving the debate further into thermodynamics and the need to make assumptions that are even more sophisticated. The extra effort that this involves is not warranted for the present purposes.

The macroscopic equations can be solved numerically with appropriate boundary conditions, to obtain the density, the velocity, the temperature and the electric field as a function of space and time in a given reactor geometry. This is the basis of the so-called fluid simulations. However, some insight can be gained by further simplifying these equations. There are two ways to do this. The first, examined in Section 2.3, is to integrate the fluid equations over the space coordinates to obtain global balance equations. This is the basis of the so-called global models that will be used extensively in this book to understand the major scaling laws in radio-frequency plasma reactors. The second, developed in Section 2.4, is to linearize the fluid equations to obtain electrodynamics properties in terms of perturbations of the steady state.

Before proceeding, the isothermal hypothesis is further examined and the electron energy relaxation length is introduced as the typical distance over which non-uniform electron temperatures can be expected.

2.2.2 Electron energy relaxation length

The isothermal approximation for charged particles will be used in most of this book. In reality, the electron and ion temperatures are not always independent of the space coordinate. It is particularly important to understand what may lead to non-uniform electron temperature, since inelastic processes are extremely sensitive to this parameter. The first condition for non-uniform electron temperature is that the electric (electromagnetic) energy is deposited non-uniformly in the electron population. This is quite often the case because a confined plasma does not necessarily experience uniform electromagnetic fields so the plasma does not absorb the energy uniformly (later chapters will show that electrons often absorb energy only near the boundaries). Nevertheless, non-uniform energy deposition does not automatically lead to non-uniform temperature, because the electron *energy relaxation length* may be large compared to the system dimensions. Under these circumstances, electrons may gain their energy in one location and subsequently share this energy with other electrons far away from this location. The rigorous calculation of the electron energy relaxation length requires kinetic theory that is beyond the scope of this book (see the calculation proposed in Lieberman and Lichtenberg [2]). Godyak [27] proposed a relatively simple expression that accounts for all electron energy loss mechanisms in a discharge:

$$\lambda_\varepsilon = \lambda_{el} \left[\frac{2m}{M} + \frac{\nu_{ee}}{\nu_m} + \frac{2}{3}\left(\frac{e\varepsilon_{exc}}{kT_e}\right)\frac{\nu_{exc}}{\nu_m} + \frac{2}{3}\left(\frac{e\varepsilon_{iz}}{kT_e}\right)\frac{\nu_{iz}}{\nu_m} + 3\frac{\nu_{iz}}{\nu_m} \right]^{-1/2}, \quad (2.37)$$

where λ_{el} is the mean free path for electron–neutral elastic collisions; the various contributions are discussed next. The first term in brackets is due to electron energy loss by elastic collisions; this leads to neutral gas heating. The second term is due to electron energy loss by electron–electron (coulomb) collisions. These were not treated above as they are usually not important in gas discharges. However, they become important in high-density RF plasmas, such as helicon plasmas studied in Chapter 8. The third and fourth terms are due to electron energy loss by inelastic collisions (ionization and excitation). Finally, the last term accounts for the kinetic energy loss at the boundaries. These various mechanisms for electron energy loss will be discussed again in the next section when introducing the power balance in global models.

Q What system parameters will mostly control the electron energy relaxation length?
A The terms in Eq. (2.37) all depend on gas pressure and composition since $\lambda_{el} = (n_g \sigma_{el})^{-1}$, and some terms depend also on the electron temperature. Molecular gases have many more inelastic processes (dissociation, vibrational and rotational excitations) than noble gases, so λ_ε will be much shorter in such gases.

Exercise 2.3: Energy relaxation length in argon For argon at 10 Pa, 300 K, with Maxwellian electrons with temperature $T_e = 2$ eV, calculate the energy relaxation length when neglecting the electron–electron collisions.

Comment: *At the low pressures typical of inductive and helicon discharges, λ_ε will be relatively large. Therefore, in many instances, although the power absorbed by electrons is localized, the electron temperature is in fact almost independent of the space coordinate. The regime when λ_ε is much larger than the system size is that of non-local electron kinetics. The first kinetic theory for the non-local regime in DC glow discharges was proposed by Bernstein and Holstein in 1954 [28], and revisited by Tsendin in 1974 [29]. It was later used in RF capacitive and inductive discharges, as described in Kolobov and Godyak [30] and Kortshagen et al. [31].*

2.3 Global particle and energy balance

The fluid equations were obtained by integrating the kinetic equations over the velocity coordinates. It is possible to further simplify the description by integrating the fluid equations over the space coordinates. Doing so establishes balance equations that govern the time variations of global (volume-averaged) quantities. To determine the discharge equilibrium, that is to calculate the mean electron density and the electron temperature for given pressure and power into the reactor, one needs to solve simultaneously two balance equations – the particle and the energy balance equations.

2.3.1 Particle balance

To obtain the particle balance equation one must integrate the fluid species conservation Eq. (2.32) over space. For the sake of simplicity, consider first an electropositive plasma between two infinite plane walls (one-dimensional geometry) placed at $x = -l/2$ and $x = l/2$; see Figure 2.5.

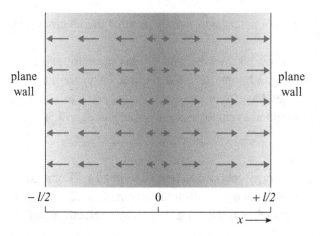

Figure 2.5 A region of plasma maintained by volume ionization between parallel walls separated by a distance l; the intensity of grey indicates plasma density and the grey arrows indicate the magnitude and direction of wall-directed fluxes where particles ultimately recombine. As will be discussed in the next chapter, the boundaries between a plasma and a surface need special attention.

Q How would the spatial average of quantities like electron density be performed over the space shown in Figure 2.5?
A The spatial average of density is its integral from $-l/2$ to $+l/2$, divided by l:

$$\bar{n}_e = \frac{1}{l}\int_{-l/2}^{l/2} n_e \mathrm{d}x. \tag{2.38}$$

The spatial integral of Eq. (2.32) can be done, term by term, treating time and space as independent coordinates:

$$\frac{\partial}{\partial t}\int_{-l/2}^{l/2} n_e \mathrm{d}x + \int_{-l/2}^{l/2}\frac{\partial (n_e u_e)}{\partial x}\mathrm{d}x = \int_{-l/2}^{l/2} n_e n_g K_{\mathrm{iz}}\mathrm{d}x. \tag{2.39}$$

The second term can be split and simplified into the fluxes at the two walls, since the flux at the centre is zero, by symmetry:

$$\int_{-l/2}^{l/2}\frac{\partial (n_e u_e)}{\partial x}\mathrm{d}x = 2\int_0^{l/2}\mathrm{d}\Gamma = 2\Gamma_{\mathrm{wall}}.$$

So the global particle balance is in effect

$$\frac{\mathrm{d}\bar{n}_e}{\mathrm{d}t} = \bar{n}_e n_g K_{\mathrm{iz}} - \frac{2\Gamma_{\mathrm{wall}}}{l}. \tag{2.40}$$

One can easily generalize this global particle balance equation to a three-dimensional vessel of volume V and total surface area A by replacing the half scale length $l/2$ with V/A (for the slab in Figure 2.5 taking a wall section of area $A_{\text{sect}} \gg l^2$ defines a volume lA_{sect} with surface area $\sim 2A_{\text{sect}}$). In Chapter 3 simple expressions will be derived for the flux at the wall as a function of the electron temperature and the electron density at the reactor centre.

> **Q** Confirm with reference to Eq. (2.27) that the particle balance Eq. (2.40) includes a strong dependence on electron temperature and hence that in the steady state ($\text{d}/\text{d}t = 0$) the electron temperature is linked to the gas pressure and the system dimensions.
>
> **A** Eq. (2.27) shows K_{iz} to be of Arrhenius (thermally activated) form. In the steady state
>
> $$K_{\text{iz}}(T_e) = \frac{\Gamma_{\text{wall}}}{\overline{n}_e} \left(\frac{1}{n_g} \frac{A}{V} \right). \qquad (2.41)$$
>
> The LHS is an exponential function of T_e. On the RHS, the wall flux is likely to be proportional to the mean electron density, so the electron density dependence vanishes; however, the gas density n_g is proportional to gas pressure. Hence the electron temperature is linked to the gas pressure and the system size.

In the case of an electronegative plasma, one would also need to consider attachment and detachment processes that capture and liberate electrons in the volume.

2.3.2 Power balance

In general in electrically sustained discharges, electrical energy is almost entirely coupled to the electrons so that consideration of the ion contribution to power absorption is not necessary to determine the discharge parameters. A global energy balance for an electrically sustained plasma can therefore be obtained by equating the power absorbed by the electrons, P_{abs}, to the rate of energy loss associated with the average lifecycle of electrons, P_{loss}, and any other loss processes.

The absorbed power term, P_{abs}, depends on the distribution of the electric field and the current density and thus on the system configuration, i.e., the reactor type. The electric field must be calculated by solving Maxwell's equation simultaneously with the fluid motion equations. This is a complicated problem and it will be shown in the following chapters how it can be simplified by using an electrical equivalent circuit model.

The rate of energy loss from the electron population, P_{loss}, is essentially independent of the system configuration. There are two ways for the electrons to dissipate the energy: (i) by undergoing collisions with the gas, transferring energy inelastically into ionization and excitation, and elastically into the thermal energy of the gas; (ii) by carrying kinetic energy to the boundaries. In the simple case of a noble gas plasma, the first contribution can be expressed as

$$P_{\text{loss,coll}} = \bar{n}_e n_g \left[K_{iz}\varepsilon_{iz} + K_{exc}\varepsilon_{exc} + \frac{3m}{M} K_{el} kT_e \right], \quad (2.42)$$

where ε_{iz} and ε_{exc} are expressed in joules to give $P_{\text{loss,coll}}$ in W m^{-3}.

> **Q** Show that the elastic contribution is based on Eqs (2.13), (2.23) and (2.31).
> **A** The loss is proportional to the number densities of the two reactants, the rate coefficient of the collision reaction (K_{el}) and the mean energy loss per reaction. From these equations, taking the maximum energy transfer fraction and the mean particle (electron) energy, the loss through elastic collisions is therefore
>
> $$\bar{n}_e \times n_g \times K_{el} \times \frac{2m}{M} \times \frac{3}{2}kT_e.$$

There are many other ways to lose energy in the case of molecular gases, for instance by dissociation of the feedstock gas or vibrational excitation; in such cases these should be included in (2.42). The second class of contribution is the kinetic energy carried to the boundaries, and lost in the boundary electrostatic field or at the wall. This, from Eq. (2.19), is written

$$P_{\text{loss,bound.}} = (2kT_e + e\Delta\phi)\Gamma_{\text{wall}} \frac{A}{V}, \quad (2.43)$$

where A is the total surface area of the boundaries and $e\Delta\phi$ is the voltage drop in the boundary sheaths. The total loss power per unit volume is therefore

$$P_{\text{loss}} = P_{\text{loss,coll}} + P_{\text{loss,bound.}} \quad (2.44)$$

Taking the case of a noble gas plasma in the steady state (d/dt = 0), Eq. (2.40) simplifies the electron power loss to

$$P_{\text{loss}} = \varepsilon_T(T_e) \Gamma_{\text{wall}} \frac{A}{V} \quad (2.45)$$

with

$$\varepsilon_T(T_e) = \varepsilon_{iz} + \frac{K_{exc}}{K_{iz}}\varepsilon_{exc} + \frac{3m}{M}\frac{K_{el}}{K_{iz}}kT_e + 2kT_e + e\Delta\phi \quad (2.46)$$

expressed in joules. Together with the transport models of Chapter 3, the energy balance links power input to a globally averaged plasma density.

Whenever there is an imbalance between the power absorbed and the power lost, there will be a change in the mean energy of the electron population. This can be derived formally from the fluid equations, including one for the energy flux – the result is

$$\frac{d}{dt}\left(\frac{3}{2}\bar{n}_e k T_e\right) = P_{abs} - P_{loss}. \tag{2.47}$$

The absorbed power is dependent on the nature of the plasma excitation and is discussed further in later chapters.

2.4 The electrodynamic perspective

Warning: This section requires an appreciation of complex numbers and the representation of trigonometric functions in terms of exponential functions with imaginary arguments. Also, when discussing plasma waves there is often a conflict in the use of symbols. Pay careful attention to the context of each symbol and any subscripts and note that in this section:

- k_B is used for Boltzmann's constant but $k = 2\pi/\lambda$, the 'wavenumber' is the magnitude of the wave vector, **k**.
- ε_0 is the permittivity of free space and $\varepsilon_p(\omega)$ is a relative permittivity.
- δ is a characteristic distance over which disturbances diminish.
- n_{ref}, n_{real} and n_{imag} are used to denote the refractive index and its complex components. Do not confuse these quantities with a particle number density n with various subscripts like 0, e, e0, i, i0 and g.
- σ_p, σ_m are conductivities NOT cross-sections.

Fluid equations may also be used to determine the plasma electrodynamics. Plasmas can be described from the view point of conductors or dielectrics. The particle motions that result from the action of an electric field on the charged components of a plasma, especially steady or low frequency (fields), can be interpreted as currents, which might be quantified by means of a conductivity. On the other hand, the displacement of charged particles, particularly at high frequency, may lead to a polarization, and one might therefore seek to describe the response in terms of a permittivity. The two approaches are equivalent and the electrodynamics of plasmas depend on the frequency domain under consideration.

Plasmas support electromagnetic waves and electrostatic waves. In the first case, as in vacuum and in dielectric media, the waves propagate through the exchange of energy between electric and magnetic fields, while in the second case the propagation is associated with the exchange of energy between an electric field and the thermal energy density of the charged components. The latter case is characterized by the fluctuating electric field being parallel to the wave vector, which points in the direction of propagation.

The response of a plasma to the electromagnetic field is usually non-linear, which makes the analysis complicated. However, when looking at small-amplitude perturbations and neglecting all second-order and higher terms, it is possible to do a linear, harmonic analysis:

$n_e \to n_{e0} + \tilde{n}_e$, where $\tilde{n}_e \ll n_{e0}$
$u_e \to u_{e0} + \tilde{u}_e$, where $\tilde{u}_e \ll v_e$

sinusoidal variations in time are expressed as $\mathbf{Re}\left[\exp i\omega t\right]$ so that $\partial/\partial t \to i\omega$
likewise, variations in space like $\mathbf{Re}\left[\exp -ikz\right]$ are simplified by $\partial/\partial z \to -ik$.

The simultaneous linearization of the fluid equations (particle motion) and of Maxwell's equations allow the determination of the plasma permittivity (or an equivalent conductivity) and the dispersion relations of various wave modes. This is done in many textbooks and we will not reproduce the calculation here, but come back to it in Section 2.4.3. When the plasma is magnetized, the medium is anisotropic and the waves behave differently depending on their direction relative to the magnetic field; we will consider this in Chapter 8. In the absence of a magnetic field, the medium is isotropic. In the following, we give a simple approach to determine the basic RF electromagnetic properties of an unmagnetized plasma.

2.4.1 Plasma conductivity and plasma permittivity

Consider a small-amplitude electromagnetic wave in the radio-frequency domain, having an electric field in the x-direction and propagating in the z-direction in an infinite plasma. The electric field may be defined using complex numbers:

$$E_x = \mathbf{Re}\left[\tilde{E}_x \exp i(\omega t - kz)\right] = \mathbf{Re}\left[\tilde{E}_x \exp i\omega(t - n_{\text{ref}}z/c)\right], \quad (2.48)$$

where n_{ref} is the refractive index to be defined by the permittivity, c is the speed of light and k is the magnitude of the wave vector in the z-direction. Let us further assume that:

– the ions do not respond to this high-frequency perturbation (see Chapter 4),
– the electron pressure gradient is not significant, effectively ignoring electron thermal energy,
– there is no steady current in the plasma so the drift speed is zero ($u_{e0} = 0$).

There is an associated oscillating magnetic field pointing in the y-direction. The two fields must be linked by Maxwell's equations and one can use Eq. (2.48) in

2.4 The electrodynamic perspective

Faraday's law followed by integration in time to specify the magnetic field. From there the Ampère–Maxwell law links the current to the electromagnetic field. It is sufficient here, however, to simply include Maxwell's concept of a displacement current arising from time-varying electric fields together with a particle current determined from the drift term of the electron fluid equations.

Within our approximations, the electron momentum conservation equation Eq. (2.33) linearizes to

$$n_{e0} m\, i\omega \tilde{u}_x = -n_{e0} e \tilde{E}_x - n_{e0} m \nu_m \tilde{u}_x. \tag{2.49}$$

The magnitudes of the perturbations in velocity and electric field amplitude are therefore related by

$$\tilde{u}_x = \frac{-e}{m(i\omega + \nu_m)} \tilde{E}_x. \tag{2.50}$$

The net current is a combination of the displacement current arising from the time-varying electric field and the conduction current due to the electron motion:

$$\tilde{J}_x = i\omega\varepsilon_0 \tilde{E}_x + n_{e0}(-e)\tilde{u}_x = i\omega\varepsilon_0 \left(1 + \frac{n_{e0} e^2}{i\omega\varepsilon_0 m(i\omega + \nu_m)}\right) \tilde{E}_x$$

$$= i\omega\varepsilon_0 \left(1 - \frac{\omega_{pe}^2}{\omega(\omega - i\nu_m)}\right) \tilde{E}_x, \tag{2.51}$$

where $\omega_{pe} \equiv (n_{e0} e^2 / m\varepsilon_0)^{1/2}$ defines a characteristic response frequency of electrons in a plasma – the so-called electron plasma frequency (see Chapter 4).

The conduction current term can be absorbed into an effective, complex permittivity for the plasma by defining

$$\varepsilon_p(\omega) = 1 - \frac{\omega_{pe}^2}{\omega(\omega - i\nu_m)} = 1 - \frac{\omega_{pe}^2}{\omega^2 + \nu_m^2} - i\frac{\nu_m}{\omega} \frac{\omega_{pe}^2}{\omega^2 + \nu_m^2}. \tag{2.52}$$

Tracking back through the calculation it can be seen that on the RHS the '1' comes from the displacement current, whereas all other contributions relate to the local particle motion. At low frequency one can simply neglect the displacement term and then define a complex conductivity, which accounts only for conduction currents:

$$\sigma_p = \frac{n_{e0} e^2}{m(i\omega + \nu_m)}. \tag{2.53}$$

Q (i) Examine the above equations to find a condition on frequency for the use of Eq. (2.53).

(ii) Use Eq. (2.53) to define a low-frequency conductivity, σ_m.

A (i) From Eq. (2.51) one can neglect the displacement term when $|\omega(\omega - i\nu_m)| \ll \omega_{pe}^2$, that is when

$$\omega(\omega^2 + \nu_m^2)^{1/2} \ll \omega_{pe}^2.$$

This is typical of many low-pressure, RF plasmas.

(ii) In the very low frequency limit, one obtains the following conductivity:

$$\sigma_m = \frac{n_{e0}e^2}{m\nu_m}. \tag{2.54}$$

Comment: See the warning about the collision frequency ν_m immediately after Eq. (2.25).

2.4.2 Plasma skin depth

The dispersion of electromagnetic waves in a plasma is equivalent to that in dielectrics in that one can define a refractive index $n_{ref}^2 = \varepsilon_p$. Given the complex nature of the relative permittivity for a plasma, one can set $n_{ref} = n_{real} + in_{imag}$, then writing $X = \nu_m/\omega$ and comparing with Eq. (2.52) gives:

$$n_{real}^2 - n_{imag}^2 = 1 - \frac{\omega_{pe}^2}{\omega^2(1 + X^2)},$$

$$2n_{real}n_{imag} = X\frac{\omega_{pe}^2}{\omega^2(1 + X^2)}.$$

Ordinary dielectrics like glass have a real refractive index that is greater than unity.

Q Infer the consequence for wave propagation of a complex refractive index by putting $n_{ref} = n_{real} + in_{imag}$ in Eq. (2.48).

A In Eq. (2.48), when $n_{imag} > 0$, waves are damped within a typical distance $\delta = c/\omega n_{imag}$. In the case where $n_{real} = 0$, there is no propagating wave and the perturbation is 'evanescent'.

2.4 The electrodynamic perspective

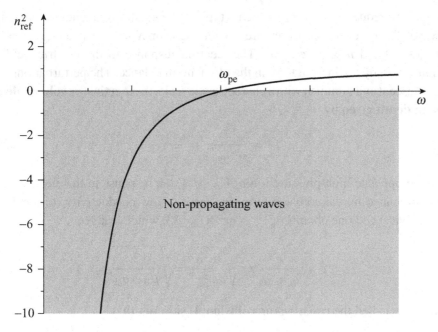

Figure 2.6 The square of the refractive index of a collisionless plasma; $n_{\text{ref}}^2 < 0$ implies an imaginary refractive index so that wave propagation is not possible and perturbations are evanescent.

There are two frequency domains that can be analysed:

1. Above the plasma frequency, $\omega > \omega_{\text{pe}}$ (typically in the GHz range), up to pressures of 100 Pa or so the collision frequency is usually very small compared to the wave frequency, so that $X \ll 1$. Then,

$$n_{\text{imag}} \simeq 0, \tag{2.55}$$

$$n_{\text{real}}^2 \simeq 1 - \frac{\omega_{\text{pe}}^2}{\omega^2}. \tag{2.56}$$

The electromagnetic waves propagate in the plasma, which behaves as a dielectric of refractive index $(1 - \omega_{\text{pe}}^2/\omega^2)^{1/2}$. The damping length is very long and the wave propagates with very weak attenuation. This frequency regime typically corresponds to microwaves with which interferometry or reflectometry can be used to determine plasma characteristics (see Chapter 10). Figure 2.6 shows the square of the refractive index. For $\omega > \omega_{\text{pe}}$ the refractive index has real values that are less than unity and thus relate to waves propagating at a phase speed greater than c, the speed of light *in vacuo* – there is no challenge to special relativity though because the energy carried by these waves travels at the group velocity, which turns out to be less than c.

2. At lower frequency, $\omega < \omega_{pe}$, which is typical of the radio-frequency domain, the waves decay. Consider a pressure low enough for $X \ll 1$. Since $\omega_{pe}^2/\omega^2 > 1$, $n_{real} \to 0$ and $n_{imag}^2 \approx \omega_{pe}^2/\omega^2$. The electron response to the electric field is dominated by the inertial term in the momentum balance. The perturbations are evanescent and diminish with a characteristic scale, the **inertial (or collisionless) skin depth** given by

$$\delta = \frac{c}{\omega n_{imag}} = \frac{c}{\omega_{pe}}. \qquad (2.57)$$

In the opposite high-pressure limit of $X \gg 1$, the response to the electric field is dominated by the collision term that leads to low conductivity (that is high resistivity) and one obtains $n_{imag}^2 \approx \omega_{pe}^2/(2\omega^2 X)$, which leads to

$$\delta = \frac{c}{\omega n_{imag}} = \sqrt{\frac{2c^2 X}{\omega_{pe}^2}} = \sqrt{\frac{2}{\mu_0 \omega \sigma_m}}, \qquad (2.58)$$

which is called the **resistive (or collisional) skin depth**.

In the present analysis it has been assumed that there is no steady magnetic field and electromagnetic waves have been shown to propagate only above the plasma frequency. It will be shown in Chapter 8 that electromagnetic waves propagate throughout the frequency spectrum in magnetized plasma.

2.4.3 Electrostatic waves

Plasmas also allow purely electrostatic disturbances to propagate through an exchange of energy between the electric field and the thermal energy density (nkT) of charged particles, with perturbations in these quantities being in the direction of the wave. The analysis will again use the linearized fluid equations for small perturbations, further simplified by the following assumptions:

– the background electron pressure gradient is negligible, though the perturbation to the electron thermal energy density plays a key rôle;
– the background electric field and gradients of particle drift speed are negligible, though the perturbation to electric field plays a key rôle.

Assuming harmonic perturbations for all quantities, e.g., $\tilde{n}_i \propto \exp[i(\omega t - kz)]$, the linearization of the particle conservation equations Eq. (2.32) leads to:

$$-i\tilde{n}_e(\omega - ku_{e0}) + in_{e0}k\tilde{u}_e = 0, \qquad (2.59)$$

$$-i\tilde{n}_i(\omega - ku_{i0}) + in_{i0}k\tilde{u}_i = 0. \qquad (2.60)$$

2.4 The electrodynamic perspective

The linearized momentum conservation equations Eq. (2.33) are

$$-imn_{e0}(\omega - ku_{e0})\tilde{u}_e = -n_{e0}q\tilde{E} - ik_BT_ek\tilde{n}_e - m\nu_m(n_{e0}\tilde{u}_e + \tilde{n}_eu_{e0}), \quad (2.61)$$

$$-iMn_{i0}(\omega - ku_{i0})\tilde{u}_i = +n_{i0}q\tilde{E} - ik_BT_ik\tilde{n}_i - M\nu_i(n_{i0}\tilde{u}_i + \tilde{n}_iu_{i0}). \quad (2.62)$$

The linearized form of Gauss's law is

$$ik\tilde{E} = \frac{q}{\varepsilon_0}(\tilde{n}_i - \tilde{n}_e). \quad (2.63)$$

It is convenient to define thermal velocities, $v_e = (k_BT_e/m)^{1/2}$ and $v_i = (k_BT_i/M)^{1/2}$, and to introduce an ion frequency analogous to ω_{pe}, namely $\omega_{pi} = (ne^2/M\varepsilon_0)^{1/2}$. Then doing the appropriate substitution and rearranging the set of equations leads to

$$[(\omega - ku_{e0})^2 - k^2v_e^2 + i\nu_m\omega - \omega_{pe}^2]\tilde{n}_e = -\omega_{pe}^2\tilde{n}_i, \quad (2.64)$$

$$[(\omega - ku_{i0})^2 - k^2v_i^2 + i\nu_i\omega - \omega_{pi}^2]\tilde{n}_i = -\omega_{pi}^2\tilde{n}_e, \quad (2.65)$$

which can be further manipulated to obtain the following dispersion relation for electrostatic (longitudinal) waves:

$$\frac{\omega_{pe}^2}{(\omega - ku_{e0})^2 - k^2v_e^2 + i\nu_m\omega} + \frac{\omega_{pi}^2}{(\omega - ku_{i0})^2 - k^2v_i^2 + i\nu_i\omega} = 1. \quad (2.66)$$

This dispersion relation has different solutions depending on the frequency domain under consideration. In most situations, one can neglect the drift velocities $u_{e0,i0}$, or consider the waves with respect to the moving fluids and then take $u_{e0,i0} = 0$. This is what we do in the following. However, counter-streaming fluids may generate instabilities (Chapter 9).

Electron plasma waves

First consider high frequencies, for which ions do not respond to the fluctuations, $\omega \gg \omega_{pi}$. Then the second term in Eq. (2.66) is negligible and the dispersion relation reads

$$\omega^2 = \omega_{pe}^2 + k^2v_e^2 - i\nu_m\omega. \quad (2.67)$$

The last term describes damping, arising in this case from elastic collisions, and may be neglected at low pressure. This relation was obtained by Bohm and Gross [240]. At large k, the electron plasma wave propagates at constant speed, v_e, and at small k they become constant-frequency waves at ω_{pe} (note that this dispersion relation has a similar form to the electromagnetic wave dispersion relation, except that the latter propagate at the speed of light at large k). The phase speed

ω/k is given by

$$v_\phi = v_e \left(1 + \frac{1}{k^2\lambda_D^2}\right)^{1/2}. \tag{2.68}$$

Ion acoustic waves

In the opposite limit of low frequencies, for which ions and electrons follow the fluctuations, one can simplify the first term by noting that $\nu_m\omega, \omega^2 \ll k^2 v_e^2$, to obtain

$$\omega^2 = \omega_{pi}^2 \left(1 + \frac{1}{k^2\lambda_D^2}\right)^{-1} + k^2 v_i^2 - i\nu_i\omega. \tag{2.69}$$

This time the wave propagates at constant speed at small k and becomes a constant-frequency wave at large k. This is more clearly seen by looking at the phase velocity, which reads

$$v_\phi = \left[\frac{k_B T_i}{M} + \frac{k_B T_e}{M}\left(\frac{1}{1+k^2\lambda_D^2}\right) - i\frac{\nu_i\omega}{k^2}\right]^{1/2}. \tag{2.70}$$

At small k and for typical conditions of low-pressure plasmas, where $T_e \gg T_i$ and $\nu_i \approx 0$, the phase velocity of the ion acoustic wave reduces to

$$v_\phi = \left[\frac{k_B T_e}{M}\right]^{1/2}. \tag{2.71}$$

This speed is also known as the Bohm speed and is of great importance in the physics of low-pressure plasmas. This speed will feature in the next chapter, when considering the criterion for the formation of non-neutral regions at plasma boundaries.

2.4.4 Ohmic heating in the plasma

The RF current density and the RF electric field in a plasma are related by a complex conductivity σ_p:

$$\tilde{J} = \sigma_p \tilde{E}. \tag{2.72}$$

It has been shown in Section 2.4.1 that when $\omega(\omega^2 + \nu_m^2)^{1/2} \ll \omega_{pe}^2$, the displacement current in the plasma can be neglected and the plasma conductivity is given by Eq. (2.53). The argument of the complex conductivity σ_p gives the phase difference between the RF current density and the RF electric field for this frequency regime:

$$\tan\theta = -\omega/\nu_m. \tag{2.73}$$

The complex exponential notation introduced above for describing sinusoidally varying quantities is useful in the analysis of linear relationships, but extra care

2.4 The electrodynamic perspective

is needed in circumstances that are non-linear, such as when dealing with power. Therefore we will now go back to using the trigonometric functions directly. The instantaneous, local, ohmic power per unit volume in a system is given by the scalar product of current density and electric field vectors. So, for example, in a simple 1-D system such as a plasma between plane parallel electrodes, passing a current $J_0 \sin \omega t$, the ohmic power dissipation per unit volume is

$$P_{v,\text{ohm}}(x, t) = J(x, t)E(x, t) \tag{2.74}$$

$$= J_0(x) \sin \omega t \, E_0(x) \sin(\omega t + \theta), \tag{2.75}$$

which includes the phase difference between current and field.

It is often convenient to express the power as a function of the RF current density, because the current density is conserved in a 1-D system and thus is independent of space. If the pressure is not too low, $\omega \ll \nu_m$ and so $\theta \approx -\omega/\nu_m$. Equation (2.75) then gives the instantaneous power dissipation as

$$P_{v,\text{ohm}}(x, t) = \frac{J_0^2}{\sigma_m(x)} \left(\frac{1 - \cos 2\omega t}{2} - \frac{\omega}{\nu_m} \sin \omega t \cos \omega t \right). \tag{2.76}$$

If the current density varies sinusoidally, at a frequency well below the electron plasma frequency, between $\pm J_0$, the time-averaged power per unit volume is

$$\overline{P}_{v,\text{ohm}}(x) = \frac{J_0^2}{2\sigma_m(x)}. \tag{2.77}$$

To compare this with a steady DC current density J_0, set $\omega t = \pi/2$, whereupon

$$P_{v,\text{ohm}}(x) = \frac{J_0^2}{\sigma_m(x)}. \tag{2.78}$$

Q Show that if there are no collisions the average power dissipation is zero.
A It can be seen from Eq. (2.73) that if there are no collisions ($\nu_m = 0$), the phase difference between the RF current and the RF electric field is $\pi/2$. Then integrating the instantaneous power, Eq. (2.74) with $\phi = \pi/2$, over one RF period gives zero, showing that there is no net energy dissipated over one RF cycle.

As ν_m/ω increases the magnitude of the phase difference decreases from $\pi/2$, which leads to power dissipation. This power dissipation transfers energy from the RF field into the electron population and in this book we will name this phenomenon 'ohmic heating' or 'collisional heating'. In low-pressure RF plasmas additional heating mechanisms exist that are 'collisionless'.

2.4.5 Plasma impedance and equivalent circuit

The results of the previous section were obtained using the current density and the electric field while treating the plasma as a conductor with complex conductivity. Exactly the same result is found if instead of Eq. (2.72), we link the current density and the electric field by a complex relative permittivity as in Eq. (2.51). In this section, the plasma will be treated as a dielectric and the whole plasma behaviour will be lumped into a single element carrying the total RF current.

The current through a medium is related to the voltage across it by its impedance. Since RF plasmas are sustained by the application of voltages in order to drive currents through a gas, it is useful to discuss the impedance of a slab of plasma in these terms. Consider first the RF impedance, Z, of a slab of dielectric material, of area A and thickness d:

$$\frac{1}{Z} = i\omega C = i\omega \frac{\varepsilon_0 \varepsilon_r A}{d}, \qquad (2.79)$$

where ε_r is the relative permittivity for the material.

Q If the dielectric material is a plasma, how could that be accommodated in Eq. (2.79)?
A If the material is a plasma its relative permittivity $\varepsilon_r \to \varepsilon_p$, Eq. (2.52).

For a slab of plasma therefore

$$\frac{1}{Z_p} = i\omega \frac{\varepsilon_0 \varepsilon_p A}{d} = i\omega \varepsilon_0 \left(1 - \frac{\omega_{pe}^2}{\omega(\omega - i\nu_m)}\right) \frac{A}{d}. \qquad (2.80)$$

Note that ε_p is a local quantity, so if Z_p is to be a global quantity this expression should be based on spatially averaged quantities. Equation (2.80) can be recast into a combination of capacitance, inductance and resistance in the following way:

$$\frac{1}{Z_p} = i\omega C_0 + \frac{1}{i\omega L_p + R_p}, \qquad (2.81)$$

where we have introduced the vacuum capacitance of the slab geometry:

$$C_0 = \frac{\varepsilon_0 A}{d}.$$

The inductance of the plasma slab, which results from the electron inertia, is

$$L_p = \frac{d}{\omega_{pe}^2 \varepsilon_0 A} = \frac{m}{ne^2} \frac{d}{A}.$$

The resistance of the plasma slab, which results from the elastic electron–neutral collisions, is

$$R_p = \nu_m L_p = \frac{m\nu_m}{ne^2} \frac{d}{A}.$$

The electrical circuit equivalent to the plasma slab is therefore composed of a capacitance in parallel with a resistance and an inductance in series. The capacitance accounts for the displacement current, which as we have seen, is generally negligible if $\omega \ll \omega_{pe}$. The equivalent circuit then reduces to a resistance and an inductance in series.

The voltage that develops across the plasma as a result of the flow of the RF current, of complex amplitude $\tilde{I}_{RF}$, is $\tilde{V}_p = Z_p \tilde{I}_{RF}$. Since Z_p is complex, there is a phase shift between current and voltage. The temporal response is linear and there are neither harmonics (multiples of ω) nor DC voltages generated across the plasma. However, note that the values of these components are functions of the plasma density, and in turn they must be functions of the amplitude of the current (or voltage, or power) in the system. The resistance and the inductance are therefore non-linear components: their values depend on the amplitude of the signals across them (this is not the usual situation in electronic circuits).

2.5 Review of Chapter 2

This chapter has covered a range of basic topics relating to plasma dynamics and equilibrium. The microscopic view of plasmas considers the constituent particles in terms of their respective time-dependent, velocity distributions in a six-dimensional phase space. Moments of the velocity distribution provide macroscopic variables such as density, fluid speed and energy. When the distribution is Maxwellian, the mean thermal speed and the mean thermal energy of a particle are

$$\bar{v} = \left(\frac{8kT}{\pi m}\right)^{1/2} \quad \text{and} \quad \bar{\varepsilon} = \frac{3}{2}kT.$$

A further result from the kinetic model is that within a Maxwellian distribution, although the net drift of particles is zero, the random thermal flux in any particular direction is

$$\Gamma_{random} = \frac{n\bar{v}}{4}.$$

In many discharge plasmas the charged particles are heavily outnumbered by the neutral gas species, so that by far the most frequent collisions of the charged particles are with gas atoms rather than with other charged particles. The most frequent collisions are elastic. Collision processes are described in terms of a mean gas-dependent cross-section ($\bar{\sigma}$) or a mean free path (λ) or a collision frequency (ν). For particles with a mean speed $\bar{v}$ in a gas of number density n_g,

$$\lambda = \frac{1}{n_g \bar{\sigma}} \quad \text{and} \quad \nu = n_g \bar{\sigma} \bar{v}.$$

Excitation and ionization collisions require a minimum, or threshold, energy (ε_{th}) to initiate the interaction. These processes then appear to be thermally activated at a rate $K \equiv <\sigma v>$ that can be expressed in the form

$$K = K_0 \exp(-e\varepsilon_{th}/kT).$$

In molecular gases the interactions of charged particles include dissociation of molecules into charged and uncharged radicals.

Integrations over the particle distributions lead to the macroscopic quantities that can be used to describe the components of plasmas in terms of fluid variables such as density (n), mean speed (u) and pressure (p). Particle mean energy is linked to the fluid temperature as above. In many situations an isothermal assumption is made, implying that the gradients in fluid temperature are weak and that heat is able to be exchanged with the fluid to maintain it at a steady temperature. When this is not the case the system should be considered to be adiabatic.

In a steady-state discharge in a low-pressure atomic gas, ionization within the volume is the dominant source of charged particles whereas recombination on bounding surfaces is the dominant loss mechanism. Under such circumstances the electron temperature of a self-sustaining plasma is controlled by a global balance between volume ionization and ion outflow at the boundaries, constrained by the gas pressure and the system size:

$$K_{iz}(T_e) = \frac{\Gamma_{wall}}{\bar{n}_e} \left(\frac{1}{n_g} \frac{A}{V} \right).$$

The input power to a self-sustaining plasma must be balanced by the power lost from the system. The power loss proceeds in proportion to the charge outflow at the boundaries and can be tracked through the lifecycle of electrons. Each electron lost from the system takes with it (i) the effective energy expended on ionization (which includes collateral losses associated with excitation), (ii) energy transferred to the neutral gas through collisions, (iii) thermal energy and potential energy deposited at the plasma boundary:

$$P_{loss} = \left[\varepsilon_{iz} + \frac{K_{exc}}{K_{iz}} \varepsilon_{exc} + \frac{3m}{M} \frac{K_{el}}{K_{iz}} kT_e + 2kT_e + e\Delta\phi \right] \Gamma_{wall} \frac{A}{V}.$$

2.5 Review of Chapter 2

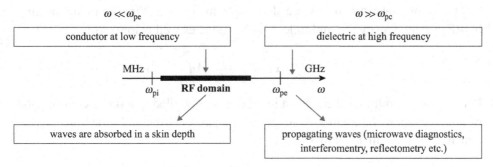

Figure 2.7 Electromagnetic properties in the different frequency ranges.

Plasmas can be viewed as being either dielectrics with a complex, frequency-dependent, relative permittivity

$$\varepsilon_p = 1 - \frac{\omega_{pe}^2}{\omega(\omega - i\nu_m)} = 1 - \frac{\omega_{pe}^2}{\omega^2 + \nu_m^2} - i\frac{\nu_m}{\omega}\frac{\omega_{pe}^2}{\omega^2 + \nu_m^2}$$

or as conductors with a complex, frequency-dependent, conductivity

$$\sigma_p = \frac{n_{e0}e^2}{m(i\omega + \nu_m)}.$$

The former naturally lends itself to high-frequency considerations while the latter suits the lower-frequency regime.

Electromagnetic waves encountering unmagnetized plasmas cannot penetrate deep into them if the wave frequency is below ω_{pe}. This will be the regime of interest for inductive discharges presented in Chapter 7. If the plasma is collisionless, the penetration is limited to the inertial skin depth

$$\delta = \frac{c}{\omega_{pe}}.$$

If the plasma is collisional, the penetration is limited to the resistive skin depth

$$\delta = \sqrt{\frac{2}{\mu_0 \omega \sigma_m}}.$$

At higher frequency, $\omega \gg \omega_{pe}$, the waves propagate and are often used as diagnostics. The diagram in Figure 2.7 summarizes the above.

Electrostatic waves propagate in plasmas but only at frequencies above ω_{pe} or below ω_{pi}. The high-frequency mode corresponds to electron plasma waves that travel with a phase speed of $(k_B T_e/m)^{1/2}$ for short wavelengths. The phase speed of long-wavelength ion acoustic waves is $(k_B T_e/M)^{1/2}$.

The mean power dissipated in a slab of plasma ($A \times d$) through ohmic heating by RF current density of amplitude J_0 at frequencies well below ω_{pe} is

$$\overline{P}_{\text{ohm}}(x) = \frac{J_0^2}{2\sigma_m(x)} Ad.$$

For the same conditions the plasma itself can be modelled as a series combination of an inductance (L_p), which accounts for electron inertia, and a resistance (R_p), which includes collisional dissipation, where

$$L_p = \frac{m}{ne^2}\frac{d}{A}, \tag{2.82}$$

$$R_p = \frac{m\nu_m}{ne^2}\frac{d}{A}. \tag{2.83}$$

3

Bounded plasma

In the previous chapter fundamental equations were established that govern the properties of low-pressure plasmas. Elementary processes such as collisions and reactions were described, and fundamental electrodynamic quantities such as the plasma conductivity and the plasma permittivity were derived. These concepts were mostly considered in the context of an infinite plasma or else were viewed as part of a global system without reference to the internal structure of the plasma volume.

Laboratory plasmas are confined. The consequence of the presence of boundaries on the structure of an electrical discharge through an *electropositive* gas will be discussed in this chapter. The basic idea to keep in mind in the discussion is that in this case charged particles are predominantly produced in the plasma volume and lost at the reactor walls. This was the basis of the global balances in the previous chapter. Conditions in the central volume may differ to some extent from those near the edge. Close to the walls a boundary layer spontaneously forms to match the ionized gaseous plasma to the solid walls; whether insulators or conductors, the walls have a major influence.

Figure 3.1 is a picture of a discharge generated between two parallel electrodes by a 13.56 MHz power supply. The discharge appears to be stratified, with regions of different properties. Light is emitted from the central region, with evidence of internal structure particularly away from the main vertical axis. There is relatively little emission from the boundary layers in front of the upper and lower electrodes. Since the emission comes from the relaxation of excited states produced originally by electron–neutral inelastic collisions, the weakness of emission is a clear sign of the markedly reduced electron density in these regions.

The purpose of this chapter is to study the stratification, shown schematically in Figure 3.2. The discussion will focus separately on two apparently distinct regions: (i) the boundary layers at the walls, which turn out to be sheaths of space charge and (ii) the plasma itself, where the net space charge is almost zero – in fact, the plasma region is usually said to be where *quasi-neutrality* prevails. When considering a

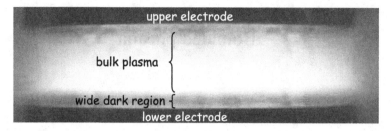

Figure 3.1 Side view of a plasma bounded by parallel-plate electrodes, showing the dark space in front of the lower electrode – note that the camera was set to view exactly along the lower electrode and does not therefore capture a clear image of the dark space adjacent to the upper electrode.

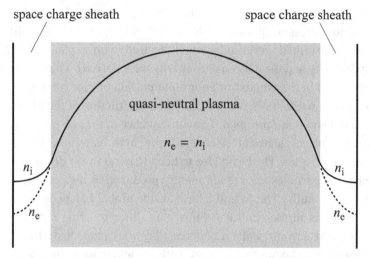

Figure 3.2 Schematic representation of a bounded plasma. The quasi-neutral plasma is separated from the walls by space charge sheaths. It is convenient to show the electrodes in a vertical orientation – in practice, gravity is unimportant and the orientation has no impact on the structure of the plasma, or of the sheaths.

DC discharge, the structure of the sheaths does not vary in time, whereas in the RF domain, the sheaths expand and contract during an RF cycle. However, in the next chapter it is shown that the fundamental properties of DC sheaths, with some modifications, apply also to RF sheaths. One of the major differences lies in the fact that under certain conditions an RF sheath makes a significant contribution to the transfer of power from the supply into the electron population of the plasma.

This chapter first explains the necessity for space charge sheaths and describes their properties as the voltage difference between a plasma and a solid boundary increases. Then the transition from the sheath to the plasma is investigated, arriving at the so-called Bohm criterion for sheath formation. Next, the three main theories of transport within the plasma region are established for the various regimes of pressure.

Plasmas of industrial interest are often more complicated than the simple, low-density, electropositive, atomic case. The sheath and transport theories are revisited in a later chapter to encompass electronegative plasmas and high-density plasmas, which both profoundly modify the transport.

> **Q** The analysis of sheaths and the plasma transport will be based on the *fluid* equations with *isothermal* electrons, and *cold* ions. Explain the meaning of the terms in italics.
>
> **A** The isothermal assumption means that temperature gradients will be ignored (see Section 2.2.2); fluid equations are based on quantities like density and drift speed, derived by averaging over particle distributions. Since $T_i \ll T_e$, thermal effects are negligible for ions – the so-called cold ion approximation.

The two fundamental equations discussed for the particle transport will be Eqs (2.32) and (2.33). When studying the sheath region, the fluid equations will be coupled to Gauss's law (in terms of the scalar electric field E) or Poisson's equation (in terms of the electrostatic potential ϕ). Finally, most of the time, one-dimensional calculations will be presented to avoid mathematical complexity and so to concentrate on the physics.

3.1 The space charge sheath region

Suppose that an object is inserted in an electropositive plasma, and that this object is not electrically connected to ground (a piece of a dielectric or a floating probe). Initially, it will collect electrons and positive ions and the corresponding current densities will be, according to (2.15),

$$J_e = -e\Gamma_e = -\frac{1}{4}en_e\bar{v}_e = -en_e\sqrt{\frac{kT_e}{2\pi m}}, \qquad (3.1)$$

$$J_i = e\Gamma_i = \frac{1}{4}en_i\bar{v}_i = en_i\sqrt{\frac{kT_i}{2\pi M}}. \qquad (3.2)$$

Since $m \ll M$ and, as already mentioned, $T_e \gg T_i$, so $J_e \gg J_i$, the object would quickly accumulate negative charge, acquiring a negative potential. It follows that electrons would then start to be repelled, reducing the electron flux, while positive ions would be accelerated towards the object. A steady state would be reached when the potential of the object is sufficiently negative for the electron flux to exactly balance that of the positive ions. Such a potential is called the *DC floating potential*;

an expression will be derived for it later in this chapter. Note that the floating potential is necessarily more negative than the potential of the plasma because $T_e \gg T_i$ and $m \ll M$. However, if the object were conducting, it could be biased by a voltage source and held at any potential with respect to the plasma, provided an appropriate current from the source were circulating through the plasma.

3.1.1 Boltzmann equilibrium and Debye length

Plasmas are DC conductors and so it is difficult to sustain large electric fields inside them. Large fields associated with charge on boundary surfaces will be localized to a narrow boundary layer, known as a space charge sheath. It will now be shown that there is a natural scale length for this space charge sheath. Suppose that the plasma adjacent to the boundary is sustained by some form of ionization. It simplifies matters if details of the process of ionization are not included here. This is equivalent to saying that the region of interest is small compared with the mean free path for ionization $(1/n_g \sigma_{iz})$ – that is something to check afterwards.

Q It is also convenient to ignore elastic collisions in the space charge region – what constraint does this put on the mean free path for elastic collisions?
A If collisions are unimportant to the modelling of the sheath region it must be narrower than the mean free path for elastic collisions. That is a more stringent condition because elastic collisions far outnumber ionization events (i.e., the mean free path between elastic collisions is much shorter than that between ionizing collisions), so this is definitely something to check afterwards.

Consider an infinite plate in the y and z-directions, in contact with an electropositive plasma having the same ion and electron density $n_{i0} = n_{e0}$. The potential of the plasma is arbitrarily set at zero and the plate is biased at a potential slightly negative with respect to this. The situation is shown schematically on the left-hand side of Figure 3.3. Since the plate has a negative surface charge and a negative potential, electrons must be repelled to some degree and their density will be reduced close to the plate.

Ions will be attracted by the negative surface charge on the plate. The electric force tends to accelerate them and thus the ion fluid speed will tend to increase on approaching the plate. According to the steady-state ion continuity equation with no local ionization, the conservation of flux requires the density to decrease as the speed increases. In the situation shown on the left of Figure 3.3, the surface charge is so small that the decrease in ion density is not yet significant – at least for the sake of the argument.

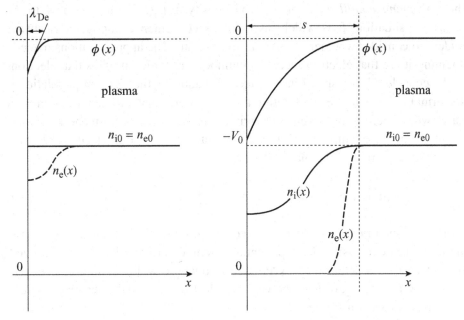

Figure 3.3 A small potential perturbation at the edge of a plasma defining the scale length (left) and a larger potential drop across a Child–Langmuir sheath (right).

Electrons, unlike ions, are very mobile because of their relatively small mass, and they will be repelled by the electric force (the electric field is directed towards the plate), with a density depletion similar to that illustrated in Figure 3.3 (left). Given that the electron mass is very small, and that collisions have been presumed to be unimportant, the inertial terms in the steady-state momentum conservation equation Eq. (2.33) are often swamped by the electric field and pressure gradient terms. The electric field is related to the gradient of potential: in one dimension, $E = -\mathrm{d}\phi/\mathrm{d}x$.

Q If the electron pressure term balances the electric force, show that the electron density establishes an exponential relationship with the local potential.

A The proposed equilibrium is between pressure gradient and electric field terms

$$kT_\mathrm{e}\frac{\mathrm{d}n_\mathrm{e}}{\mathrm{d}x} = -n_\mathrm{e}eE = n_\mathrm{e}e\frac{\mathrm{d}\phi}{\mathrm{d}x},$$

which integrates to the following important, exponential relationship:

$$n_\mathrm{e}(x) = n_{\mathrm{e}0}\exp\left(\frac{e\phi(x)}{kT_\mathrm{e}}\right). \qquad (3.3)$$

The *Boltzmann equilibrium* for electron density, Eq. (3.3), applies only to the isothermal situation. It is widely used for its convenience, though it is always prudent to consider whether or not it is appropriate. The approximations made in obtaining it are that electron inertia is unimportant (which implies that electrons can be considered to respond instantaneously) and that the electron population is isothermal (which implies that within the system of interest there are no times or places where there is any significant variation in the mean electron energy).

Within the approximation of a small potential perturbation ($|e\phi| \ll kT_e$), the space charge in front of the plate is therefore

$$e(n_i - n_e) = en_{e0}\left[1 - \exp\left(\frac{e\phi}{kT_e}\right)\right] \simeq -\frac{e^2 n_{e0}\, \phi(x)}{kT_e}, \qquad (3.4)$$

where quasi-neutrality has been invoked at the edge of the plasma ($n_{i0} = n_{e0}$) and in the last expression the exponential potential variation has been linearized. The potential ϕ is determined by Poisson's equation, which is a combination of Gauss's law and the equivalence between the electric field and the negative potential gradient, so that in one dimension

$$\frac{(n_i - n_e)e}{\varepsilon_0} = \frac{dE}{dx} = -\frac{d^2\phi}{dx^2}. \qquad (3.5)$$

Poisson's equation relates space charge to potential. Using Eq. (3.4),

$$\frac{d^2\phi}{dx^2} = \frac{e^2 n_{e0}\, \phi}{\varepsilon_0 kT_e}.$$

The appropriate solution of this linearized equation should satisfy $\phi = 0$ when $x \to \infty$, so

$$\phi(x) = \phi_0 \exp\left(-\frac{x}{\lambda_{\mathrm{De}}}\right), \qquad (3.6)$$

where

$$\lambda_{\mathrm{De}} = \sqrt{\frac{\varepsilon_0 kT_e}{n_{e0} e^2}} \qquad (3.7)$$

is the scale length of the space charge region, commonly known as the Debye length. The small negative potential perturbation imposed at the plate relaxes exponentially within a typical distance λ_{De} as the free charges in the plasma become distributed so as to screen the electrostatic potential.

Exercise 3.1: Debye length Calculate the Debye length for a plasma in which the electron density is $n_{e0} = 1.0 \times 10^{16}$ m^{-3} and $kT_e/e = 2.0$ V.

3.1 The space charge sheath region

Note that Eq. (3.7) may also be written

$$\lambda_{De} = \frac{v_e}{\omega_{pe}}, \quad (3.8)$$

where $v_e = (kT_e/m_e)^{1/2}$ is the electron thermal speed (not the mean speed $\bar{v}_e$) and ω_{pe} is the electron plasma frequency. In the previous chapter, it was shown that electromagnetic waves in a non-magnetized plasma diminish with a characteristic skin depth δ. Here, electrostatic fields are shown to be screened with a characteristic length λ_{De}.

Q Show that $\lambda_{De}/\delta = v_e/c$ and hence deduce that electromagnetic waves penetrate further into plasmas than electrostatic perturbations.

A Since $\delta = c/\omega_{pe}$ (from Eq. (2.57)), substituting $\omega_{pe} = c/\delta$ in Eq. (3.8) gives the required result. Since the electron thermal speed is much smaller than the speed of light, $\lambda_{De} \ll \delta$ and so electrostatic perturbations penetrate less far than electromagnetic waves.

3.1.2 The ion matrix model

In the preceding analysis it was assumed that $|e\phi| \ll kT_e$ and that the ion density remained unchanged from that in the plasma. The constraint on the potential is now lifted but the constant ion density will be retained, so that throughout the sheath the ions provide a uniform space charge, as if there were no acceleration of the ion fluid. This is a convenient approximation that leads to a simple analysis; in fact, a stationary matrix of ions might arise transiently if somehow the electrons were swept away by the rapid application of a negative potential on the wall in Figure 3.3 or if collisions were so frequent that ions were not accelerated in the sheath.

First consider what happens to the space charge of electrons when the local potential becomes large and negative: $e\phi \ll -kT_e$. The electrons will be strongly repelled.

Q According to the Boltzmann relation, what value of $e\phi/kT_e$ would reduce the electron density to 1% of its initial value?

A Using Eq. (3.3), the problem amounts to solving $\exp(e\phi(x)/kT_e) = 0.01$. Taking natural logarithms: $e\phi(x)/kT_e = \ln(0.01) = -4.6$.

Comment: In laboratory plasmas the electron temperature is typically in the range 1–5 eV, so a retarding potential of a few volts can substantially reduce the electron density.

So if $e\phi \ll -kT_e$, the electron space charge in the sheath can be completely neglected. The potential in the sheath must then satisfy Poisson's equation Eq. (3.5) with constant ion density and zero electron density:

$$\frac{d^2\phi}{dx^2} = -\frac{e\,n_{i0}}{\varepsilon_0}.$$

This can easily be integrated twice from the wall where $x = 0$ through the ion space charge to the boundary with the plasma:

$$\phi(x) = -\frac{e\,n_{i0}}{\varepsilon_0}\left(\frac{x^2}{2} + C_1 x + C_2\right).$$

Two boundary conditions must be supplied to determine the constants C_1 and C_2. Since the plasma is a conductor it is reasonable from a sheath point of view to set the electric field, $-d\phi/dx$, to zero at the boundary with the plasma $x = s$: that requires $C_1 = -s$. The second condition is simply that the potential at $x = s$ is zero; that is, the plasma boundary is taken as the reference for the potential. That requires $C_2 = s^2/2$, whereupon

$$\phi(x) = -\frac{e\,n_{i0}}{2\varepsilon_0}(x-s)^2. \tag{3.9}$$

This electron-free 'ion matrix model' is the simplest model of a space charge sheath. It has two major shortcomings, as its description implies – electrons are excluded from the model and the ions do not 'flow' through the sheath, accelerated by the sheath field. Nevertheless, it gives an initial means of estimating the size of space charge sheaths. If the plate potential $\phi(0) = -V_0$ with respect to the plasma, then

$$V_0 = \frac{e\,n_{i0}}{2\varepsilon_0}s^2; \tag{3.10}$$

note that V_0 is the magnitude of the potential across the sheath and that the potential has been defined so that in the sheath $\phi < 0$. Since the net space charge would be lower with both accelerated ion flow and electrons, this model underestimates the sheath thickness for a given sheath potential.

Exercise 3.2: Ion matrix model Divide both sides of Eq. (3.10) by kT_e/e and rearrange it to show that an ion matrix sheath with 200 V across it, at the edge of a plasma with electron temperature 2 eV, will be about 14 Debye lengths thick (at the plasma boundary $n_{i0} = n_{e0}$).

3.1.3 The Child–Langmuir law

In this section ion flow will be included in the sheath model, but electron space charge will again be neglected on the grounds that the model will be restricted to regions of large negative potential: $e\phi \ll -kT_e$. To obtain the ion density variation as a function of the potential variation, simultaneously continuity and momentum equations for the ions, Eqs (2.32) and (2.33), must be considered.

In the interest of simplicity, the ion fluid will be presumed to be cold so until further notice, $T_i \to 0$ and at any point in space all ions move at the fluid speed $u(x)$ (the so-called mono-energetic ion assumption). Ions will also be presumed to be singly charged.

The low-pressure (collisionless) case

Consider first the low-pressure limit in which the friction force is negligible ($Mu\nu_m \ll eE$). In the steady state, the electric force acting on the cold ion fluid is balanced by the remaining inertial term. In one dimension:

$$n_i M u \frac{du}{dx} = n_i e E.$$

Q Combine this ion momentum equation with $E = -d\phi/dx$ to confirm that the ion energy is conserved.

A The combined equation is

$$Mu \frac{du}{dx} = -e \frac{d\phi}{dx},$$

which integrates to

$$\left(\frac{1}{2}Mu^2 + e\phi\right) = \text{constant}.$$

So the total ion energy (kinetic plus potential) is conserved.

If the ions are at rest where the potential is zero, it follows that

$$\frac{1}{2}Mu(x)^2 + e\phi(x) = 0, \qquad (3.11)$$

and that $e\phi \leq 0$. To this can be added the steady-state ion continuity equation, which balances the divergence of the ion flux with production and loss of ions, but it has already been stated that ionization in the sheath region can usually be discounted and recombination is predominantly a surface process. That means the continuity equation simply requires that the ion flux has no divergence and therefore remains constant. In terms of the ion current density,

$$J_i = en_i(x)u(x). \tag{3.12}$$

Note that although it is more consistent than the ion matrix model, there is a logical inconsistency in this model, since Eq. (3.11) gives the ion speed on entering the sheath ($\phi = 0$) as zero, yet Eq. (3.12) has the speed–density product non-zero throughout, which requires infinite density at the plasma boundary. Fortunately, there are no serious implications for the calculation developed here because ions are accelerated to very large speeds in the sheath. Later, a more careful consideration will show that ions do not in fact enter the space charge region with zero speed, nor even with the ion thermal speed.

The combination of Eqs (3.11) and (3.12) allows the positive ion density to be expressed as a function of the potential,

$$n_i(x) = \frac{J_i}{e}\left(-\frac{2e\phi(x)}{M}\right)^{-1/2}. \tag{3.13}$$

Substituting this density into Poisson's equation, Eq. (3.5), (with $n_e = 0$) leads to the following differential equation for the potential:

$$\frac{d^2\phi}{dx^2} = -\frac{J_i}{\varepsilon_0}\left(-\frac{2e\phi(x)}{M}\right)^{-1/2}. \tag{3.14}$$

Note that since

$$\frac{d}{dx}\left(\frac{d\phi}{dx}\right)^2 = 2\frac{d\phi}{dx}\frac{d^2\phi}{dx^2},$$

Eq. (3.14) can be changed into a form that can be formally integrated by multiplying first by $2d\phi/dx$ and then integrating from a general point $x = x_1$ up to the plasma boundary at $x = s$. It is convenient to use the notation ϕ' to represent the derivative of $\phi(x)$ with respect to x:

$$(\phi'(s))^2 - (\phi'(x_1))^2 = 4\frac{J_i}{\varepsilon_0}\left(\frac{2e}{M}\right)^{-1/2}\left[(-\phi(s))^{1/2} - (-\phi(x_1))^{1/2}\right]. \tag{3.15}$$

At the plasma boundary ($x_1 = s$) two conditions can be set as for the ion matrix model, namely that the field and the potential are both zero: $\phi(s) = \phi'(s) = 0$. Using these conditions, one can integrate again from $x_1 = 0$, where again $\phi(0) = -V_0$ is the potential at the surface, to $x_1 = s$ where the potential has been defined to be zero; this gives

$$V_0^{3/4} = \frac{3}{2}\left(\frac{J_i}{\varepsilon_0}\right)^{1/2}\left(\frac{2e}{M}\right)^{-1/4} s. \tag{3.16}$$

3.1 The space charge sheath region

This relation is known as the Child–Langmuir law. It was first derived as the voltage–current relationship for a fixed space occupied by electron space charge in a thermionic diode [32,33]. In the present case the sheath size s is not independently fixed but is related to the magnitude of the potential between the plasma boundary and the surface, V_0, and to the ion flux crossing the sheath J_i/e (which will be determined in Section 3.2). Equation (3.16) shows that for a given positive ion current extracted from the plasma, the sheath thickness increases as the 3/4 power of the voltage applied to an electrode.

Exercise 3.3: Child–Langmuir model Rearrange Eq. (3.16) to show that, according to the low-pressure Child–Langmuir model, a sheath with 200 V across it at the edge of a plasma with electron temperature 2 eV will be about 25 Debye lengths thick if the ion current density is $J_i = n_{e0}e\sqrt{kT_e/M}$.

The high-pressure (fully collisional) case

In the low-pressure case the energy of the ion fluid within the sheath is not transferred to the gas, so the conservation of energy could be applied to the ion motion. This is not always a good approximation. Taking the example of argon gas, the ion–neutral mean free path is about $\lambda_i \approx 4$ mm at a pressure $P \approx 1$ Pa, from Eq. (2.30). The sheath size may easily be greater than λ_i, even at relatively low pressure. In this situation, energy conservation is not useful in determining the ion motion. Including the collision term in the cold ion momentum equation Eq. (2.33) leads to a high-pressure limit in which the electric force is entirely balanced by collisions, that is the friction force. In the one-dimensional steady-state case it reads

$$n_i M u \nu_i = e n_i E, \qquad (3.17)$$

where the pressure term is neglected because ions are cold. It will be assumed that the mobility is constant on the basis of the ion collision frequency being fixed by ions moving slowly between collisions at their thermal speed $\bar{v}_i \gg u$, so that

$$\nu_i = \bar{v}_i/\lambda_i. \qquad (3.18)$$

The ion fluid drift speed is then proportional to the electric field

$$u = \frac{e}{M\nu_i}E = \mu_i E, \qquad (3.19)$$

where $\mu_i \equiv e/(M\nu_i)$ is called the ion mobility. Even though there are collisions, it is still assumed that ionization and recombination do not occur in the sheath so the ion flux remains constant and $J_i = en_i u$ throughout. Poisson's equation can now be written for the high-pressure limit:

$$\phi''(x) = -\frac{J_i}{\varepsilon_0 \mu_i \phi'(x)}, \tag{3.20}$$

where $\phi''(x)$ is the second derivative of $\phi(x)$ with respect to x. Note that as before there is an inconsistency in this model since at the plasma boundary, where the electric field (and hence the ion speed) is presumed to be zero, the ion space charge must become infinite to maintain a finite positive ion current.

Q Show that in the high-pressure case

$$V_0 = \sqrt{\frac{8\,s^3 J_i}{9\,\varepsilon_0 \mu_i}}. \tag{3.21}$$

A Multiplying Eq. (3.20) by $\phi'(x)$ and integrating with the same boundary conditions as before:

$$\left(\phi'(x_1)\right)^2 = \frac{2 J_i}{\varepsilon_0 \mu_i}(s - x_1).$$

Integrating $\phi'(x_1)$ from $x = 0$ where $\phi(0) = -V_0$ to the plasma boundary leads to Eq. (3.21).

Benilov [34] refers to this case as the Mott–Gurney law, noting its origin in semiconductor physics. This approximation is not satisfactory in most high-pressure sheaths because the ion fluid speed always exceeds the thermal speed. In the next section the mobility is taken to be a function of the drift speed, giving a more useful representation for many plasma processing discharges (in particular in etching discharges).

The intermediate-pressure (collisional) case

An intermediate regime, between free-flow (collisionless) motion and fully collisional motion, has been introduced. This regime is called the variable mobility regime [24]. It mainly consists of considering the ion fluid speed instead of the ion thermal speed in determining the motion between collisions. The mobility then becomes a function of the fluid speed and is written, in the one-dimensional case, as

$$\mu_i = \frac{2e\lambda_i}{\pi M |u|}, \tag{3.22}$$

where λ_i is the ion–neutral mean free path. Following the same integration procedure as in the fully collisional case above, a new Child–Langmuir law

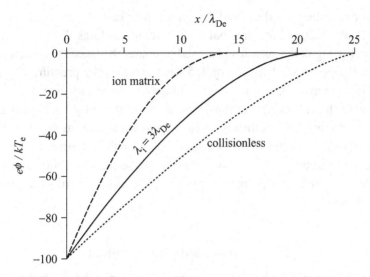

Figure 3.4 Normalized potential profiles for space charge sheaths according to various models. Collisions tend to slow the ions, increasing their contribution to the space charge and so narrowing the sheath – the intermediate-pressure model has been used to calculate the collisional case. In each case the parameters have been set to match the examples in the text.

expression is obtained, which can be written in normalized form as

$$\frac{s}{\lambda_{De}} = \left(\frac{8}{9\pi}\frac{\lambda_i}{\lambda_{De}}\right)^{1/5}\left(\frac{n_{e0}e\sqrt{kT_e/M}}{J_i}\right)^{2/5}\left(\frac{5}{3}\frac{eV_0}{kT_e}\right)^{3/5}. \qquad (3.23)$$

This relation will be termed the 'collisional' Child–Langmuir law in this text.

Exercise 3.4: The intermediate-pressure Child–Langmuir law Show that according to the intermediate-pressure (collisional) model, a sheath with 200 V across it, at the edge of a plasma with electron temperature 2 eV, will be a little over 20 Debye lengths thick if the ion current density is such that $J_i = n_{e0}e\sqrt{kT_e/M}$, which is no more than a convenient choice of scale factor, and the mean free path for ions is such that $\lambda_i = 3\lambda_{De}$.

Review

Figure 3.4 compares the ion matrix model with the two versions of the Child–Langmuir models. The reduction of the ion space charge in the non-matrix models makes the sheath thicker. The more collisions there are, the less the ion space charge density falls, so that a highly collisional sheath tends towards the ion matrix assumption of constant ion density.

The three variations of the Child–Langmuir law, Eqs (3.16), (3.21) and (3.23), present the same general features but with different scalings. For a given positive ion current flowing in the sheath and a fixed pressure, the sheath size increases with voltage (to the power of 3/4, 2/3 or 3/5 depending on the pressure regime). For a given positive ion current and a fixed voltage V_0, the sheath size decreases with pressure, with the following scalings: $s \propto p^{-1/3}$ in the fully collisional case and $s \propto p^{-1/5}$ in the partially collisional case. What remains unsatisfactory, however, for all three models is the 'undetermined' value of the positive ion current and the problem of sustaining a finite ion current at the plasma boundary, where all three Child–Langmuir models presume that the ion speed is zero – the next section addresses these issues.

3.2 The plasma/sheath transition

When the voltage across a sheath is very large compared to the electron temperature $(eV_0 \gg kT_e)$, the electron density is virtually zero in most of the sheath. However, in the vicinity of the plasma/sheath transition, the electron density must be comparable to the ion density. To study the transition, the space charge of both electrons and ions must be included in Poisson's equation; this will be done now.

3.2.1 The Bohm criterion: the transition from sheath to plasma

Consider the situation shown in Figure 3.5, which represents the boundary region between a quasi-neutral plasma in a low-pressure gas and a space charge sheath; as before, the ions will be assumed to be cold. The plasma/sheath transition takes place at $x = s$, where $\phi = 0$ and $n_{is} = n_{es} = n_s$. In the ion matrix and Child–Langmuir models the electric field at the plasma/sheath boundary is also set to zero, but that is not consistent with a non-zero ion current. A more cautious approach is simply to suggest that the field is *nearly*, but not exactly, zero: i.e., $\phi' \approx 0$.

Applying the isothermal assumption allows the electron density to be linked to the local potential by the Boltzmann relation (Eq. 3.3)

$$n_e(x) = n_s \exp\left(\frac{e\phi(x)}{kT_e}\right). \tag{3.24}$$

For the cold ions, the low-pressure continuity and momentum equations, neglecting the effects of ionization and momentum transfer collisions, are

$$n_i(x)u(x) = n_s u_s \tag{3.25}$$

$$\frac{1}{2}Mu(x)^2 + e\phi(x) = \frac{1}{2}Mu_s^2, \tag{3.26}$$

3.2 The plasma/sheath transition

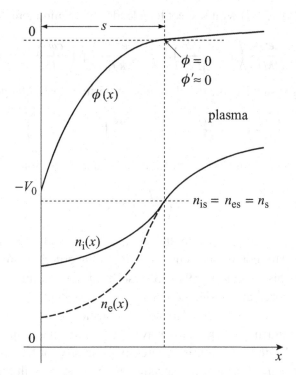

Figure 3.5 Densities and potential around the plasma/sheath transition region.

where u_s is the positive ion fluid speed at the plasma/sheath transition, building in the fact that the ion fluid is not at rest at this point. The aim now is to find a specification for that speed. Substituting the ion speed u from Eq. (3.26) into Eq. (3.25) gives an expression for the positive ion density as a function of the potential:

$$n_i(x) = n_s \left(1 - \frac{2e\phi(x)}{Mu_s^2}\right)^{-1/2}.$$

The net space charge in the sheath is therefore

$$\rho = en_s \left[\left(1 - \frac{2e\phi}{Mu_s^2}\right)^{-1/2} - \exp\left(\frac{e\phi}{kT_e}\right)\right]. \tag{3.27}$$

Looking at Figure 3.5, it is clear that the potential falls on passing from the plasma into the sheath (x decreasing), so $d\phi/dx > 0$, while the net space charge becomes positive, as x decreases, so $d\rho/dx < 0$; even though both ion and electron densities decrease, the latter must fall more rapidly. Thus, for $\phi < 0$,

$$\frac{d\rho}{d\phi} < 0.$$

Differentiating Eq. (3.27) with respect to ϕ leads to a requirement that

$$\frac{e^2 n_s}{M u_s}\left(1 - \frac{2e\phi}{M u_s}\right)^{-3/2} < \frac{e^2 n_s}{kT_e} \exp\left(\frac{e\phi}{kT_e}\right).$$

Expanding this for small ϕ, the development of positive space charge requires that

$$\frac{e^2 n_s}{M u_s}\left(1 + \frac{3e\phi}{M u_s} \cdots\right) < \frac{e^2 n_s}{kT_e}\left(1 + \frac{e\phi}{kT_e} \cdots\right)$$

and for $\phi < 0$ this inequality is satisfied if at the boundary the ion speed is such that

$$u_s = \left(\frac{kT_e}{M}\right)^{1/2}; \tag{3.28}$$

the leading terms then cancel, but the next order will always ensure that the condition is met. This condition is known as the Bohm criterion, after the physicist who established this relation in 1949 [35], so from now on this quantity will be called the Bohm speed and designated u_B.

Note that the Bohm speed is equal to the ion acoustic speed that was derived in the context of low-frequency electrostatic waves in the previous chapter (Eq. (2.71)). In order for the sheath to form, it is apparently necessary for the ions to reach the 'sound' speed $u = u_B$. It is as if the sheath were coincident with the formation of a shock in the boundary-directed flow of the ion fluid [36].

> **Q** What potential difference would ions have to fall through without collision, starting from rest, to acquire the Bohm speed?
>
> **A** Equation (3.11) describes the consequence of collisionless ion acceleration – replacing $\phi(x)$ in that equation by $\Delta\phi$, and using Eq. (3.28) for the final ion speed, the potential drop required to achieve the Bohm speed is given by
>
> $$e\Delta\phi = -\frac{1}{2} M \left(\frac{kT_e}{M}\right).$$

In order for a sheath to form at the boundary, the potential variation between the centre and the edge of a collisionless plasma therefore does not need to be more than $kT_e/2e$. Actually, this is only true if all ions are created at rest, in the centre. In practice, ionization takes place throughout the volume. In that case the centre-to-edge potential will have to be higher to compensate for some ions being much slower than others.

> **Q** What is likely to be the effect of introducing momentum transfer collisions into the ion motion on the centre-to-edge potential difference?

> **A** Since collisions will tend to slow the ion fluid, in order to achieve the Bohm speed at the boundary, the potential between centre and edge will again have to become larger than $kT_e/2e$.

The ion flux at the plasma boundary

When the sheath is free of ionization, the ion flux leaving the plasma, which is $\Gamma_i = n_s u_B$ according to the above, is also equal to the positive ion flux arriving at the wall:

$$\Gamma_{\text{wall}} = n_s u_B. \tag{3.29}$$

The flux leaving the discharge volume is a key quantity for global models (see Eqs (2.40), (2.45) and (2.47)). Also, $e\Gamma_i$ gives the ion current density entering the space charge sheath region, which is required for the calculation of the sheath size using the Child–Langmuir model.

So far, the ion flux to the boundary is expressed as a function of the density at the sheath edge n_s. In the collisionless ionization-free case, the potential in the centre must be $kT_e/2e$ higher than at the edge, and it follows from the Boltzmann relation that the density in the centre n_0 must be higher than that at the edge:

$$\frac{n_s}{n_0} = \exp\left(-\frac{1}{2}\right) \approx 0.6. \tag{3.30}$$

In the more realistic case where ions undergo collisions with neutrals, and where ionization within the plasma is included, the potential between centre and edge will be greater and consequently the density drop will be more pronounced. Examining these cases will be the topic of Section 3.3.

The potential of a floating surface

Now that an expression for the ion flux crossing into the sheath has been quantified, it is possible to find the potential difference across the sheath that would retard the electron flux to such an extent that the net current is exactly zero. This will be the steady potential that any insulating, or electrically isolated, surface would 'naturally' acquire when exposed to a region of plasma. This floating potential was introduced at the start of Section 3.1.

The ion flux to the surface has just been calculated. The electron flux was discussed in the previous chapter (Eq. (2.17)) – when a surface retards a Maxwellian distribution of electrons of temperature T_e so only those with energy greater than $e\Delta\phi$ can reach it, the flux arriving at the surface is

$$\Gamma_e = \frac{n_s \bar{v}_e}{4} \exp\left(-\frac{e\Delta\phi}{kT_e}\right). \tag{3.31}$$

Q Starting from

$$\Gamma_e = \Gamma_i$$

show that the floating potential of a surface exposed to a region of plasma with cold ions of mass M and Maxwellian electrons (temperature T_e) is

$$V_f = \frac{kT_e}{e} \frac{1}{2} \ln\left(\frac{2\pi m}{M}\right) \quad (3.32)$$

with respect to the potential at the plasma/sheath boundary.

A Substituting for the electron and ion fluxes from Eqs (3.31) and (3.29), with $\Delta\phi = -V_f$, gives

$$\frac{n_s \bar{v}_e}{4} \exp\left(\frac{eV_f}{kT_e}\right) = n_s u_B,$$

which gives the required result on taking logarithms. Note that $V_f < 0$, since $m \ll M$.

Exercise 3.5: Ion flux and energy at a floating surface For a plasma with Maxwellian electrons characterized by a temperature of 2 eV and a central density of 10^{16} m^{-3}, in argon at a pressure for which the mean free path for charge exchange collisions is $\lambda_i \sim 10$ mm, estimate the ion flux and the ion energy flux (power density) arriving at a floating surface.

3.2.2 The transition from plasma to sheath

The transition from quasi-neutral plasma to space charge sheath can also be viewed starting from the plasma. Instead of separate electron and ion densities, a single plasma density n is considered:

$$n_e = n_i = n. \quad (3.33)$$

This quasi-neutrality condition, also called the plasma approximation, will be used again in the next section when considering transport properties within the plasma. In addition to the assumption of quasi-neutrality, collisions will be included through a drag term in the ion momentum equation of the ion fluid. Ionizing collisions will be considered in the next section, but for the present case it must be supposed that the plasma is created somewhere further 'upstream'. A steady flux of ions enters the region of interest. The ion fluid motion is determined by (i) an equation of continuity that sets the divergence of the flux to zero (no volume sources or

3.2 The plasma/sheath transition

sinks)

$$(nu)' = 0 \tag{3.34}$$

and (ii) a force balance

$$nMuu' = neE - nMu\nu_i, \tag{3.35}$$

where ν_i is the collision frequency for momentum transfer between ions and the background gas. The inclusion of the electric field in the quasi-neutral plasma needs some comment. From the sheath perspective the field vanishes at the plasma boundary. However, within the plasma a small field is required to reduce the flow of electrons, and to sustain the flow of ions, to the plasma boundary. In this way, volume production and surface loss can be kept in balance. The next section will quantify the electrical potential profile in the plasma region that self-consistently achieves this balance. The presence of the electric field in the plasma means that the plasma is not exactly neutral, which is why it is described as quasi-neutral, meaning that ion and electron densities are almost, but not exactly, equal. Finally, the electron force balance is assumed to be dominated by the electric field ($E = -\phi'$) and the pressure gradient, so that

$$-ne\phi' = -kT_e n'. \tag{3.36}$$

This is equivalent to the Boltzmann equilibrium. Using Eq. (3.36), Eqs (3.34) and (3.35) can be rewritten:

$$un' + nu' = 0, \tag{3.37}$$

$$nuu' = -u_B^2 n' - nu\nu_i. \tag{3.38}$$

These can be manipulated further to express the density and velocity gradients in the following way:

$$u' = \frac{u^2 \nu_i}{u_B^2 - u^2}, \tag{3.39}$$

$$n' = \frac{-nu\nu_i}{u_B^2 - u^2}. \tag{3.40}$$

It is evident that both gradients (n', u') become singular at $u = u_B$, which suggests that the quasi-neutral solution (i.e., a solution forcing $n_e = n_i$) breaks down when ions reach the Bohm speed. See Figure 3.6. This time starting in the plasma, it appears that the boundary with the sheath occurs when positive ions reach the Bohm speed. Notice that in Figure 3.5 the boundary is at $x = 0$ and in the plasma ($x > 0$) the ion speed is directed to the boundary ($u < 0$). Then, as Eq. (3.40) shows, the density increases away from the boundary ($n' > 0$), since $u^2 < u_B^2$.

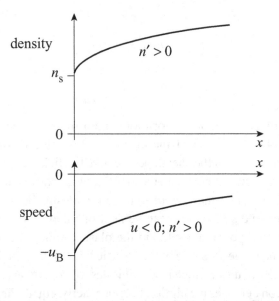

Figure 3.6 Schematic plasma density and ion fluid speed profiles on the plasma side ($x > 0$) of a plasma/sheath transition from a 1-D quasi-neutral plasma model with elastic ion–neutral collisions. Such a model cannot resolve the sheath and matches singularities in the density and speed profiles to the boundary at $x = 0$.

3.3 The plasma region: transport models

After studying the sheath and the plasma/sheath transition, we move further away from the boundary and now consider charged particle transport within the plasma itself, where the plasma is produced by the ionizing collisions of electrons with the background gas.

The transport within the plasma depends strongly on the pressure regime of interest, which determines the relative importance of the various terms in the continuity equation and the force balance equations, Eqs (2.32) and (2.33), specifically for the ion fluid. The temperature of all particles will be considered to be uniform. The gas temperature and that of the ion fluid will be supposed to be close to that of the walls that confine the plasma. However, it is easy to show that the value of the electron temperature may not be arbitrarily chosen – it was shown in Section 2.3 that a global particle balance leads to

$$K_{\text{iz}}(T_{\text{e}}) = \frac{\Gamma_{\text{wall}}}{\bar{n}_{\text{e}}} \left(\frac{1}{n_{\text{g}}} \frac{A}{V} \right), \tag{3.41}$$

where Γ_{wall} is the flux leaving the plasma, Eq. (3.29), and $\bar{n}_{\text{e}}$ is the plasma density averaged across space defined by the volume V and boundary area A. Writing the density profile in the form $n(x) = n_0 h(x)$ for a 1-D symmetrical plasma region, where $h(x)$ is the form of the density profile, the mean plasma density is

$$\bar{n}_e = n_0 \frac{2}{l} \int_0^{l/2} h(x) dx.$$

In this section, the form of the density profile, $h(x)$, will be discussed in the different pressure regimes. There are three models traditionally used to describe the charged particle transport in plasma discharges. They correspond to the three pressure regimes that were discussed in the context of the Child–Langmuir law. The difference between these models arises from the assumptions made to treat the transport of ions. The first to be considered is a collisionless solution at low pressure, obtained in the early years of discharge physics by Tonks and Langmuir [37]. Then, the opposite limit of high pressure, in which ions are fully collisional, will be studied. This regime was described by Schottky [38], a few years before the low-pressure analysis of Tonks and Langmuir. Finally, the intermediate pressure regime of Godyak and Maximov [39] is discussed. In each case the goal will be to obtain an expression for the edge-to-centre density ratio, which will be given the symbol h_l, where the subscript (l for linear) indicates an axial 1-D model.

Thus, Eq. (3.41) sets the electron temperature through the rate coefficient K_{iz}, in terms of the Bohm speed u_B, gas density n_g, the system dimensions A/V and the edge-to-centre density ratio n_s/n_0 (from Eq. (3.29)). It will be shown in due course that the electron temperature depends solely on the product of the gas pressure and the system size.

3.3.1 Low-pressure models

The sheath models did not need ionization to sustain the space charge because ions and electrons were supposed to penetrate the region from an 'upstream' source. In the case of the bulk volume of a plasma, ionization must be included to sustain it in the presence of wall-directed fluxes. In one dimension, the ion continuity equation with ionization is

$$(nu)' = n n_g K_{iz}, \quad (3.42)$$

where n_g is the neutral gas density and K_{iz} is the ionization rate coefficient defined by Eq. (2.27). The collisionless fluid model is examined first; the model of Tonks and Langmuir will follow later. The ion momentum equation can be written as being predominantly a balance between inertial and electrical forces if one considers a steady, collisionless flow of the ion fluid:

$$nMu u' = neE. \quad (3.43)$$

The total neglect of collisions from the momentum conservation equation is artificial, since the fluid model requires all ions at any point to have the same speed – strictly, the momentum required for newly generated ions to get up to speed should be included. Nevertheless, the simplicity of the model makes it amenable to analysis, from which some insight can be gained. To complete the set of equations, there is the Boltzmann relation in the form of Eq. (3.36).

To solve these equations it is noted that the integration of Eq. (3.43) from the centre ($x = 0$) into the positive half-plane, with $\phi(0) = 0$ and $u(0) = 0$, gives (remembering that $E = -\phi'$):

$$u = \left(\frac{2e}{M}\right)^{1/2}(-\phi)^{1/2} \tag{3.44}$$

and with Eqs (3.42) and (3.36):

$$\frac{e}{kT_e}\phi' u - \frac{e}{Mu}\phi' = n_g K_{iz}. \tag{3.45}$$

Substituting Eq. (3.44) into Eq. (3.45) and integrating with the condition $\phi(0) = 0$, yields

$$\frac{-2e}{3kT_e}(-\phi)^{3/2} + (-\phi)^{1/2} = n_g K_{iz}\left(\frac{M}{2e}\right)^{1/2} x, \tag{3.46}$$

which is the equation for the potential. Using Eq. (3.44) again translates this into an equation for the ion speed as a function of position:

$$u - \frac{u^3}{3u_B^2} = n_g K_{iz} x. \tag{3.47}$$

Equation (3.47) is a third-order algebraic equation which can be solved analytically to obtain $u(x)$, and in turn $\phi(x)$ and $n(x)$.

In Section 3.2.2 the speed and density derivatives were shown to become singular when $u = u_B$, i.e., where the sheath forms. Equation (3.47) now locates the plasma/sheath transition (i.e., the position at which $u = u_B$) at

$$x_s = \frac{2}{3}\frac{u_B}{n_g K_{iz}}. \tag{3.48}$$

Note that if the sheath size is small compared to the plasma size, $x_s \approx l/2$. Equation (3.48) determines the electron temperature then – a very similar result was found from the global particle balance in the previous chapter. Equation (3.44) gives the potential drop between the centre, where $u = 0$, and the plasma/sheath transition, where $u = u_B$, to be $\phi(x_s) - \phi(0) = -kT_e/2e$. Consequently, from the Boltzmann relation, the collisionless edge-to-centre density ratio is $h_l \equiv n_s/n_0 = 0.6$ (as obtained at the end of Section 3.2.1).

3.3 The plasma region: transport models

Letting $\xi \equiv 2x/3x_s$, the solution of Eq. (3.47) is

$$u(\xi) = 2u_B \cos\left\{\frac{1}{3}\left(4\pi - \arctan\left(\frac{4}{9\xi^2} - 1\right)^{1/2}\right)\right\}, \quad (3.49)$$

which in turn determines the potential and the density as a function of the position through the two relations

$$\phi(\xi) = -\frac{M}{2e}u^2(\xi), \quad (3.50)$$

$$n(\xi) = n_0 \exp\left(\frac{e\phi(\xi)}{kT_e}\right). \quad (3.51)$$

In the above derivation, the ion momentum equation did not include a term to account for the reduction of momentum due to newly generated ions being born at rest but having instantaneously to catch up with the ion fluid flow. To deal with this the analysis should be repeated with Eq. (3.43) replaced by

$$nMu\,u' = neE - nMn_g K_{iz} u; \quad (3.52)$$

a complete analytical solution is not particularly simple [40, 41], though further insight is possible without it.

Q Substituting Eqs (3.42) and (3.36) into Eq. (3.52) leads to

$$(nu^2)' = -u_B^2 n'.$$

Integrate from the centre to the edge to show that in this case $h_l = 0.5$ and determine the boundary potential.

A At the centre $u = 0$ and $n = n_0$ while at the edge $u = u_B$ and $n = n_s$. The integration simplifies because

$$\int y' dx = \int dy.$$

So

$$[n_s u_B^2 - 0] = -u_B^2 [n_0 - n_s].$$

That means

$$h_l = 0.5 \quad (3.53)$$

$$\phi_s = -\ln 2 \left[\frac{kT_e}{e}\right]. \quad (3.54)$$

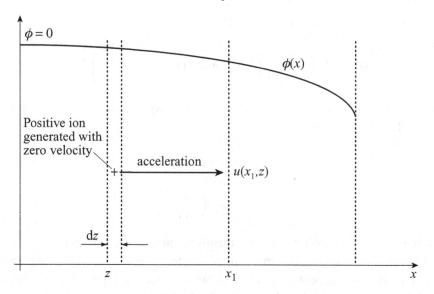

Figure 3.7 Schematic of how the Tonks–Langmuir model accounts for the distribution of ion speed at any point in terms of the acceleration of ions born at all points 'upstream' in the potential profile.

The new solution is therefore quite similar to the previous one, but the collisional drag leads to a larger potential drop between the centre and the sheath edge and consequently to a smaller h_1.

Tonks and Langmuir [37], however, took a more general approach avoiding the above somewhat contrived fluid model that requires instantaneous acceleration of new-born ions up to the local fluid speed. Their solution treats the ions kinetically; see Figure 3.7. The supposition is that as in the first part of this section the motion of ions is collisionless so energy is conserved within the ion population. However, ions only gain energy by falling freely in the electric field from their point of generation. That means that ions generated at rest at one particular position only contribute to the flux downstream from this position. There is no longer a single ion fluid speed but a distribution of speeds reflecting the distribution of places where ions are generated. As shown in Figure 3.7, the ion density at x is the result of ions generated at all positions between zero and x. The elementary contribution to density at x from ions born at z is therefore a function of the rate of generation in a slab of width dz at position z, with $z < x$:

$$\mathrm{d}n = \frac{n(z) n_g K_{iz} \, \mathrm{d}z}{u(x, z)}, \tag{3.55}$$

where $u(z, x)$ is the speed at which ions generated at position z arrive at position x. This speed follows from the energy conservation equation, and is

$$u(x, z) = \left(\frac{2e}{M}\right)^{1/2} (\phi(z) - \phi(x))^{1/2}. \quad (3.56)$$

The plasma density at position x is obtained by doing the sum (the integral) of all contributions of slabs dz between $z = 0$ and $z = x$:

$$n(x) = \left(\frac{M}{2e}\right)^{1/2} \int_0^x \frac{n(z) n_g K_{iz} dz}{(\phi(z) - \phi(x))^{1/2}}. \quad (3.57)$$

The calculation is not complete because $n(z)$ is not yet specified. To do this, quasi-neutrality is assumed so the plasma density follows the electron Boltzmann equilibrium of Eq. (3.3) with $n_{e0} = n_0$. This leads to an integral equation for the potential:

$$\exp\left(\frac{e\phi(\xi)}{kT_e}\right) = \left(\frac{kT_e}{2e}\right)^{1/2} \int_0^\xi \frac{\exp(e\phi(\xi_1)/kT_e) d\xi_1}{(\phi(\xi_1) - \phi(\xi))^{1/2}} \quad (3.58)$$

where $\xi \equiv n_g K_{iz} x / u_B$ and $\xi_1 \equiv n_g K_{iz} z / u_B$. Tonks and Langmuir originally found a power series solution for this equation. Later, Thompson and Harrison [42] found a closed-form solution in terms of Dawson functions. The solution allows the evaluation of the following important quantities, to be compared to the fluid solutions:

$$h_l = 0.425, \quad (3.59)$$

$$\phi_s = -0.854 \left[\frac{kT_e}{e}\right]. \quad (3.60)$$

Figure 3.8 compares the three density profiles determined in this section on plasma transport in the low-pressure (collisionless) limit. The solid line is the solution of Tonks and Langmuir, the dashed line is the fluid solution with the ionization term in the momentum equation and the dash-dotted line is the fluid solution neglecting the ionization term in the momentum equation.

3.3.2 High-pressure model

The opposite pressure limit was considered in 1924 by Schottky [38], in a radial model of the positive column in DC glow discharges, for which the ion–neutral mean free path was significantly shorter than the column diameter, $\lambda_i \ll r_0$. In this case the ion motion is collisional, so energy is not conserved within the ion fluid and the ion drag force must play an important role. Collisions keep the ion fluid speed small or comparable with the ion thermal speed in the major part of the

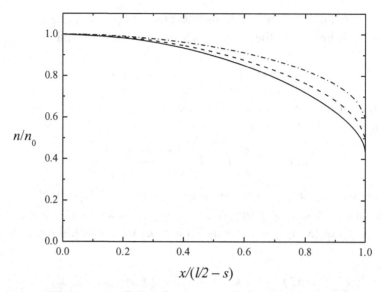

Figure 3.8 Density profile in the low-pressure regime: the solid line is the solution by Tonks and Langmuir, the dashed line is the fluid solution with the ionization term in the momentum equation and the dash-dotted line is the fluid solution neglecting the ionization term in the momentum equation.

discharge. The thermal motion of the electrons is even more dominant over any drift motion. If the situation is still dominated by production of charged particles in the volume and loss of charged particles by recombination at the walls, the electric field must be such that the electron and ion fluxes at any position within the plasma are the same, otherwise quasi-neutrality would not be maintained. Since both charges appear to diffuse together in this collision-dominated motion at the same density, n, and the same speed, u, the expression *ambipolar diffusion* is used to describe the process. The isothermal force balance from the momentum equation for each charged species is

$$0 = -neE - kT_e n' - nmu\nu_e$$
$$0 = neE - kT_i n' - nMu\nu_i.$$

These can each be rearranged as expressions, respectively, for the flux in terms of electrons and ions:

$$nu = -n\mu_e E - D_e n' \qquad (3.61)$$
$$nu = n\mu_i E - D_i n', \qquad (3.62)$$

where diffusion coefficients have been introduced to describe thermally driven motion: $D_e = kT_e/m\nu_e$ and $D_i = kT_i/M\nu_i$; the field-driven motion is described in terms of the mobilities $\mu_e = e/m\nu_e$ and $\mu_i = e/M\nu_i$.

3.3 The plasma region: transport models

There are three quantities to be determined: n, u and E and the third equation, as before, comes from a continuity equation that includes the ionization source:

$$(nu)' = n\, n_g\, K_{iz}. \tag{3.63}$$

Using Eqs (3.61) and (3.62), the electric field that maintains the equality of the electron and ion fluxes is

$$E = -\frac{D_e - D_i}{\mu_e + \mu_i}\frac{n'}{n}. \tag{3.64}$$

Substituting this expression back into Eq. (3.61), the ambipolar flux $\Gamma = nu$ can be written as an effective ambipolar diffusion flux

$$\Gamma = -D_a n', \tag{3.65}$$

where an ambipolar diffusion coefficient has been defined as

$$D_a = \frac{\mu_i D_e + \mu_e D_i}{\mu_i + \mu_e}. \tag{3.66}$$

Q Show that in the usual situation of $\mu_i \ll \mu_e$ and $T_i \ll T_e$,

$$D_a \approx \frac{kT_e}{M v_i} \tag{3.67}$$

and that

$$\Gamma \ll n\mu_e E,\ D_e n'.$$

A Putting $\mu_i \ll \mu_e$ in Eq. (3.66),

$$D_a \approx \frac{\mu_i}{\mu_e} D_e + D_i$$

$$\approx \mu_i \frac{kT_e}{e} + D_i$$

$$\approx \mu_i \frac{kT_e}{e} + \mu_i \frac{kT_i}{e}.$$

The first result follows from inserting $T_i \ll T_e$ and $\mu_i = e/M v_i$. Thus, $D_a \gg D_e$, so on combining Eqs (3.67) and (3.61), it becomes apparent that the ambipolar drift must be the difference between two much larger effects, the electric field drift and thermal diffusion, that are not quite exactly balanced. *Comment:* Equation (3.67) can be obtained directly if one supposes that the electrons are in Boltzmann equilibrium and the ions are cold, $T_i = 0$.

It appears that ions 'diffuse' faster than their own free diffusion coefficient (if no electrons were present) and electrons diffuse slower than their own free diffusion coefficient, i.e., $D_i < D_a < D_e$. This is due to the fact that the ambipolar electric field acts differently on electrons and ions: it accelerates the ions out of the discharge while it confines the electrons.

To find the density profile, Eq. (3.65) is combined with Eq. (3.63) to obtain a second-order differential equation:

$$n'' = -\beta^2 n, \qquad (3.68)$$

where $\beta^2 = n_g K_{iz}/D_a$. This equation has a solution that is a linear combination of sine and cosine functions. Considering the typical case in which the plasma is confined between two electrodes placed at $x = \pm l/2$, a symmetrical solution is appropriate, the simplest of which is

$$n(x) = n_0 \cos \beta x. \qquad (3.69)$$

This in turn gives

$$\Gamma(x) = -D_a n'(x) \qquad (3.70)$$
$$= D_a n_0 \beta \sin \beta x. \qquad (3.71)$$

A suitable boundary condition would be to set the plasma density to zero at the vessel walls at $x = \pm l/2$, neglecting for the moment that a thin sheath may in practice separate the quasi-neutral plasma from the wall:

$$\beta = \left(\frac{n_g K_{iz}}{D_a}\right)^{1/2} \approx \frac{\pi}{l}. \qquad (3.72)$$

This is the boundary condition used by Schottky, and Eq. (3.72) is sometimes called the Schottky condition.

Q Equation (3.72) can be said to determine the electron temperature – explain how this is so.
A Since $\beta = (n_g K_{iz}/D_a)^{1/2}$ and D_a, u_B and K_{iz} are all functions of T_e, then Eq. (3.72) links the system size, the gas density and the electron temperature (cf. Eq. (3.48)).

In the absence of any more certain knowledge, a sheath edge might be set at the place where the ion fluid reaches the Bohm speed ($x = l/2 - s$). The edge-to-centre density ratio, h_l, that characterizes the density profile would then be derived from the flux at the sheath edge:

$$n_s u_B = D_a n_0 \beta \sin \beta(l/2 - s). \tag{3.73}$$

Thus

$$h_1 = \frac{n_s}{n_0} = \frac{\beta D_a}{u_B} \sin\left[\beta\left(\frac{l}{2} - s\right)\right]. \tag{3.74}$$

Under circumstances where the sheath size is small, $s \ll l/2$, Eq. (3.72) applies and

$$h_1 \approx \frac{\pi D_a}{l u_B} = \pi \frac{u_B}{\bar{v}_i} \frac{\lambda_i}{l}. \tag{3.75}$$

In the 'high-pressure' regime $\lambda_i \ll l$, though somewhat offset by the $u_B/\bar{v}_i$ ratio that scales with the square root of the electron-to-ion temperature, h_1 will tend to be small, scaling with p^{-1}. Franklin [43] has shown from a full solution of the fluid equations that above about 10 Pa between plates separated by about 3 cm, the ions do not in fact reach the Bohm speed, which places an upper pressure limit on the validity of including a sheath in the model.

The original Schottky model sets the electron density to zero at the wall. Looking back at Eq. (3.64), it is clear that since $D_e \neq D_i$ and the density gradient is finite, then the electric field becomes infinite at the boundary, even though there is no room for a sheath. In fact, with $n_e = 0$ at the plasma boundary the Debye length is infinite and so if there were any sheath it would be infinitely thick. In addition, the finite particle flux at the boundary means that infinite particle speeds must accompany zero particle density. This condition is therefore unsatisfactory.

3.3.3 Intermediate pressure

Many processing discharges operate in an intermediate-pressure regime for which neither Tonks–Langmuir nor Schottky solutions are appropriate. This regime is such that the ion–neutral mean free path is smaller, but not significantly smaller, than the typical discharge size $\lambda_i \leq l$. The fluid conservation equations for ions and electrons are essentially those of the Schottky model, except that the collision frequency of ions is not taken as a constant but is supposed to depend on the ion fluid speed. This reflects the fact that the ion thermal motion no longer dominates over the drift motion and so ions move between collisions essentially at the fluid speed with relatively little influence of the thermal speed. Under these circumstances, it turns out that the collision frequency is related to the mean free path by

$$v_i = \frac{\pi}{2} \frac{u_i}{\lambda_i}. \tag{3.76}$$

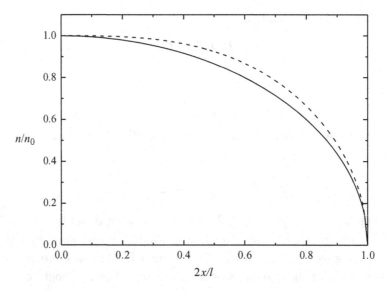

Figure 3.9 Density profiles according to Raimbault et al. [44] (dashed line) and to the heuristic solution of Lieberman and Lichtenberg [2] (solid line).

This form replaces the high-pressure form used earlier, $v_{e,i} = \bar{v}_{e,i}/\lambda_{e,i}$. A similar argument was used in discussing the intermediate-pressure regime of the Child–Langmuir sheath model (cf. Eq. (3.22)). That means that for intermediate pressure the mobility of ions, for example, should be expressed as

$$\mu_i = \frac{e}{M v_i(u)} = \frac{2e\lambda_i}{\pi M |u(x)|}, \tag{3.77}$$

and therefore varies in space. Putting this form of mobility into Eqs (3.61) and (3.62) in combination with Eq. (3.63) does not lead to a linear differential equation for any one of the variables n, u and ϕ as it did in the high-pressure case.

A solution for the variable mobility model has been given by Godyak [39]. The exact solution is somewhat cumbersome, and the density profile is only obtained in an implicit fashion. Recently, Raimbault et al. [44] obtained a more general solution and included the effect of neutral depletion on the transport, as will be seen in Chapter 9. Lieberman and Lichtenberg [2] have shown that the density profile in this intermediate-pressure regime is fairly well approximated by the formula

$$n(x) = n_0 \left[1 - \left(\frac{2x}{l} \right)^2 \right]^{1/2}, \tag{3.78}$$

as shown in Figure 3.9, where the density profiles are compared.

3.3 The plasma region: transport models

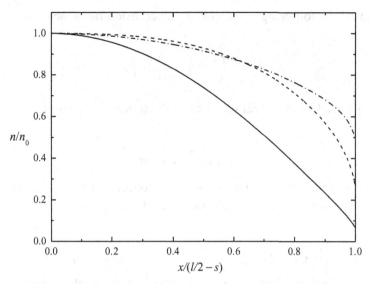

Figure 3.10 Density profiles for the different models with $l = 0.03$ m in argon: the dash-dotted line is for the Tonks–Langmuir model ($p < 1$ Pa), the dashed line is for the Godyak model ($p = 3$ Pa), the solid line is for the Schottky model ($p = 30$ Pa).

As discussed above, the appropriate boundary condition should be that the ions reach the sound speed at the sheath edge, that is at $x = l/2 - s$. With this boundary condition, one obtains the expression for the edge-to-centre density ratio, i.e., the h_l factor. Raimbault et al. have shown that this factor is

$$h_l \approx 0.877 \left(\frac{l}{2\lambda_i}\right)^{-1/2}. \qquad (3.79)$$

Figure 3.10 compares the density profile for the Tonks–Langmuir model, the Schottky model and the Godyak (variable mobility) model. The Tonks–Langmuir curve has been obtained in the collisionless limit, for a 0.03 m vessel this would obtain for room temperature argon below about 1 Pa, while the Schottky model curve has been obtained for a pressure of 30 Pa. The Godyak model curve has been obtained for a pressure of 3 Pa.

The general trend is that at low pressure the density is flatter in the central region, and falls more abruptly near the edge. It is interesting to note that the intermediate profile is flatter than the Tonks–Langmuir profile near the discharge centre, but falls more near the edge with a smaller h_l factor. This is partially due to the fact that within the Godyak model, the friction force is too small in the centre where $u < \bar{v}_i$. To correct this effect, and simultaneously introduce a smooth transition from Godyak to Schottky as the pressure increases, Chabert et al. [45]

have proposed the following expression of the collision frequency:

$$\nu_i = \frac{\overline{v}_i}{\lambda_i}\left(1 + \left(\frac{\pi}{2}\frac{u_i}{\overline{v}_i}\right)^2\right)^{1/2}, \qquad (3.80)$$

which can be used in numerical solutions of the transport equations.

3.4 Review of Chapter 3

This section summarizes the key results and concepts obtained in this chapter, which are necessary to follow the discussion of the next chapters.

3.4.1 Positive ion flux leaving the plasma

In the previous chapter it was shown that the highest level of simplification to describe the equilibrium in a confined plasma (sustained by an unspecific external power source) is to evaluate simultaneously the particle balance, given by Eq. (2.40), and the energy balance, given by Eq. (2.47). Solving these two equations requires an expression for the flux of charged particles leaving the plasma, as a function of the average density, or as discussed, as a function of the plasma density in the discharge centre. The transport models described in this chapter allowed the finding of an expression suitable for various pressure ranges. It has been found that in planar geometry, the flux leaving the plasma is given by

$$\Gamma = h_1 n_0 u_B, \qquad (3.81)$$

where n_0 is the plasma density at the discharge centre and h_1 is the edge-to-centre density ratio, which depends on the pressure regime under consideration. The three pressure regimes can be joined heuristically to obtain

$$h_1 \approx 0.86\left[3 + \frac{1}{2}\frac{l}{\lambda_i} + \frac{1}{5}\frac{T_i}{T_e}\left(\frac{l}{\lambda_i}\right)^2\right]^{-1/2}, \qquad (3.82)$$

where l is the electrode separation and λ_i is the ion–neutral mean free path. The first term is the dominant term at low pressure when $\lambda_i \gg l$. In the intermediate-pressure regime, the second term (from the Godyak solution) dominates, and at higher pressure, when $\lambda_i \ll l$, the last term dominates. Note that this latter depends on the electron temperature, which is also a function of pressure (see the next section).

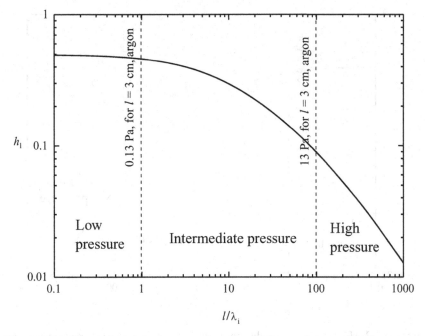

Figure 3.11 Heuristic axial edge-to-centre density factor, $h_l \equiv n_s/n_0$, from equation (3.82) with $T_i/T_e = 0.02$. For a 3 cm gap in argon, the two dashed lines correspond to 0.13 Pa and 13 Pa respectively.

Similar factors have been obtained in cylindrical geometry [2, 39]. The heuristic radial edge-to-centre density factor in that case is

$$h_{r0} \approx 0.8 \left[4 + \frac{r_0}{\lambda_i} + \frac{T_i}{T_e} \left(\frac{r_0}{\lambda_i} \right)^2 \right]^{-1/2}, \quad (3.83)$$

where r_0 is the radius of the cylinder. In this case, the flux at the radial wall is simply given by $\Gamma = h_{r0} n_0 u_B$.

The h factors are plotted in Figure 3.11 (for axial h_l) and Figure 3.12 (for radial h_{r0}) as a function of l/λ_i and r_0/λ_i, respectively. The two dashed lines approximately separate the different models valid in different pressure regimes. For the axial case, the plate separation (gap) is 3 cm, typical of etching reactors. In this case, for argon gas, the intermediate-pressure (variable mobility) model is valid between 0.13 Pa (1 mTorr) and 13 Pa (100 mTorr), which is the typical pressure window of plasma etching reactors. For the radial case, a radius of 6 cm is typical of the source tube of helicon or cylindrical inductive reactors (see Chapters 7 and 8). Then, again for argon gas, the transition from low to intermediate pressure occurs at 0.065 Pa (0.5 mTorr). In most cases, plasma processing reactors like cylindrical inductive

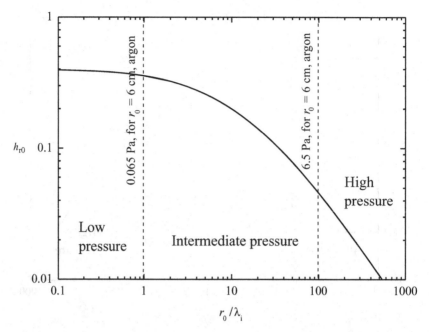

Figure 3.12 Heuristic radial edge-to-centre density factor, $h_{r0} \equiv n_s/n_0$, from Eq. (3.83) with $T_i/T_e = 0.02$. For a 6 cm radius in argon, the two dashed lines correspond to 0.065 Pa and 6.5 Pa, respectively.

reactors will operate above this threshold, i.e., in the intermediate-pressure regime. Plasma thrusters may operate below 0.065 Pa, but the plasma is often magnetized by means of an external static magnetic field and the transport theories described above are not valid. The effect of the magnetic field on transport will be discussed in Chapter 9.

Finally, in the context of electronegative plasmas and/or high-density plasmas these h factors must be revisited (Chapter 9).

3.4.2 Electron temperature

An important result of the previous chapter is the fact that, to a first approximation, the electron temperature is independent of the plasma density (and thus of the power deposited in the plasma) – it only depends on the product of the gas pressure by the typical reactor dimensions. In the steady-state, one-dimensional case of interest here, Eq. (2.41) can now be written

$$\bar{n}_e n_g K_{iz} l = 2 h_1 n_0 u_B, \qquad (3.84)$$

where the LHS represents the production of particles by ionization within the plasma volume, and the RHS represents the flux of particles to the two electrodes.

Q Show that the mean density of the Schottky model is $\bar{n}_e = 2n_0/\pi$.
A This is effectively a straightforward integration of the cosine profile between zero and $\pi/2$:

$$\frac{\bar{n}_e}{n_0} = \frac{2}{l}\int_0^{l/2} \cos(\pi x/l)\,\mathrm{d}x = \frac{2}{\pi}.$$

Note that while the density profile is a cosine at high pressure, it becomes flatter at lower pressure so the following inequality must hold:

$$\frac{2}{\pi} \leq \frac{\bar{n}_e}{n_0} \leq 1. \tag{3.85}$$

This shows that the mean electron density is quantitatively never more than 36% less than the central electron density. Thus, for the purposes of estimation, it is acceptable to set $\bar{n}_e = n_0$ in Eqs (3.84). Then, using Eqs (2.27) and (2.28) for the ionization rate, one can isolate the strong temperature dependence in Eq. (3.84):

$$\frac{kT_e}{e} = \varepsilon_{iz}\left[\ln\left(\frac{l n_g K_{iz0}}{2h_1 u_B}\right)\right]^{-1}; \tag{3.86}$$

notice that the error in setting $\bar{n}_e = n_0$ is substantially lessened by the logarithm. However, Eq. (3.86) is not an explicit formula for the electron temperature because K_{iz0}, u_B and h_1 also depend weakly on T_e. Therefore, a typical procedure is to use $kT_e/e = 3$ V to estimate the value for u_B and h_1, and then to calculate kT_e/e with Eq. (3.86) and iterate a few times. Figure 3.13 shows a calculation of the electron temperature as a function of pressure for an argon plasma generated between two plates spaced by $l = 3$ cm. The solid line is a self-consistent iterative calculation, that is T_e variations are included in h_1 and u_B. The dashed line is a non-self-consistent calculation using fixed values: $u_B = 2500$ m s^{-1} and $T_i/T_e = 0.02$. The data for argon are from Table 2.1.

In a finite cylinder, the particle balance may be written

$$n_g K_{iz}\pi r_0^2 l = u_B(\pi r_0^2 h_1 + 2\pi r_0 l h_{r0}), \tag{3.87}$$

where r_0 is the cylinder radius and l is the cylinder length. One could now rework the temperature calculation starting from this balance.

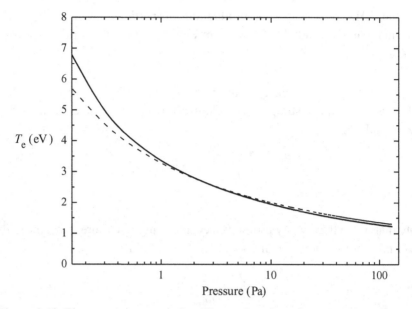

Figure 3.13 Electron temperature (in eV) as a function of pressure for an argon plasma generated between two plates spaced by $l = 3$ cm. The solid line is a self-consistent calculation, that is T_e variations are included in h_l and u_B. The dashed line is a non-self-consistent calculation using Eq. (3.86) with fixed values: $u_B = 2500$ m s^{-1} and $T_i/T_e = 0.02$.

3.4.3 Floating potential and scalings for the sheath thickness

This chapter began by establishing the necessity to form sheaths. It was shown that if any surface is electrically isolated (for example a piece of a dielectric or a floating probe), then a sheath forms adjacent to it to collect ions while repelling sufficient electrons to obtain zero net current. Invoking an exact flux balance

$$\frac{1}{4} n_s \bar{v}_e \exp\left(\frac{eV_f}{kT_e}\right) = n_s u_B, \tag{3.88}$$

where

$$V_f = \frac{kT_e}{2e} \ln\left(\frac{2\pi m}{M}\right) \tag{3.89}$$

is the potential of an isolated surface with respect to the plasma/sheath boundary. For an argon plasma, $V_f \approx -4.7 kT_e/e$. Since the electron temperature is typically around 3 eV, the potential drop across a floating sheath is typically around 15 V. One of the questions addressed in the following chapter is how this result is modified in the presence of RF components of potential.

3.4 Review of Chapter 3

The simplest description of the sheath is the ion matrix model. It neglects the effect of ion acceleration in the sheath and sets the electron space charge to zero. The sheath thickness is simply related to the difference in potential across the sheath:

$$\frac{s}{\lambda_{De}} = \sqrt{\frac{2eV_0}{kT_e}}. \tag{3.90}$$

The model underestimates the thickness of steady sheaths but it is particularly appropriate for the transient response of a sheath to a negative voltage step.

If a large, steady (negative) potential is applied across a sheath, the Child–Langmuir model relates the voltage across the sheath, V_0, to the sheath thickness, s, in terms of the electron temperature and Debye length (and so the electron density). Again, three pressure regimes were identified, but each is shifted to higher pressure than the corresponding transport regime because sheaths are usually much thinner than the plasma dimension. The ion current in each case is due to the Bohm flux; the plasma density is presumed to be in the range 10^{15}–10^{16} m^{-3}.

Collisionless:

$$\frac{\lambda_i}{\lambda_{De}} > 10 \quad \text{typically } p \lesssim 30\,\text{Pa in argon}$$

$$\frac{s}{\lambda_{De}} = \left(\frac{4\sqrt{2}}{9}\right)^{1/2} \left(\frac{eV_0}{kT_e}\right)^{3/4}. \tag{3.91}$$

Intermediate pressure:

$$\sqrt{\frac{T_i}{T_e}} < \frac{\lambda_i}{\lambda_{De}} < 10 \quad \text{typically } 30\,\text{Pa} < p \lesssim 4\,\text{kPa in argon}$$

$$\frac{s}{\lambda_{De}} = \left(\frac{8}{9\pi}\frac{\lambda_i}{\lambda_{De}}\right)^{1/5} \left(\frac{5\,eV_0}{3\,kT_e}\right)^{3/5}. \tag{3.92}$$

High pressure:

$$\frac{\lambda_i}{\lambda_{De}} < \sqrt{\frac{T_i}{T_e}} \quad \text{typically } p \gtrsim 4\,\text{kPa in argon}$$

$$\frac{s}{\lambda_{De}} = \left(\frac{9}{8}\frac{\omega_{pi}}{\nu_i}\right)^{1/3} \left(\frac{eV_0}{kT_e}\right)^{2/3}. \tag{3.93}$$

It appears that for given voltage and plasma density, the sheath tends to shrink with increasing pressure since $\lambda_i \propto p^{-1}$.

4
Radio-frequency sheaths

So far in this book on radio-frequency plasmas the properties of plasmas have been investigated in the absence of periodic time-dependent parameters or boundary conditions, therefore effectively in a DC steady state. In this chapter the restriction to DC conditions will be relaxed to prepare the ground for the discussion of plasmas that are sustained by radio-frequency (RF) power supplies. Although quantities such as electric fields and potentials then become a combination of steady and periodic values, there are many useful situations that appear to be (RF) steady states when viewed over many cycles – all relevant quantities exhibit coherent oscillations and identical conditions are reproduced within each cycle. When the plasma is sustained by a combination of volume ionization and surface loss, and the response of ions is restricted by their inertia, as is the case in many RF plasmas, the density structure of the plasma shows barely any temporal modulation. The ion space charge in sheath regions is similarly robust. That is, the density profile of the plasma and that of the ions in the sheath remain steady. However, because the electrons are much more mobile, they are able to respond virtually instantaneously, thereby changing the spatial extent of sheaths and quasi-neutral plasmas. The potential profile is related to the spatial distribution of charges through Gauss's law, and this will change in line with applied potentials and consequent rapid redistribution of electrons. In view of the fact that the plasma remains quasi-neutral, any rapid spatial change in potential occurs in the space charge sheath rather than in the plasma.

The discussions in this chapter relate to conditions in the region of a single sheath that is subjected to RF-modulated boundary conditions. Such a situation arises in RF plasmas when a substrate is supported on an independent electrode, separate from any structures involved in plasma generation. In some circumstances the RF excitation of the plasma leads to RF fluctuations of the plasma potential, or it may be that the substrate electrode is itself connected to an independent RF source. In either case there is a component of RF potential across the sheath at the substrate. A similar circumstance can arise around an electrostatic probe inserted in an RF plasma.

The first step will be to reflect on how ions would respond to changing acceleration and also on what might be retained from the original DC steady-state descriptions in the previous chapter. Then certain cases will be analysed to provide quantitative models. In later chapters, these models will be extended to cover the whole of a bounded plasma, sustained by RF fields.

4.1 Response times

4.1.1 RF modulation of a DC sheath

The first situation to be considered is that of a plasma confined by a vessel which will be taken as being at ground potential so that there is some well-defined reference. Suppose too that the plasma is sustained by some external agency that does not require any current to flow to the vessel walls. A space charge sheath will form between the vessel surface and the main body of the plasma comprising a net positive space charge, self-organized in such a way that there is no net current through the sheath – the potential of the wall with respect to that of the plasma will therefore be V_f, as given in Eq. (3.32). Equivalently, one can say that the plasma naturally floats with respect to a grounded vessel at a potential $V_p = -V_f$:

$$V_p = \frac{kT_e}{e} \frac{1}{2} \ln\left(\frac{M}{2\pi m}\right).$$

Next consider an isolated section of the wall, forming an independent electrode. Again, any currents necessary for sustaining the plasma do not pass through this electrode in any significant quantity. That isolated surface in turn floats with respect to the plasma and because it is adjacent to the same plasma, of electron temperature T_e, it will find itself at ground potential, though isolated from it – if it were otherwise a net current would be drawn from the plasma. Suppose next that the electrode is connected to ground via a large capacitor maintaining its DC isolation but now allowing RF currents to cross its sheath. The potential across the capacitor will be whatever it needs to be to ensure that the electrode draws no net current, and it has already been established that this occurs with both sides of the capacitor at ground potential. Figure 4.1 summarizes the above, with the inclusion of a voltage source that at this stage is switched off: $V_1 = 0$. It will be shown later in this chapter that the consequence of $V_1 \neq 0$ is the development of a steady (DC) voltage across the capacitor.

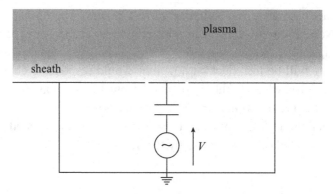

Figure 4.1 A capacitively coupled electrode adjacent to a plasma excited by some unspecified external means; $V = V_1 \sin \omega t$.

4.1.2 Characteristic frequencies

It is helpful to establish the speed with which charged particles can respond to changes in electric fields. It has already been shown that the highest macroscopic electric fields are associated with space charge sheaths. The transport models in Chapter 3 showed that in quasi-neutral plasma the potential changes by about kT_e/e over a distance comparable with the length scale of the vessel, whereas in the sheath the same change in potential occurs over much shorter distances that are comparable with the Debye length, λ_{De}. It follows that the strength of the electric field in the sheath region can be estimated to be greater than, or of the order of, $kT_e/e\lambda_{De}$. The response time for an electron in the vicinity of the sheath might therefore be judged to be the time that a decelerating field of this magnitude would take to slow a thermal electron to rest, or equivalently, starting from rest, the time taken for an electron to reach the thermal speed. In that case, starting from the equation of motion,

$$m\frac{dv}{dt} = -e\frac{kT_e/e}{\lambda_{De}},$$

the time for the speed of the electron to reach $v_e = \sqrt{kT_e/m}$ from rest in a constant electric field, that is the response time τ_e, is found by integrating; the minus sign just gives the direction of the motion, so the response time is:

$$\tau_e = \frac{m\lambda_{De}}{kT_e}\sqrt{kT_e/m} = \frac{\lambda_{De}}{\sqrt{kT_e/m}} = \omega_{pe}^{-1}. \tag{4.1}$$

This is also the characteristic time that is deduced from considering a typical electron ($v = v_e$) travelling through the Debye distance. In Section 2.4 the frequency ω_{pe} was introduced as a characteristic frequency in connection with electromagnetic

4.1 Response times

and electrostatic waves in plasmas – here it is seen to be associated with the response of electrons to the large electrostatic environment of the sheath/plasma boundary.

For ions, a similar argument can be applied but, in the vicinity of the sheath, the important speed to consider is the Bohm speed ($\sqrt{kT_e/M}$) rather than the ion thermal speed. The ion response time in or near the sheath is then

$$\tau_i = \frac{M\lambda_{De}}{kT_e}\sqrt{kT_e/M} = \frac{\lambda_{De}}{\sqrt{kT_e/M}} = \omega_{pi}^{-1}. \quad (4.2)$$

This is also roughly the time it takes an ion to cross the first Debye length of the sheath, having entered at the Bohm speed.

Q Show that the electron and ion plasma frequencies defined in Eqs (4.1) and (4.2) are consistent with

$$\omega_{pe} = \sqrt{\frac{ne^2}{m\varepsilon_0}} \quad \text{and} \quad \omega_{pi} = \sqrt{\frac{ne^2}{M\varepsilon_0}}. \quad (4.3)$$

A From Eq. (4.1)

$$\omega_{pe} = \sqrt{\frac{kT_e ne^2}{m\varepsilon_0 kT_e}} = \sqrt{\frac{ne^2}{m\varepsilon_0}}. \quad (4.4)$$

Equations (4.1) and (4.2) have the same form with m and M interchanged, so the ion plasma frequency expression is also confirmed.

Exercise 4.1: Response times Calculate the ion and electron response times near the boundary of an argon plasma where the charged particle density is 10^{16} m^{-3} and compare them with the period of a 13.56 MHz sine wave.

To further appreciate the significance of plasma frequencies in time-varying sheaths, consider the motion of an ion in an electric field that is of the same order of magnitude as a typical sheath field but which is varying sinusoidally at angular frequency ω. In that case we can model the ion motion as follows:

$$\frac{d^2x}{dt^2} = \frac{e}{M}\frac{kT_e}{e\lambda_{De}}\sin(\omega t),$$

with x the position of the ion. The solution will have the form

$$x = -x_0 \sin \omega t,$$

so

$$x_0\omega^2 = \frac{e}{M}\frac{kT_e}{e\lambda_{De}}.$$

The amplitude of the oscillation thus has the following frequency scaling for oscillations in a field of this characteristic strength:

$$\frac{x_0}{\lambda_{De}} = \frac{\omega_{pi}^2}{\omega^2}. \tag{4.5}$$

Thus, at modulation frequencies greater than the ion plasma frequency, the oscillation amplitude of an ion would be less than the Debye distance. The displacement of electrons scales similarly with ω_{pe} in place of ω_{pi}. The following questions explore in general how a sheath responds to sinusoidal modulation. More detailed analysis then follows in the rest of the chapter.

Q Suppose that in Figure 4.1 the voltage source provides an output voltage $V_1 \sin \omega t$. Outline the nature of the response of the plasma/sheath/electrode system in each of the following cases: (a) $\omega < \omega_{pi}$, (b) $\omega_{pi} < \omega < \omega_{pe}$ and (c) $\omega_{pe} < \omega$; assume that the plasma remains steady so the ion current entering the sheath can be presumed to remain constant.

A (a) $\omega < \omega_{pi}$: There will be slow oscillations in the sheath potential and thickness since Eq. (4.5) suggests large displacements of ions at low frequency.
(b) $\omega_{pi} < \omega < \omega_{pe}$: More rapid oscillatory changes may leave the ions barely disturbed while the electrons are swept back and forth, as the potential across the sheath varies.
(c) $\omega_{pe} < \omega$: In this frequency regime neither ions nor electrons are able to react to the changing potential.
Comment: See further details in the next section.

Q Based on the argument leading up to Eq. (4.5), state how the amplitude of oscillations is likely to scale with the amplitude of an RF electric field.

A Setting the RF field to be α times the characteristic sheath field of $kT_e/e\lambda_{De}$ would introduce a factor of α in Eq. (4.5), so the field dependence of the oscillation amplitude is linear.
Comment: Models of the sheath region are required to give a scaling with RF voltage.

Exercise 4.2: Characteristic frequency Evaluate the ion plasma frequency for a plasma with 10^{16} ions m^{-3} when the ion mass is 1 amu (hydrogen), 18 amu (water) and 40 amu (argon).

4.1.3 Frequency domains

Low-frequency domain ($\omega \ll \omega_{pi}$)

According to Eq. (4.2) ions moving at the Bohm speed would traverse a region that is 1 Debye length wide in a time of ω_{pi}^{-1}. In the very low-frequency regime therefore ions (and electrons) are able to respond faster than an applied alternating voltage would change conditions at an electrode surface. The time for ions to cross the sheath is much shorter than the oscillation period. Very low-frequency oscillations in the sheath potential can therefore be treated as a series of quasi-static states – that is, a DC sheath model might be applied at any instant. The current and voltage might then be estimated from an ion matrix model or one of the Child–Langmuir models discussed in the previous chapter, though allowance may have to be made for the space charge of electrons when the magnitude of the instantaneous potential is low. Whenever the potential across the sheath is equal to the floating potential (Section 3.2.1), the electron and ion current will cancel and the net current in the external circuit will be zero. At all other potentials there is an imbalance of charge arriving at the surface and currents must flow through the external circuit and back to the plasma through the ground electrode. If the ground electrode is sufficiently large compared with the isolated one, then the return electrode will be able to accommodate the return current through a similar but less significant fluctuation of the potential between it and the plasma (see also Section 10.2).

Intermediate-frequency domain ($\omega \lesssim \omega_{pi}$)

As the RF frequency approaches the ion plasma frequency the transit time of ions across a sheath becomes comparable with the RF period. Their incomplete interaction with the varying sheath field under these circumstances complicates the ion dynamics. The energy gained by an ion in crossing the sheath now depends on the phase and frequency of the RF modulation and therefore the ion energy distribution function may be adjusted through these control of parameters – see Section 4.2.3.

Higher-frequency domain ($\omega_{pi} \ll \omega < \omega_{pe}$)

Oscillations at frequencies above the ion plasma frequency tend to leave the ions barely disturbed while the electrons are swept back and forth, as the potential across the sheath varies. In this regime electrons near the plasma/sheath interface are still able virtually instantaneously to redistribute in response to changes in the charge on the capacitor driven by the applied voltage. This is the situation at the boundary in many of the RF-excited plasmas that are discussed in this book.

Q Anticipate the effect on the ions of collisions with the background gas.
A Collisions with background gas will transfer momentum and energy out of the ion motion. The energy distribution at the surface will therefore be modified. *Comment: Collisions were included in the DC models of Chapter 3.*

Beyond the electron plasma frequency ($\omega_{pe} < \omega$)

In this regime even the electrons are unable to keep up with changes induced by the applied voltage. There is insufficient time to maintain the instantaneous quasi-neutrality of the plasma bulk. Under these circumstances an electrostatic disturbance can be launched from the electrode into the plasma as a wave of charge inequality propagating according to the dispersion relation for electron plasma waves, introduced in Section 2.4.3.

4.2 Ion dynamics

4.2.1 Ion motion in a steady sheath

Following on from the consideration of ions oscillating in RF fields, it is instructive to model the motion of ions in a collisionless sheath, first under steady conditions using one of the DC sheath models in Chapter 3 and then in the presence of RF modulation of the sheath potential.

The general equation of motion for a singly charged ion is

$$\frac{d^2x}{dt^2} = \frac{e}{M}E(x).$$

Note that since we are examining the motion of a typical ion, x denotes its position as well as the general spatial coordinate.

In a steady (DC) sheath the electric field is independent of time. The simplest space charge model (Section 3.1.2) assumes constant ion density ($n_i = n_0$ = constant) and no electrons ($n_e = 0$). This can be described as an ion matrix, step model in view of the fixed ion density and the sudden transition in electron density at the boundary. In that case the electric field varies linearly with distance. Starting from zero field at the plasma boundary ($x = s$), on moving into the sheath ($x < s$)

$$E = \frac{n_0 e}{\varepsilon_0}(x - s), \tag{4.6}$$

so for the ions moving in the $-x$ direction

$$\frac{d^2x}{dt^2} = \omega_{pi}^2(x - s).$$

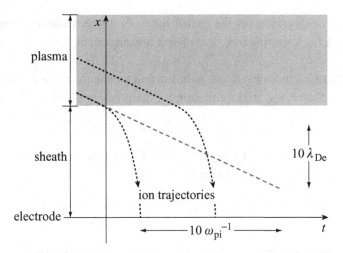

Figure 4.2 Schematic ion trajectories (dashed lines) in a steady sheath: ions approach the sheath boundary close to the Bohm speed, being significantly accelerated when inside the sheath.

Ion motion in a *steady* sheath is not expected to be oscillatory, so an appropriate solution has the form

$$x = s - x_0 \sinh(\omega_{\text{pi}} t), \qquad (4.7)$$

so that

$$v = -x_0 \omega_{\text{pi}} \cosh(\omega_{\text{pi}} t), \qquad (4.8)$$

with $-x_0 \omega_{\text{pi}}$ being the speed of the ions as they enter the sheath at $x = s$. This model is not self-consistent since the ion density is taken as constant even though ions are accelerating. Under the conditions of any step model there is no lower limit on the ion speed at the boundary (since the space charge is always positive), but one might propose matching this crude sheath model to a quasi-neutral plasma that expels ions at the Bohm speed. Applying a Bohm speed boundary condition to a single integration of the motion equation sets the scale length of the ion motion as $x_0 = u_B/\omega_{\text{pi}} = \lambda_{\text{De}}$.

Figure 4.2 shows the trajectories of two ions, the first of which is described directly by Eq. (4.7) and the second one is shifted in time, crossing into the sheath at some later time. In this figure notice how the ions approach from the plasma at the Bohm speed ($u_B = \lambda_{\text{De}} \omega_{\text{pi}}$, so the initial trajectory traverses $10\,\lambda_{\text{De}}$ in $10\,\omega_{\text{pi}}^{-1}$). Note also that the rapid steepening of the curves towards $x = 0$ indicates that the transit time is dominated by the slower motion near the plasma sheath boundary.

Q Derive an expression for the transit time of ions crossing a strongly biased ($eV_0 \gg kT_e$), electron-free, ion matrix sheath, having entered at the Bohm speed.

A The time taken by an ion to cross the ion matrix sheath, τ_{IM}, is found by setting $x = 0$ and $x_0 = \lambda_{De}$ in the equation for the position of an ion, Eq. (4.7):

$$0 = s - \lambda_{De} \sinh(\omega_{pi} \tau_{IM}).$$

Next, combining this with Eq. (A.3.1) and then applying the limiting approximation $2\sinh z \to \exp z$ for large z gives

$$\tau_{IM} \approx \omega_{pi}^{-1} \ln(2\sqrt{2eV_0/kT_e}). \tag{4.9}$$

The ion transit time is a few times ω_{pi}^{-1}. It is dominated by the time it takes to cross the first Debye length or so when the ion speed is about u_B, after which ions are rapidly accelerated so that the dependence on the potential across the sheath is weak.

For an ion moving without collisions in a steady field its kinetic energy change is derived directly from the local potential

$$\Delta w = eE\Delta x.$$

So, direct integration of the electric field in Eq. (4.6) gives the energy gained by an ion as it moves in the steady field from the sheath/plasma boundary at $x = s$ to $x = 0$ as

$$\int_s^0 eE\,dx = \frac{1}{2}\frac{n_0 e^2}{\varepsilon_0}s^2.$$

Not surprisingly in this steady case the same result can be deduced from the observation that after falling through the total sheath potential, $V_0 = n_0 e/2\varepsilon_0 s^2$ (Eq. (3.10)), the kinetic energy of an ion will be increased by eV_0. Other distributions of space charge would change the spatial variation of the electric field but this would not affect the net transfer of electrical potential into kinetic energy.

It must be remembered that as a representation of reality, the ion matrix, step model is strictly limited to transients that occur faster than ions can respond (though it also approximates aspects of a highly collisional sheath). Its application here to study aspects of ion motion in a sheath is purely for the purpose of illustrating the general principles in circumstances where the mathematical analysis is tractable. More realistic models in general require more sophisticated analysis or numerical methods.

4.2 Ion dynamics

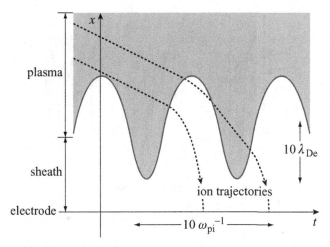

Figure 4.3 Schematic ion trajectories in a temporally modulated sheath ($\omega \sim \omega_{pi}$): ions approach the sheath boundary close to the Bohm speed, being significantly accelerated when inside the sheath.

4.2.2 Ion motion in an RF sheath

When the sheath potential fluctuates at around the ion plasma frequency then it is important to consider the effect of the ion transit time and the phase at which an ion approaches the space charge region. It has already been shown that the transit time $\tau_{IM} \approx \omega_{pi}^{-1}$. Under these circumstances some of the ions will fall into the sheath when it has a relatively small but fast growing potential across it. Other ions, having entered the sheath when the potential difference was rising more slowly close to its highest, may then be overtaken by the plasma boundary during the phase when the sheath width decreases. These latter ions then re-enter the sheath when the potential difference is lower. Figure 4.3 illustrates these different trajectories. When $\omega \sim \omega_{pi}$, the speed of the oscillating plasma boundary becomes comparable with the ion Bohm speed.

The ion trajectories through sheaths modulated by frequencies higher than ω_{pi} become increasingly less responsive to the sheath motion. The case for $\omega \sim 5\,\omega_{pi}$ is shown in Figure 4.4. Once an ion comes within the range of the moving sheath its speed begins to increase, driven effectively by the time-averaged electric field in the region swept by the plasma/sheath boundary.

4.2.3 Ion energy distribution function

For the collisionless fluid models of electropositive plasmas discussed so far, in the absence of temporal variations, ions at any position all have the same speed

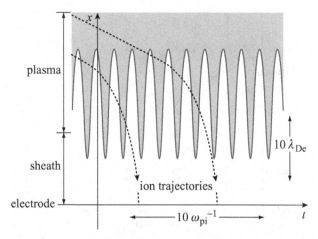

Figure 4.4 Schematic ion trajectories in a temporally modulated sheath ($\omega \sim 5\,\omega_{pi}$): ions approach the sheath boundary close to the Bohm speed, being significantly accelerated when inside the sheath and being repeatedly overtaken by the oscillating sheath.

and energy whether in the plasma or the sheath – the ion energy is simply a function of position. It has also been tacitly assumed that even when collisions with the background gas are significant, the ion fluid remains mono-energetic. In this section the effects of plasma structure, sheath modulation and collisions in the sheath will be considered as corrections to the model to account for the distribution of ion energies that arrive at surfaces immersed in real RF plasmas.

Ion energy distributions emerging from a DC plasma

In the Tonks–Langmuir model of the plasma (Section 3.3.1) the mono-energetic assumption for the ion fluid was not imposed. Instead, this model considers the ion population at any point to be comprised of ions that have fallen freely, starting at rest, from various points upstream – ions are being generated everywhere in proportion to the local electron density. So at the plasma/sheath boundary there will be a distribution of ion speeds (and hence ion energy). Ions born close to the boundary have barely begun to move, whereas the greatest speed and energy corresponds to ions having fallen through the entire potential difference between the centre and the edge ($0.854 kT_e/e$).

An expression for the ion energy distribution function (IEDF) at the plasma/sheath boundary can be extracted from the model equations of Section 3.3.1 [46]. The result is plotted in Figure 4.5. After entering the sheath there will be virtually no addition to the distribution as the electron number density falls rapidly

4.2 Ion dynamics

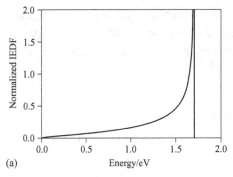
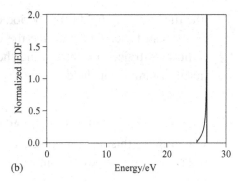

Figure 4.5 The ion energy distribution function from a Tonks–Langmuir model with $kT_e/e = 2$ V (a) at the plasma/sheath boundary – the maximum ion energy corresponds with ions falling through $0.854kT_e/e$ and (b) after collisionless acceleration through a 25 V DC sheath – the entire distribution is shifted up the energy axis by the potential across the sheath.

and so the entire distribution is shifted by the change in electrostatic potential, and the IEDF maintains its shape (the height and width are unchanged). At a surface where the potential is V_s below that of the plasma/sheath boundary the IEDF is spread between eV_s and $eV_s + 0.854kT_e$. As Figure 4.5 indicates, after crossing a DC sheath with several kT_e/e across it, the distribution tends to become virtually mono-energetic.

Q What will be the effect of elastic collisions (i) in the plasma and (ii) in the sheath?

A (i) Collisions will spread the distribution to lower energy so that the peak in Figure 4.5 will be lowered while the numbers of particles at lower energies are increased, maintaining the same area under the curve. (ii) As the distribution is being shifted towards higher energy by the sheath potential, collisions will be spreading the distribution back down towards zero energy, again maintaining the same area under the curve.
Comment: In the sheath, ions readily gain sufficient energy to excite and ionize the background gas – ionization will add new particles to the distribution.

Q In a steady-state sheath, the conservation of energy ensures a transfer of electrical potential into the kinetic energy of ions at any point in space. Suggest how to determine the energy of an ion when the sheath field is changing.

A When the electric field changes in time the energy of any particular ion at a given point in space is no longer related to the local electrical potential. In that case the ion energy will depend on the detail of the trajectory through the time-varying field if the period of the oscillation is comparable with the ion transit time. To determine the energy of an ion it will be necessary to follow its trajectory, integrating the energy it gains through motion in the local instantaneous field.

Low-frequency sinusoidal modulation

Warning: In this section be careful to distinguish between particle energy, w, and angular frequency, ω (omega).

For simplicity consider first the case of mono-energetic ions crossing into a sheath, across which there is a large sinusoidal modulation of potential. The voltage across the sheath changes most slowly through the phases of maximum and minimum. Thus if the period of the modulation is much larger than the ion transit time, at the surface there will be markedly more ions having energies corresponding to the maximum and minimum potentials compared with the intermediate range. This is illustrated in Figure 4.6. The shape of this idealized distribution is derived as follows. Ions arrive at the plasma boundary at a steady rate dN/dt. The potential difference across the sheath is presumed to comprise a steady component, V_0 (which will be considered further in Section 4.3), and a sinusoidal component with amplitude $V_1 \leq V_0$:

$$w = e(V_0 + V_1 \sin \omega t) + M u_B^2/2.$$

This equation conveniently links energy and time. Then, defining the ion energy distribution function at any energy w as the number of ions with energy in the range w to $w + dw$:

$$f_i(w) = \frac{dN}{dw} = \frac{dN/dt}{dw/dt}$$
$$= \frac{dN/dt}{eV_1 \omega \cos \omega t}.$$

For a high modulation voltage, $V_1 \gg kT_e/e$, it is scarcely necessary to consider a more realistic distribution of ions leaving the plasmas as in the previous section. However, in the case of low modulation the mono-energetic behaviour needs to be convolved with the distribution that arrives at the plasma/sheath boundary (Figure 4.5).

Q How would the low-frequency modulation of the IEDF be affected by (i) a modulation waveform that had narrow peaks and broad troughs and (ii) collisions?

A (i) An asymmetric waveform that spends more time through the minimum potential region, and least time through the high potential region, will lead to an asymmetric IEDF with the lower peak more emphasized than the upper peak.
(ii) Collisions will scatter some ions down to a lower energy, reducing the peaks and spreading the distribution down to zero energy.

Intermediate and high-frequency sinusoidal modulation

When the modulation frequency is increased to around the characteristic response frequency of the ions, they no longer respond fully to the RF fields. A full solution of the ion dynamics requires numerical methods that relate the instantaneous values of the potential across the sheath, the width of the sheath and the charged particle currents [47].

At intermediate frequency the ions cross the sheath in a time that is comparable with the RF period. When the pressure is high enough for ion–neutral collisions to occur within the sheath, the ion trajectory depends not only on the phase at which the ion enters the sheath but also the phase at which a collision occurs. In certain conditions the resulting IEDF contains many pairs of peaks in the cascade down to lower energy, corresponding with the periodic field structure in the sheath.

As one might anticipate, increasing the modulation frequency narrows the IEDF, effectively drawing the high and low-energy peaks closer together. At sufficiently high frequency ($\omega \gg \omega_{pi}$), the peaks effectively merge to give a mono-modal IEDF. Under these circumstances ions spend several RF cycles in the region that is swept by the sheath/plasma boundary. Figure 4.4 shows schematically how ion trajectories develop over several cycles of high-frequency oscillations. Notice that the ions are never slowed as they pass out of the bulk plasma onto the electrode (or other) surfaces. Furthermore, the ion motion in this regime is effectively governed by the average fields so that the trajectories closely resemble those of a steady sheath (Figure 4.2). A review of various models of IEDFs is given in [48].

Q The IEDF at a surface where the sheath is subjected to an RF modulation is shown in Figure 4.7; comment on the modulation frequency and the ion–neutral collision frequency, ν_{i-n}.
A In the collisionless and weakly collisional regimes there is a single, 'high'-energy peak so it appears that $\omega \gg \omega_{pi}$; as the pressure increases the distribution is spread to lower energies suggesting that $\omega_{pi} \sim \nu_{i-n}$ at 4 Pa.

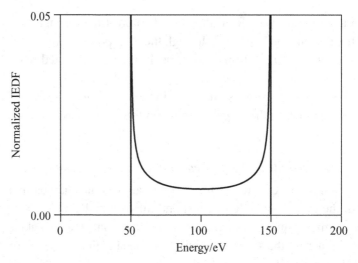

Figure 4.6 The normalized ion energy distribution (IEDF) at a surface that is biased 100 V (DC) below the potential of the plasma and modulated with a 50 V amplitude sinusoidal potential – singly charged ions, in the absence of collisions, $\omega \ll \omega_{pi}$.

Q The IEDF at a surface where the sheath is subjected to an intermediate frequency modulation $V_1 \sin \omega t$ will have a bimodal structure for heavy ions. How would you expect the energy separation between the peaks (Δw) to scale with the V_1, ω and M, the ion mass.

A The electric field in the sheath will increase with V_1 (though not necessarily in strict proportion) and the extent of the ion response will decrease as ω/ω_{pi} increases, so one might tentatively propose

$$\Delta w \propto V_1/\sqrt{M}\omega.$$

Comment: This does indeed appear to be the case [48].

4.3 Electron dynamics

In the previous section it was supposed that the instantaneous voltage across an RF-modulated sheath would have a DC component of potential. This is generally the case and in this section it will be shown that the non-linearity of the electron dynamics is responsible for rectifying the RF voltage.

In an RF sheath the current collected by any electrode has, in addition to the electron and ion (particle) currents, a contribution from the time-varying electric field. This extra component of RF current increases in proportion with the frequency

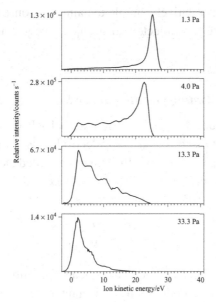

Figure 4.7 Argon ion energy distributions at the grounded electrode of a 13.56 MHz plasma with low power input (~2 W) in a CCP between 10 cm diameter plates, 2.5 cm apart [49].

since it depends on the rate of change of electric field. The electric field at an electrode is directly related to its surface charge. In a plasma environment that surface charge is itself matched by the space charge in the sheath, since the electric field falls rapidly in crossing from the sheath (where it is of $\sim V_{\text{sheath}}/\lambda_{\text{De}}$) into the quasi-neutral plasma (where it is $\sim kT_{\text{e}}/el$). This means that the displacement of charge at the sheath/plasma boundary (say n_0 electrons pushed back at a speed u_s) is directly linked to the electric field at the electrode. The associated conduction current in the plasma equals the displacement current at the electrode:

$$n_s e u_s = \varepsilon_0 \left. \frac{dE}{dt} \right|_{\text{electrode}}. \qquad (4.10)$$

When the sheath motion is periodic, the displacement current averages to zero, whereas particle currents do not necessarily do so, though conditions must be such that the net particle current is zero for an isolated surface. During continuous RF modulation of the potential between a surface and the plasma, the instantaneous electron flux arriving at the surface will be modulated through the dependence of the Boltzmann retardation factor on the instantaneous potential:

$$\Gamma_{\text{e}} = \frac{1}{4} n_s \bar{v}_{\text{e}} \exp\left(\frac{eV(t)}{kT_{\text{e}}}\right). \qquad (4.11)$$

This section applies this relatively simple dynamic response to the high-frequency regime when a surface is subjected to a capacitively coupled RF modulation (Figure 4.1).

4.3.1 Floating potential under RF bias ($\omega \gg \omega_{pi}$)

With reference to Figure 4.1, before any external RF oscillation is started, the electron and ion fluxes must be equal as the electrode is isolated by the capacitance – the Bohm flux of ions is balanced by a Boltzmann-retarded thermal flux of electrons. The voltage difference between the electrode surface and the plasma boundary will be the floating potential defined in the previous chapter:

$$V_{f_{DC}} = \frac{kT_e}{2e} \ln\left(\frac{2\pi m}{M}\right), \qquad (4.12)$$

where the subscript has been extended to indicate the ambient DC condition.

During continuous RF modulation the instantaneous electron flux arriving at the surface will be modulated through the dependence of the Boltzmann retardation factor on the instantaneous potential, whereas that of the ions will remain unchanged. Since there can be no steady current passing through the capacitor, conditions must settle down such that the electron flux averaged over the RF cycle equals the steady ion flux, otherwise the net current would not be zero. The displacement current automatically averages to zero, so it does not need to be included here. Therefore

$$< \frac{1}{4} n_s \bar{v}_e \exp\left(\frac{e(V_1 \sin \omega t + V_{f_{RF}})}{kT_e}\right) > = n_s u_B, \qquad (4.13)$$

where the potential $V_{f_{RF}}$ relates to floating with RF bias, and is the yet to be determined 'RF floating potential' and the angled brackets imply averaging over one period ($2\pi/\omega$):

$$< f(t) > = (\omega/2\pi) \int_0^{2\pi/\omega} f(t) dt.$$

Since in the present case $f(t)$ includes an exponential function, it is useful to know that

$$(\omega/2\pi) \int_0^{2\pi/\omega} \exp(a \sin \omega t) dt = I_0(a) \qquad (4.14)$$

in which I_0 is the zero-order modified Bessel function which has the form shown in Figure 4.8; for small arguments the function tends to unity and for large arguments it becomes close to a pure exponential.

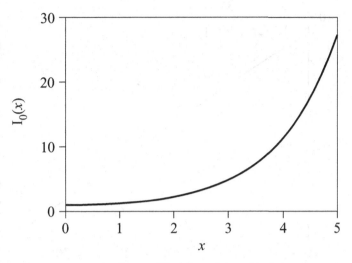

Figure 4.8 The modified Bessel function $I_0(x)$.

Q Show that the presence of an RF voltage across a sheath shifts its floating potential by $-(kT_e/e)\ln(I_0(eV_1/kT_e))$ (which is known as the RF self-bias).

A Starting from Eq. (4.13), the time-averaged flux balance requires that

$$\frac{1}{4}n_s\bar{v}_e \exp\left(\frac{e(V_{f_{RF}})}{kT_e}\right)(\omega/2\pi)\int_0^{2\pi/\omega}\exp\left(\frac{eV_1\sin(\omega t)}{kT_e}\right)dt = nu_B.$$

Expressing the integral in terms of the modified Bessel function:

$$\frac{1}{4}n_s\bar{v}_e \exp\left(\frac{e(V_{f_{RF}})}{kT_e}\right)I_0\left(\frac{eV_1}{kT_e}\right) = nu_B.$$

Inserting the usual expressions for $\bar{v}_e$ and u_B, rearranging and taking logarithms now gives the floating potential under RF-biased conditions:

$$V_{f_{RF}} = \frac{kT_e}{e}\left[\frac{1}{2}\ln\left(\frac{2\pi m}{M}\right) - \ln I_0\left(\frac{eV_1}{kT_e}\right)\right]. \quad (4.15)$$

Comparison with Eq. (4.12) shows the additional shift in floating potential.

A reference electrode without RF bias would adopt a potential given by Eq. (4.12), so that the potential difference between the biased and unbiased electrode amounts to just the second term in Eq. (4.15). This term is therefore often called the 'self-bias voltage'. This relationship between the RF self-bias and the amplitude of the applied RF voltage is shown in Figure 4.9. For $V_1 < kT_e/e$ the floating potential is close to the DC value, but for $V_1 \gg kT_e/e$ the magnitude of the floating potential gets ever closer to the RF amplitude. Figure 4.10 shows

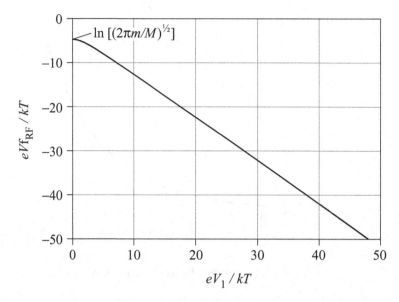

Figure 4.9 Floating potential for a sinusoidally modulated sheath in an argon plasma (40 amu); the floating potential approaches the RF amplitude V_1 at high bias while at very low bias the RF self-bias becomes negligible and the floating potential approaches the DC value.

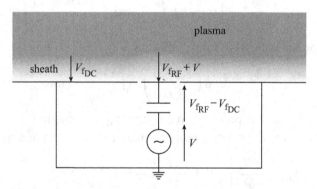

Figure 4.10 Distribution of DC and RF potentials assuming that the area of the RF-driven electrode is small compared with the ground area and that external capacitance is sufficiently large for it to have negligible RF impedance compared with the electrode sheath; $V = V_1 \sin \omega t$.

how the DC and RF potentials are primarily distributed across the various elements when the RF electrode is small and the external capacitance is large. In terms of Kirchoff's laws, the original voltage sources are the external RF and the normal floating potential at the ground sheath. The RF voltage almost entirely appears inverted across the adjacent sheath, accompanied by a DC component, the

4.3 Electron dynamics

self-bias, that adds to the normal floating potential; the self-bias voltage must also appear inverted across the external capacitor so that there is no net voltage around a loop from the plasma, round the external circuit and back to the plasma.

The RF impedance of the sheath has not been required for this discussion of the self-bias but it will be considered later after a fuller model of the sheath has been outlined.

4.3.2 Ion flux and energy onto an RF electrode

On the basis of the above discussion estimates can be made of the thickness of the RF sheath. More importantly, estimates can also be made of the flux of ion kinetic energy arriving at a plane RF-biased electrode. Suppose that the plasma in Figure 4.10 is formed in argon at 1 Pa (ion–neutral mean free path about 4 mm), has a mean charged particle density of 10^{16} m^{-3} and an electron temperature of 2 eV. Suppose also that the RF potential applied to the isolated electrode has an amplitude of 50 V at a frequency of 13.56 MHz.

Exercise 4.3: Collisionality of RF sheaths On the basis of the DC components of voltage only, use the ion matrix model to estimate the width of the sheath at the surface of the RF-biased electrode in Figure 4.10 given the above conditions and suggest whether or not ions are likely to cross this space without collision.

Ions that cross the sheath without collisions arrive at the electrode with kinetic energy equivalent to the electrostatic potential energy. The ion plasma frequency compared with the RF frequency is

$$\frac{\omega_{pi}}{\omega} = \frac{\sqrt{\left(10^{16} \times [1.6 \times 10^{-19}]^2\right) / \left(1.67 \times 10^{-27} \times 40 \times 8.9 \times 10^{-12}\right)}}{2\pi \times 13.56 \times 10^6} \sim \frac{1}{4}.$$

So the ions will barely respond to the RF fluctuations and will reach the electrode having an energy close to the mean sheath potential $E_{\text{ion}} \sim -eV_{\text{fRF}}$.

Exercise 4.4: Ion energy flux Estimate the ion energy flux onto the surface of the RF-biased electrode in Figure 4.10 for a collisionless plasma of central density 10^{16} m^{-3} and electron temperature 2 eV.

4.3.3 RF self-bias with more than one frequency component

If they are completely independent, which is to say that there is no phase synchronization between them, additional RF components each produce additional self-bias in proportion to $\ln(I_0(a_i))$, where a_i is the amplitude of the ith component.

When there is a strict phase relationship then the self-bias depends on that phase difference, especially if the frequencies are not too widely separated (such as a fundamental and its second harmonic). For phase-locked harmonic waveforms the self-bias is generally less than would be achieved if the effects of each simply contributed independently.

Q Calculate the self-bias of an isolated electrode for a square wave modulation, amplitude V_{sq}, period τ.

A Modifying the integral in Eq. (4.13):

$$\frac{1}{4}n\bar{v}_e \exp\left(\frac{e(V_{\text{fRFsq}})}{kT_e}\right)(1/\tau)\left[\int_0^{\tau/2} \exp\left(\frac{eV_{\text{sq}}}{kT_e}\right) dt \right.$$

$$\left. + \int_{\tau/2}^{\tau} \exp\left(-\frac{eV_{\text{sq}}}{kT_e}\right) dt \right] = nu_B,$$

which simplifies to

$$V_{\text{fRFsq}} = \frac{kT_e}{e}\left[\frac{1}{2}\ln\left(\frac{2\pi m}{M}\right) - \ln\left(\cosh\left(\frac{eV_{\text{sq}}}{kT_e}\right)\right)\right].$$

Exercise 4.5: Tailored IEDF The form of the IEDF depends on various factors. Suggest how these might be used to manipulate the precise shape of the IEDF to achieve a narrow distribution at a specific energy on a biased, insulating substrate.

4.4 Analytical models of (high-frequency) RF sheaths

This section investigates the structure and impedance of an RF sheath in the frequency range $\omega_{\text{pi}} \ll \omega \ll \omega_{\text{pe}}$. It is necessary to build models that include the distribution of space charge and the displacement currents (associated with time-varying fields) as well as the particle currents. A simple analysis treats the space charge using a constant ion density ($n = n_0$), as in the DC ion matrix model; this case is sometimes also called a 'homogeneous' sheath model. More realistic models include ion motion resulting in a non-uniform (or 'inhomogeneous') distribution of ion space charge.

4.4.1 Equivalent circuit of an RF sheath

There are three parallel contributions to the current that crosses an RF-modulated sheath formed at an electrode of area A, namely the ion, electron and displacement currents:

$$I_{\text{RF}} = -n_0 e u_B A + \frac{n_0 e \bar{v}_e}{4} A \exp\frac{-eV_{\text{sh}}}{kT_e} + I_d. \tag{4.16}$$

4.4 Analytical models of (high-frequency) RF sheaths

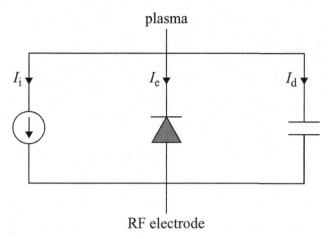

Figure 4.11 A schematic circuit model of an RF sheath featuring a constant current source (ion current), a reverse-biased diode (electron current) and a non-linear capacitor (displacement current). The ion source and diode will each dissipate power in proportion to the current through it and the component of in-phase voltage across it.

In the high-frequency range the steady Bohm flux of ions in circuit terms has the appearance of a constant current source. The electron current is controlled by a Boltzmann exponential function of the voltage across the sheath, which gives it the same current–voltage characteristic as a reverse-biased diode. Since a capacitor is the archetypal component through which a displacement current flows, this provides the third parallel element of the equivalent circuit shown in Figure 4.11. This third element though is no ordinary capacitor, as the capacitance changes in response to the RF modulation – the dielectric (sheath) thickness depends on the amplitude of the RF modulation. Therefore it is necessary next to consider models for the behaviour of the space charge in an RF sheath so that this non-linear element can be specified.

4.4.2 Constant ion density models of RF sheaths

The first case to examine is that of a constant ion density (ion matrix or homogeneous) sheath with a superimposed RF modulation. From the perspective of the electrons, the ions therefore appear to form a fixed matrix of space charge, which has a constant density, as in the DC case. Although this model is not self-consistent, its simplicity allows rapid insight without the burden of numerical methods. Figure 4.12 sets out the geometry and confirms the definition of key quantities in the analysis. The task is to find expressions for the current, sheath potential and sheath width that arise in response to the application of an RF modulation of the sheath.

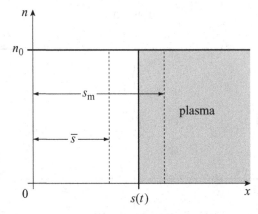

Figure 4.12 An RF ion matrix sheath; the ion density is constant throughout at n_0 and $s(t)$ is the instantaneous sheath width, so that $n_e = n_0$ for $x > s(t)$; s_m is the maximum sheath extent and $\bar{s}$ is the mean sheath width.

Gauss's law in the sheath relates the electric field gradient to the net space charge:

$$\frac{\partial E}{\partial x} = \frac{n_0 e}{\varepsilon_0}\left(1 - \exp\frac{e\phi}{kT_e}\right), \tag{4.17}$$

in which Φ is the potential with respect to that in the plasma where $n_e = n_0$. The electron space charge can be ignored in Gauss's law wherever $|e\phi/kT_e| \gg 1$. It is reasonable to continue the analysis on the basis that this holds everywhere, presuming the total voltage across the sheath to be so large that the region where $e\phi/kT_e \gg 1$ does not hold is small in extent. Then, using $E = 0$ at $x = s(t)$, that is at the plasma/sheath boundary, Eq. (4.17) can be simply integrated to give the electric field at any point in the sheath as

$$E(x,t) = \frac{n_0 e}{\varepsilon_0}(x - s(t)). \tag{4.18}$$

The total voltage across the sheath (with respect to the electrode), at any instant, is found by integrating the field from the electrode at $x = 0$, to the place where electron and ion number densities are equal:

$$\begin{aligned}
V_{sh}(t) &= -\int_0^{s(t)} E(x,t)\,dx \\
&= -\left[\frac{n_0 e}{\varepsilon_0}\left(\frac{x^2}{2} - s(t)x\right)\right]_0^{s(t)} \\
&= \frac{n_0 e}{2\varepsilon_0}s(t)^2.
\end{aligned} \tag{4.19}$$

4.4 Analytical models of (high-frequency) RF sheaths

Equation (4.18) also shows that Maxwell's displacement current in the sheath is everywhere given by

$$\varepsilon_0 \frac{\partial E}{\partial t} = -n_0 e \frac{ds}{dt}.$$

This also conveniently shows that the displacement current in the sheath is continued across the sheath/plasma boundary as a conduction current carried by the motion of plasma electrons ($-n_0 e$), moving with the speed that defines the boundary motion (ds/dt). The current in the external circuit as it enters the sheath at the electrode can be viewed as the sum of three contributions, namely the displacement, ion and electron currents:

$$J(t) = -n_0 e \frac{ds}{dt} - n_0 e u_B + \frac{n_0 e \bar{v}_e}{4} \exp \frac{-eV_{sh}}{kT_e}, \quad (4.20)$$

where the signs are consistent with Figure 4.12 and the electron current has been determined in terms of the flux reaching the electrode surface. The potential difference V_{sh} measures the instantaneous voltage across the sheath and is referenced to the electrode rather than to the plasma; it turns out that this makes the analysis simpler on this occasion. Note that although the electron space charge in the sheath has been neglected, the electron current cannot be ignored so easily, owing to the high thermal speed of electrons.

To establish the current, voltage and sheath width, Eqs (4.19) and (4.20) must be solved together with a third equation involving at least one of these quantities. Two cases are considered – voltage-driven sheath modulation and current-driven sheath modulation.

Voltage drive

The non-linearity of the sheath has already been shown to rectify applied RF voltage, leading to self-bias, so that the voltage-driven sheath with zero mean current has an instantaneous bias *with respect to the electrode surface* given by

$$V_{sh}(t) = V_1 \cos \omega t - V_{f_{RF}}. \quad (4.21)$$

Then, using Eq. (4.19), the sheath edge motion is found to be

$$s(t) = [V_1 \cos \omega t - V_{f_{RF}}]^{1/2} \left(\frac{2\varepsilon_0}{n_0 e}\right)^{1/2} \quad (4.22)$$

and the mean sheath width is

$$\bar{s} = \left[2 \left|\frac{eV_{f_{RF}}}{kT_e}\right|\right]^{1/2} \lambda_{De}.$$

Inserting this expression for $s(t)$ into Eq. (4.20) for the total current density gives

$$J(t) = +(2\varepsilon_0 n_0 e)^{1/2} \frac{V_1 \sin(\omega t)}{\left[V_1 \cos \omega t - V_{f_{RF}}\right]^{1/2}} - n_0 e u_B$$
$$+ \frac{n_0 e \bar{v}_e}{4} \exp \frac{-e\left(V_1 \cos \omega t - V_{f_{RF}}\right)}{kT_e}. \quad (4.23)$$

The first term on the RHS averages to zero owing to its trigonometric components, as it must do since it represents periodic displacement current. The second and third terms cancel since $V_{f_{RF}}$ was defined in Section 4.3.1 precisely to ensure that they would, so that $<J(t)> = 0$. Note that although the current has been constrained to have no DC component, Eq. (4.23) contains frequency components at higher harmonics – see Figure 4.13.

Thus the voltage-driven RF sheath in the ion matrix approximation has voltage, width and current described, respectively, by Eqs (4.21), (4.22) and (4.23). Since the current is not a pure sinusoid and $|J(t)|$ is not directly proportional to $\tilde{V}$, the circuit concept of a steady complex impedance is not applicable – a circuit element based on the current–voltage relationship of a sheath will be non-linear.

Current drive

An alternative way to drive an RF modulation is to specify the current injected into the plasma through a capacitively coupled electrode such as that in Figure 4.1. A pure sinusoidal current might be thought to be the most appropriate choice, but for an analytical solution a different one must be made. By imposing a total current that has the following form:

$$J(t) = -J_0 \sin \omega t - n_0 e u_B + \frac{n_0 e \bar{v}_e}{4} \exp \frac{-eV_{\text{sh}}(t)}{kT_e}, \quad (4.24)$$

a simple expression can be obtained for the rate of change of sheath width:

$$n_0 e \frac{ds}{dt} = J_0 \sin \omega t.$$

It then follows that the sheath width is subjected to a purely sinusoidal variation of amplitude

$$s_0 = \frac{J_0}{n_0 e \omega}, \quad (4.25)$$

4.4 Analytical models of (high-frequency) RF sheaths

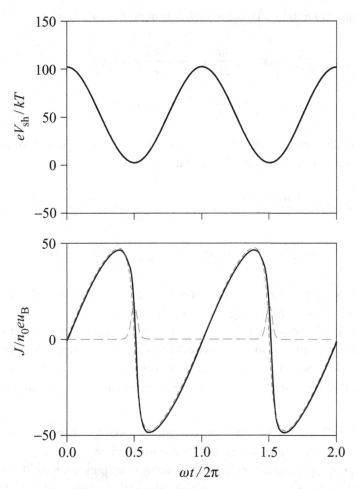

Figure 4.13 Normalized sheath voltage, eV_{sh}/kT_e, and normalized currents during two cycles of voltage-driven RF modulation of a constant density ion matrix sheath. Total current J/n_0eu_B is the solid line; the displacement current component is shown as a grey short-dashed line and the electron current is shown as a grey long-dashed line; the ion current is a steady component of magnitude -1. For the case shown $n_0 = 10^{16}$ m^{-3}, $kT_e/e = 2$ V, $V_1 = 100$ V. Compare with Figure 4.14.

such that

$$s(t) = \bar{s} - s_0 \cos \omega t. \tag{4.26}$$

The mean sheath width $\bar{s}$ is to be determined by the requirement for capacitive coupling, which is that

$$< J(t) = 0 >.$$

Inserting this $s(t)$ into the basic ion matrix model, Eq. (4.19), gives

$$V_{\text{sh}}(t) = \frac{n_0 e}{2\varepsilon_0} \left(\bar{s} - \frac{J_0}{n_0 e \omega} \cos \omega t \right)^2,$$

$$= \frac{n_0 e}{2\varepsilon_0} (\bar{s} - s_0 \cos \omega t)^2,$$

which can be expanded to give the DC and time-varying components of the voltage that necessarily appears across the sheath:

$$V_{\text{sh}}(t) = \frac{n_0 e \bar{s}^2}{2\varepsilon_0} - \frac{\bar{s} J_0}{\varepsilon_0 \omega} \cos \omega t + \frac{J_0^2}{2\varepsilon_0 n_0 e \omega^2} \left(\frac{1 + \cos 2\omega t}{2} \right), \quad (4.27)$$

$$= \frac{n_0 e}{2\varepsilon_0} \left[\bar{s}^2 - 2\bar{s} s_0 \cos \omega t + s_0^2 \left(\frac{1 + \cos 2\omega t}{2} \right) \right]. \quad (4.28)$$

The time-varying part has a second harmonic term in addition to that at the drive frequency. Now using this voltage in Eq. (4.24) and setting the average total current to zero effectively constrains $\bar{s}$. The mean electron conduction current cannot be simplified in terms of the modified Bessel function (as in Section 4.3.1) owing to the appearance of the second harmonic term, so it is necessary to resort to a numerical iteration to find the value of $\bar{s}$ that leads to zero net current. See Figure 4.14.

> **Q** Compare Figure 4.14 with the current and voltage behaviour for a pure capacitance and comment on the sheath impedance.
> **A** For a pure capacitance the current and voltage are purely sinusoidal and out of phase by 90°; at first sight this looks similar to the case in the figure. However, the sheath voltage is not a pure sinusoid (and even has a DC component) and $|V_{\text{sh}}|$ is not directly proportional to J_0, so the circuit concept of a steady complex impedance does not properly fit a single, ion matrix sheath. The sheath impedance is both non-linear and time-dependent.

The current-driven RF sheath in the ion matrix approximation has current and voltage described, respectively, by Eqs (4.24) and (4.28). Its width must in general be found by iteration. If $J_0 \gg n_0 e u_B$ the current is only significantly non-sinusoidal during the times near the voltage minimum, when there is a pulse of electron current; this pulse is discernible in Figure 4.14. Numerical analysis shows that as J_0 increases above about $0.3 \times n_0 e \bar{v}_e/4$, the departure of the current from a simple sinusoid becomes less than 10%; the mean sheath thickness $\bar{s}$ is then within 20%

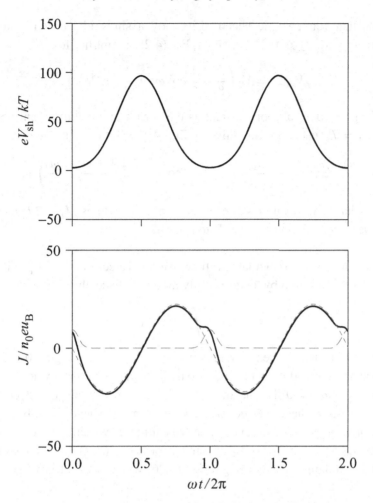

Figure 4.14 Normalized sheath voltage, eV_{sh}/kT_e, and normalized currents during two cycles of current-driven RF modulation of a constant density ion matrix sheath. Total current $J/n_0 eu_B$ is the solid line; the displacement current component is shown as a grey short-dashed line and the electron current is shown as a grey long-dashed line; the ion current is a steady component of magnitude -1. For the case shown $n_0 = 10^{16}$ m^{-3}, $kT_e/e = 2$ V, $J_0 = 75$ A m^{-2}. Compare with Figure 4.13.

of s_0. So, for large external, sinusoidal currents the sheath motion can be presumed also to be sinusoidal, but in that case the sheath must be presumed to vanish for one instant in each cycle, which means at that moment the potential difference between the electrode and the plasma is zero, allowing the ion charge built up during that cycle to be neutralized – this simplification will be used in the next chapter.

Q Show that for a single sheath driven by a sinusoidal current waveform $J_0 \sin \omega t$ with $J_0 \gg 0.3 \times n_0 e \bar{v}_e/4$, Eq. (4.28) simplifies to

$$V_{\rm sh}(t) = V_0 \left(\frac{3}{8} - \frac{1}{2} \cos \omega t + \frac{1}{8} \cos 2\omega t \right). \qquad (4.29)$$

A If $V_{\rm sh}$ goes to zero at one instant as the sheath momentarily vanishes, then $\bar{s} = s_0 = J_0/n_0 e \omega$. Putting this into Eq. (4.28) and rearranging gives

$$V_{\rm sh}(t) = \frac{J_0^2}{2\varepsilon_0 n_0 e \omega^2} \left(1 - 2 \cos \omega t + \left(\frac{1 + \cos 2\omega t}{2} \right) \right),$$

which has a maximum value at $\omega t = \pi$ that defines $V_0 = 2J_0^2/\varepsilon_0 n_0 e \omega^2$. Further manipulation then leads to Eq. (4.29).

It is also interesting to calculate the time-averaged electron density profile within the region swept through by a periodically modulated sheath:

$$\bar{n}_e(x) = \frac{\omega}{2\pi} \int_0^{2\pi/\omega} n_e(x, t) dt.$$

The high-current-driven sheath sweeps back and forth in the region $0 < x \le 2s_0$; the motion is sinusoidal so one need only consider half the cycle. According to Eq. (4.26) at $\omega t = 0$ the sheath starts with zero width and the electron front reaches right up to the electrode, while at $\omega t = \pi$ the sheath is fully expanded. Consequently, in the first half cycle, at any position x_1 within the region swept by the sheath, $n_e(x, t) = n_0$ for $s(t) \ge x_1$ and elsewhere $n_e(x, t) = 0$ – the switchover from n_0 to zero happens, according to Eq. (4.26) with $\bar{s} = s_0$, at x_1 when

$$\omega t_1 = \arccos \left[\frac{s_0 - x_1}{s_0} \right]. \qquad (4.30)$$

Therefore

$$\bar{n}_e(x_1) = \frac{\omega}{\pi} n_0 \int_0^{t_1} dt$$

$$= \frac{n_0}{\pi} \arccos \left[\frac{s_0 - x_1}{s_0} \right].$$

Taking the average over space, it turns out that

$$\frac{1}{2s_0} \int_0^{2s_0} \bar{n}_e(x_1) dx_1 = \frac{n_0}{2} \qquad (4.31)$$

shows that globally, the spatially and temporally averaged electron density in the region swept by the sheath is half that in the plasma bulk. Though perhaps not

4.4.3 Child–Langmuir models of RF sheaths ($\omega \gg \omega_{pi}$)

The previous section treated the RF sheath as an extension of the DC ion matrix model with particle currents. The model is not self-consistent as the ion space charge is not allowed to respond to the local electric field. In this section that restriction is lifted and the ion space charge is now determined through motion in the mean electric field. From the perspective of the electrons, the ions therefore again appear to form a fixed matrix of space charge, though now that charge is not uniformly distributed in space. Because the ion density is non-uniform, this scenario is also sometimes described as the basis of the 'inhomogeneous' model of an RF sheath.

Q Under what circumstances is it justified to suppose that the ions respond only to the *mean* local electric field in the sheath?

A It was shown at the start of this chapter that the ion transit time across the sheath is comparable with the reciprocal of the ion plasma frequency, so when $\omega/\omega_{pi} \gg 1$ the ions take several RF cycles to cross the sheath; within one RF period the ion speed cannot be significantly altered. Therefore in this frequency regime the ion motion is determined by the mean field.

It has been shown in Chapter 3 that taking a more consistent account of the ion space charge in a steady sheath leads to the Child–Langmuir relationship between the current, the thickness of the sheath and the potential difference across it. That model will now be extended to analyse an RF sheath. The density of the ion fluid in the RF sheath is presumed to be determined by acceleration in the *mean* (time-averaged) field, with ions entering the RF-modulated 'sheath' region at the Bohm speed. The equations to be solved are

$$\frac{\partial E(x,t)}{\partial x} = \frac{e}{\varepsilon_0}[n_i(x) - n_e(x,t)], \tag{4.32}$$

$$\frac{\partial \phi(x,t)}{\partial x} = -E(x,t), \tag{4.33}$$

$$n_i(x) = n_0 \left(1 - \frac{2e}{kT_e} <\phi(x,t)>\right)^{-1/2}, \tag{4.34}$$

$$n_e(x,t) = n_0 \exp\left(\frac{e\phi(x,t)}{kT_e}\right), \tag{4.35}$$

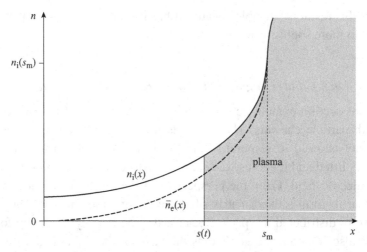

Figure 4.15 Charged particle densities in the vicinity of an RF Child–Langmuir sheath. The solid line shows the profile of steady ion density; instantaneous electron density is represented by the grey area and is delimited by $x = s(t)$. The mean electron density is shown by the broken line. Ions enter the ion sheath ($x = s_m$) at the Bohm speed and fall freely in the mean potential.

with boundary conditions that set the potential to zero where the electron and ion densities are equal and that prescribe either the voltage across the sheath or the total current through it. These are already too complicated to solve analytically, and an obvious first simplification is to use again the notion that the electron density steps discontinuously to zero at the boundary of the oscillating sheath structure. That replaces Eq. (4.35) with

$$n_e(x, t) = n_i(x) \quad \text{for } x \geq s(t)$$
$$= 0 \quad \text{for } x < s(t),$$

which is a reasonable approximation when the total voltage across the region is much larger than kT_e/e. This situation is illustrated in Figure 4.15, which also shows the resulting average electron density. There are two important things to note. First, as before, from the ion perspective there is a clear and fixed boundary between a neutral plasma and a non-neutral sheath that is located at $x = s_m$. On the sheath side of the sheath/plasma boundary the ions see an average electron density. Second, from the electron perspective the boundary between a neutral plasma and an instantaneous space charge sheath, $x = s(t)$, moves back and forth, with the density at this boundary therefore being strongly modulated through the RF cycle. Electrons are in fact swept back and forth during the RF oscillation in such a way that they exactly neutralize the ion space charge on the plasma side of the moving sheath/plasma boundary, $x > s(t)$. As before, the displacement of electrons at

4.4 Analytical models of (high-frequency) RF sheaths

the edge of the instantaneous boundary of the non-neutral region defines the RF current.

Voltage drive

In this case it is supposed that the potential difference across the sheath varies sinusoidally on top of a mean value (chosen such that the net particle current is zero):

$$V_{sh} = V_1 \cos \omega t - V_{fRF}.$$

The instantaneous sheath thickness, $s(t)$, must then be calculated from the distribution of ion space charge in the mean potential and the applied voltage waveform. But the mean potential at any point can only be calculated once the instantaneous potential is known at that point throughout the RF cycle. Thus an iterated numerical solution or some clever mathematical trick is required. Figure 4.15 is a result from an iterated solution.

Another, less rigorous approach is to make a crude adaptation of the result for a DC Child–Langmuir sheath. In an RF-modulated sheath the ion space charge is not as effective as it is in a DC sheath because of the periodic neutralization by the electrons driven by the RF potential. A crude estimate of the spatial extent of an RF Child–Langmuir sheath can be obtained (Eq. (3.16)) by artificially decreasing the effectiveness of the ion space charge at all points, which can be achieved by introducing a fraction, α, to scale it down, leading to a relationship between the mean sheath voltage and the maximum sheath width:

$$s_m = \frac{2}{3} \left(\frac{\varepsilon_0}{\alpha J_i} \right)^{1/2} \left(\frac{2e}{M} \right)^{1/4} (-V_{fRF})^{3/4}, \tag{4.36}$$

where $J_i = n_i(s_m) e\, u_B$ and $0 < \alpha < 1$; for smaller applied voltages α will be closer to zero and for larger voltages, α will be closer to unity.

One could then proceed to apply this idea to the spatial distribution of ion space charge. Then the instantaneous RF displacement current that would be associated with this RF sheath can be deduced by taking the electron density at the position of the moving sheath edge (which equals the local ion density) and the instantaneous speed of the sheath edge to form the current density $n_i(s) e\, ds/dt$. The current must be periodic but it is not expected to be a simple sinusoidal current because the speed of the sheath edge and the ion density at the sheath edge both vary with position. Furthermore, as in the homogeneous ion matrix case, the electron particle current should be included for completeness if the total current in the external circuit is to be determined.

In modelling a capacitively coupled RF discharge (see Chapter 5) it is necessary to treat two sheaths in series, one each side of the plasma. For this case it turns out to be more convenient to analyse the current-driven situation, so the voltage-driven RF sheath will not be examined further.

Current drive

For a current-driven sheath there are four steps in the reasoning that relates the sinusoidal RF current to the mean sheath potential and the mean sheath size.

(i) First, as in the voltage-driven case, where n_s represents the boundary density in any DC formula that is developed into an RF model, it should perhaps be replaced by αn_s, with $0 < \alpha < 1$ to account for the periodic compensating effect of electron space charge.

(ii) Next, the time-varying electric field in the sheath is linked to the displacement current by

$$J_0 \sin \omega t = \varepsilon_0 \frac{\partial \tilde{E}}{\partial t}.$$

The time-varying electric field will then certainly have a harmonic component ($\cos \omega t$), but in the simplest cases the field does not change sign so there must be a DC component of the field that scales with E. So this DC (or mean) field $\overline{E} = \,<E>$ can also be expected to be proportional to J_0/ω. Then, higher frequency and/or lower current will lead to lower mean fields. (Under some circumstances, in fact, the sheath field does transiently change sign during the RF cycle [51–55], but this sheath field reversal is so brief that it does not invalidate the general argument here.)

(iii) It can be shown from the normal Child–Langmuir relationship, which applies when ions fall freely through the sheath, that the potential across the space charge is proportional to the fourth power of the field that it produces.

> **Q** Look back to Chapter 3 and identify where in the development of the collisionless Child–Langmuir sheath the potential can be seen to scale with the fourth power of the electric field.
>
> **A** Equation (3.15) with appropriate sheath edge conditions ($\phi'(s) = 0$ and $\phi(s) = 0$) shows that
>
> $$(\phi'(x_1))^2 = 4\frac{J_i}{\varepsilon_0}\left(\frac{2e}{M}\right)^{-1/2}(-\phi(x_1))^{1/2}.$$
>
> Thus the fourth-power relationship between field ($-\phi'$) and potential is confirmed.

4.4 Analytical models of (high-frequency) RF sheaths

Supposing that this strong non-linearity applies equally to the RF field, it can be anticipated that the amplitude of the potential across an RF sheath is

$$V_0 \propto \left(\frac{J_0}{\omega}\right)^4. \tag{4.37}$$

(iv) The final step is to adapt the DC version of the Child–Langmuir model (Eq. (3.16)) to relate the width of the region swept through by the sheath to this mean sheath potential. The ion current is still the Bohm current, but the factor of α must be included to allow for the partial compensation of the ion space charge by the periodic ingress of electrons. The RF version of the Child–Langmuir model for the mean sheath width s_m is therefore anticipated to take the form

$$s_m = \frac{(J_0/\omega)^3}{6(\alpha n_s)^2 \, \varepsilon_0 k T_e}. \tag{4.38}$$

A more thorough analysis by Lieberman [56] confirms the scaling of potential with the current (Eq. (4.37)) and shows that for a high current drive $6\alpha^2 = 12/5$, which suggests that the effectiveness of the ion space charge is given by $\alpha \sim 0.63$, through a combination of rearrangement in the mean field and periodic screening by electrons as the sheath contracts and expands. Lieberman [56], neglecting the particle currents, goes on to show that the instantaneous voltage across a sheath driven by a sinusoidal RF current contains contributions up to the fourth harmonic, owing to the non-linearity of the sheath. So, the more realistic modelling of ions in equilibrium with the mean field leads to a richer spectrum than the fundamental and second harmonic found in the case of constant ion density. The full analysis leads to an expression for the voltage across a high-current-driven, inhomogeneous sheath in terms of a voltage scaling parameter, which is given here for later reference:

$$H = \frac{1}{\pi \varepsilon_0 k T_e n_s} \left(\frac{J_0}{\omega}\right)^2, \tag{4.39}$$

with amplitudes of the first four Fourier amplitudes in the voltage waveform being

$$\frac{e\overline{V}}{kT_e} = \frac{3}{4}\pi H + \frac{9}{32}\pi^2 H^2,$$

$$\frac{eV_\omega}{kT_e} = \pi H + 0.34\pi^2 H^2,$$

$$\frac{eV_{2\omega}}{kT_e} = \frac{1}{4}\pi H + \frac{1}{24}\pi^2 H^2,$$

$$\frac{eV_{3\omega}}{kT_e} = -0.014\pi^2 H^2.$$

The scaling parameter can also be expressed as $\pi H \approx (s_m/\lambda_{De})^2$ – sheath modulations of several Debye lengths are expected, so in practice the analysis is chiefly of interest when $H > 1$ and then the harmonics make significant contributions.

4.5 Summary of important results

- There are characteristic response frequencies for electrons ω_{pe} and ions ω_{pi} in and adjacent to a plasma. These formulas depend on the square root of particle number density over particle mass. Ions tend to cross space charge regions in times that are a few ω_{pi}^{-1}.
- The typical domain of RF plasmas is the range $\omega_{pi} < \omega < \omega_{pe}$.
- In the RF domain only the lightest ions respond to the full RF cycle; most ions effectively experience time-averaged fields.
- Ion energy distributions at surfaces reflect the form of the plasma upstream of the sheath, fluctuations in RF fields in the sheath relative to ω_{pi}, and collisions. If $\omega < \omega_{pi}$, in the absence of collisions, the distributions tend to be bimodal. Collisions scatter ions down to lower energy and for $\omega \approx \omega_{pi}$ multiply peaked distributions can occur.
- RF modulation of sheaths shifts the conventional floating potential more negative. At large RF modulation the shift is almost in proportion to the amplitude of the RF potential.
- RF sheaths look like a parallel combination of a current source (ion current), a reverse biased diode (electron current) and a non-linear capacitance (displacement current). Models of RF sheaths can be set up considering the modulation to be forced by an RF voltage or an RF current. Owing to the non-linearity of the sheath, sinusoidal voltages lead to currents with higher harmonics whereas a pure sinusoidal current leads to voltage waveforms with higher harmonics. In modelling RF sheaths one can use different models for the ion space charge. The simplest is a fixed ion density model that ignores all ion dynamics. A more realistic model allows ions to respond only to the mean fields, and this represents the situation where $\omega_{pi} < \omega$. (One can also include ion motion to account for the dynamics of lighter ions, but this was not detailed in this chapter.)

5

Single-frequency capacitively coupled plasmas

Capacitively coupled plasmas (CCPs) have been used for several decades for the etching and deposition of thin films. CCPs consist of two parallel electrodes, typically of radius $r_0 \sim 0.2$ m separated by $l \sim 3 - 5$ cm, and biased by a radio-frequency power supply, typically operating at 13.56 MHz. A plasma forms between the electrodes, from which it is separated by space charge sheaths, the thicknesses of which vary at the excitation frequency – see Chapter 4.

A very important aspect of the following discussion is the fact that in many industrial systems the neutral gas pressure may be below 10 Pa. This has two major consequences for the physics of such plasmas. Firstly, collisionless processes then play a significant role in the transfer of energy to the electron population from electromagnetic fields. The usual collisional (joule) heating term is too small to explain the high electron densities observed. Secondly, the electron energy relaxation length may then become larger than the physical dimensions of the discharge (see Chapter 2). This introduces non-local effects into the plasma kinetics and electrodynamics.

Q Compare the vacuum wavelength of electromagnetic radiation and the collisionless skin depth of a plasma of density $n_e \leq 10^{16}$ m^{-3} at $f = 13.56$ MHz with the characteristic system dimensions.

A The vacuum wavelength, λ, is $c/f \sim 22$ m $\gg r_0$ and the collisionless skin depth, δ, is $c/\omega_{pe} > 0.05 \gtrsim l$.

Comment: In this regime, the discharge can be modelled in the electrostatic approximation, and all the physics can be associated with a single dimension, perpendicular to the electrodes.

The first chapters of this book have explored the fundamental mechanisms governing weakly ionized bounded plasmas, including collisions and reactions, electromagnetic properties and transport of charged particles to the boundaries and the boundaries themselves. In particular, it has been shown that in the simplest quantitative analysis, the so-called global models, low-pressure plasma reactors can be described in terms of two equations, namely the particle and energy balance equations, Eqs (2.40) and (2.47). These equations require expressions for the generation of charged particles by volume reactions and for the loss of particles through fluxes of charged particles to the wall. These were discussed in Chapters 2 and 3. The energy equation (Eq. (2.47)) also requires an expression for the power absorbed by electrons from the external supply. This has not so far been discussed, and it will be central to the purpose of this and the three following chapters.

In RF discharges the absorbed power depends strongly upon the way in which energy is coupled from the external power supply into the electric fields in which electrons are accelerated. There are three ways (modes) to couple the energy provided by the RF generator to the electrons: the electrostatic (E) mode, the evanescent electromagnetic (H) mode, and the propagating wave (W) mode. CCPs generally operate in the E-mode, although it will be shown in the next chapter that this is not necessarily the case at very high frequency. Inductively coupled plasmas (ICPs), described in Chapter 7, generally operate in the H-mode, although they may operate in the E-mode at low power. Finally, helicon plasmas generally operate in the W-mode but may also operate in the E and H-modes. The physics of these various modes, along with the transitions between modes, will be described in this and the following chapters.

An equivalent-circuit description of the discharge is useful for the calculation of the absorbed power term in the global models. Equivalent-circuit models account for the plasma and sheaths in terms of impedances that link instantaneous voltages and currents. These impedances are derived by integrating the local electromagnetic fields over plasma and sheath regions. The RF electric and magnetic fields in the analysis are thereby replaced by RF voltages and currents – quantities that can also be conveniently compared with direct measurements. Sections 5.1 and 5.2 develop appropriate expressions for these impedances.

The global model of the symmetrical single-frequency capacitively coupled discharge is presented in Section 5.3. In this section the most important scaling laws of capacitive discharges are discussed. Section 5.4 focuses on other interesting phenomena, such as asymmetric discharges, higher pressure regimes and series resonances. Finally, Section 5.5 summarizes the important results of this chapter and underlines the principal limitation of CCPs, namely that the flux and the energy of ions bombarding the electrodes cannot be independently controlled. This topic is then taken up in Chapter 6.

5.1 Constant ion density, current-driven model

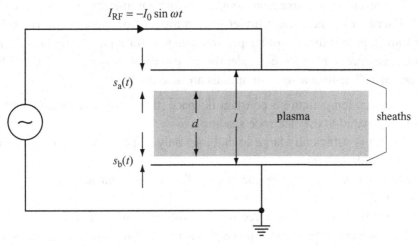

Figure 5.1 Schematic of the symmetrical capacitive discharge model. The definition of the current is chosen to simplify the analysis.

5.1 A constant ion density, current-driven symmetrical model

A basic model of a symmetrical capacitively coupled plasma was introduced in the 1970s by Godyak [39, 57]. This model is sometimes called the homogeneous model. A more sophisticated ('inhomogeneous') model has also been introduced by Godyak [39] and revisited by Lieberman [2, 56, 58]. The inhomogeneous model is based on Child–Langmuir models of the RF sheaths (see Chapter 4); it will be discussed briefly later in this chapter.

Consider the situation shown schematically in Figure 5.1, where there is a quasi-neutral plasma that extends over a distance d, separated from the electrodes by two time-varying sheaths of thicknesses $s_a(t)$ and $s_b(t)$. While the sheath 'a' expands, during one half RF cycle, sheath 'b' contracts, and *vice-versa*.

Q Considering the distributions of current and voltage between the electrodes, suggest why the voltage that appears across the electrodes is a less attractive control parameter than the current for the purposes of a simple model.

A At any instant the current is continuous across all planes parallel to the electrodes; the voltage is distributed across two distinct space charge sheaths and the plasma, none of which is certain to exhibit a linear response. Thus, specifying a sinusoidal current in the external circuit ensures sinusoidal current within the plasma, whereas an externally imposed voltage between the electrodes is no guarantee that the sheaths or the plasma will conveniently experience sinusoidal differences of potential.

The RF current is a convenient control (input) parameter – the sheaths will then be described by the current-driven sheath model from Chapter 4. The other input parameters are: the power supply frequency f (or equivalently $\omega = 2\pi f$), the neutral gas pressure p (or equivalently the gas density $n_g = p/kT_g$), and the electrode gap l. The following assumptions are also made:

(i) The electron temperature is constant in space (the energy relaxation length is large compared to the electrode spacing).
(ii) Ion inertia is sufficiently large so that ions only respond to the time-averaged electric field; $\omega \gg \omega_{pi}$.
(iii) Electron inertia is negligible. Electrons follow the instantaneous electric field; $\omega \ll \omega_{pe}$.
(iv) The system is divided into three regions: the quasi-neutral plasma, in which $n_e = n_i = n_0$ (constant) and E is (nearly) zero; two sheaths ('a' and 'b'), where $n_e = 0$ and $n_i = n_0$, so $E \neq 0$.
(v) The electrostatic regime applies, which requires $\lambda \gg R$ and $\delta \gg l$, such that the voltage between the electrodes is independent of their radius R.

In the next sections, the voltages across the sheaths and across the plasma are calculated for a given RF current, $I_{RF} = -I_0 \sin \omega t$. This leads to a definition of the impedance of each of these elements and an equivalent circuit model of the capacitive discharge. The results obtained in Chapter 4 on RF sheaths will be developed for a combination of two RF sheaths connected in series by a plasma.

Warning: Do not confuse the modified Bessel function, $I_0(x)$, from the previous chapter with the amplitude of an RF current, I_0, in this chapter.

5.1.1 Electric field and potential between the electrodes

A schematic of the constant ion density model is shown in Figure 5.2; $s_a(t)$ is the instantaneous position of the boundary between sheath 'a' and the plasma, $\bar{s}$ is its time-averaged position, and s_m is its maximum extent. Similar notation applies to sheath 'b'.

Although it was shown in Chapter 3 that it is not generally appropriate to set $E \equiv 0$ throughout the plasma, the electric field in a quasi-neutral plasma is extremely small (virtually zero) *compared with* that in the space charge sheaths.

Q By considering the continuity of the RF current, show that in the low-pressure regime ($\omega \gg \nu_m$) the electric field within the plasma is much smaller than that in the sheath when $\omega \ll \omega_{pe}$.

A The RF current crosses the sheath region (sh) predominantly as displacement current associated with the time variation of the electric field, whereas in

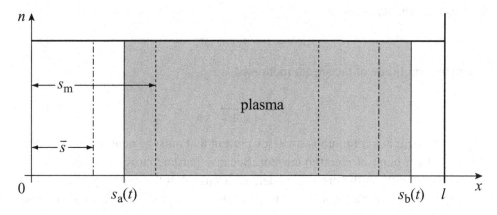

Figure 5.2 Schematic of the sheath dynamics for the constant ion density model.

the plasma (p) the electron fluid has a finite, non-zero conductivity σ and the displacement current is negligible. The continuity of current therefore requires that

$$\varepsilon_0 \frac{\partial E_{sh}}{\partial t} = \sigma E_p,$$

which, given the sinusoidal nature of the current, requires that

$$\left|\frac{E_{sh}}{E_p}\right| \sim \frac{\sigma}{\varepsilon_0 \omega}.$$

Using Eq. (2.53) for the conductivity and taking the low-pressure limit leads to the conclusion that, since $\omega \ll \omega_{pe}$,

$$|E_{sh}/E_p| \gg 1.$$

Therefore, when considering the sheath regions, the electric field in the plasma can be neglected. The electric field in sheath 'a' is given by Eq. (4.18) with $s(t) = s_a(t)$:

$$E(x, t) = \frac{n_0 e}{\varepsilon_0}(x - s_a(t)); \tag{5.1}$$

in the interval $0 \le x \le s_a$ the electric field is directed towards the electrode.

From the current-driven model, the motion of the sheath is given by Eq. (4.26). In the high-current regime the particle current crossing the sheath is negligible, so sinusoidal sheath motion results from an externally imposed sinusoidal current, of amplitude I_0, distributed uniformly over an electrode of area A (i.e.,

$I_0 \gg 0.3 \times An_0 e\bar{v}_e/4$). Then

$$s_a(t) = \bar{s} - s_0 \cos \omega t, \tag{5.2}$$

where the amplitude of the sheath motion is

$$s_0 = \frac{I_0}{n_0 e A \omega}. \tag{5.3}$$

In practice there is a continuous small ion current that must be neutralized over one RF cycle by a burst of electron current. Because particle currents have not been included explicitly in the present model, as discussed at the end of Section 4.4.2, $s_a(t)$ must be zero for an instant (so that the sheath vanishes and electrons escape to the electrode), which requires that

$$\bar{s} = s_0.$$

In this model, as the sheaths grow and contract in anti-phase, the plasma dimension remains constant at $l - 2s_0$. The motion of the plasma/sheath boundary close to electrode 'b' is therefore

$$s_b(t) = l - s_0(1 + \cos \omega t). \tag{5.4}$$

In driving a sinusoidal current between the electrodes a potential difference will appear between them, but the form of the potential in this model is determined by the non-linear impedances of the two sheaths. For definiteness, electrode 'b' at $x = l$ will be set at ground potential and the as yet unknown potential at $x = 0$ will be written $V_{ab}(t)$; an expression for this voltage will be found in due course.

In the region $0 \leq x \leq s_a(t)$, that is in sheath 'a', the potential $\phi(x, t)$ between the electrode and the plasma can be calculated by integrating Eq. (5.1) from zero, with the boundary condition that $\phi(0, t) = V_{ab}(t)$:

$$E_a(x, t) = \frac{n_0 e}{\varepsilon_0} [x - s_a(t)], \tag{5.5}$$

$$\phi(x, t) = -\frac{n_0 e}{\varepsilon_0} \left[\frac{x^2}{2} - s_a(t)x\right] + V_{ab}(t). \tag{5.6}$$

The electric field varies linearly in space while the potential is quadratic. In the plasma region the electric field is zero (assumption (iv)) and the potential is independent of x. Since the potential must be continuous at the plasma/sheath boundary its value in the plasma is $\phi(s_a(t), t)$, so throughout the plasma, $s_a(t) < x < s_b(t)$,

$$\phi_p(t) = +\frac{n_0 e}{2\varepsilon_0} s_a(t)^2 + V_{ab}(t). \tag{5.7}$$

5.1 Constant ion density, current-driven model

In the region $s_b(t) \leq x \leq l$, that is in sheath 'b', the potential between the plasma at $x = s_b(t)$ and the electrode is obtained by integrating Eq. (4.18) with $s(t) = s_b(t)$, imposing the boundary condition that $\phi(s_b(t), t) = \phi_p(t)$:

$$E_b(x,t) = \frac{n_0 e}{\varepsilon_0} [x - s_b(t)], \tag{5.8}$$

$$\Phi(x,t) = -\frac{n_0 e}{\varepsilon_0} \left[\frac{x^2}{2} - s_b(t)x + \frac{s_b(t)^2}{2} \right] + \phi_p(t). \tag{5.9}$$

Finally the potential difference between the electrodes, $V_{ab}(t)$, can be determined by requiring that the potential at $x = l$ is zero. In that case Eq. (5.9) gives

$$\phi_p(t) = +\frac{n_0 e}{2\varepsilon_0} (l - s_b(t))^2. \tag{5.10}$$

Using Eqs (5.2) and (5.4) for the positions of the two sheath/plasma boundaries, Eqs (5.7) and (5.10) combine to give

$$V_{ab}(t) = \frac{n_0 e}{2\varepsilon_0} s_0^2 (1 + \cos \omega t)^2 - \frac{n_0 e}{2\varepsilon_0} s_0^2 (1 - \cos \omega t)^2$$

$$= V_0 \cos \omega t, \tag{5.11}$$

with

$$V_0 = 2n_0 e s_0^2 / \varepsilon_0. \tag{5.12}$$

So where's the non-linearity?

The result in Eq. (5.11) is remarkable. Space charge sheaths are inherently non-linear, so the imposing of a sinusoidal current passing through the space charge sheaths and the plasma would not be expected in general to be associated with a sinusoidal voltage. Yet the constant ion density model is a special case for which a symmetrical sheath/plasma/sheath system is linear, in the sense that a sinusoidal current is associated with a sinusoidal voltage.

Non-linearity is still present within the system as the profiles of potential between the electrodes reveal. Figure 5.3 shows profiles at different phases of the RF cycle. At the instant when $\omega t = 0$, sheath 'a' is collapsed and sheath 'b' is at its maximum expansion, $s_b = l - 2s_0$; there is no voltage difference between the plasma and the powered electrode 'a'. At $\omega t = \pi$, sheath 'a' is fully expanded and sheath 'b' has collapsed. At any other time during the RF cycle, the potential in the plasma is greater than the potential of each electrode. This has to be the case, otherwise electrons would be lost so quickly that a plasma could not be sustained.

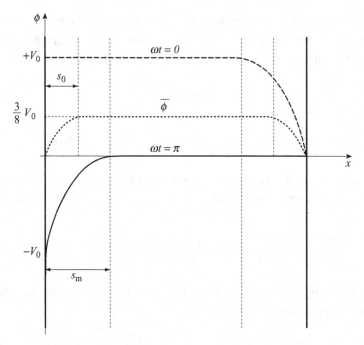

Figure 5.3 Potential $\phi(x, t)$ between the electrodes for two different times during the RF cycle, along with the time-averaged value.

Each sheath has a non-linear response to the sinusoidal current (as shown in Section 4.4.2). That at electrode 'a' leads to a difference of potential between that of the plasma (Eq. (5.7)) and that of the electrode (Eq. (5.11)):

$$\phi_p(t) - V_0 \cos \omega t = \frac{V_0}{4}(1 - \cos \omega t)^2$$

$$= V_0 \left[\frac{3}{8} - \frac{1}{2}\cos \omega t + \frac{1}{8}\cos 2\omega t\right]. \qquad (5.13)$$

The potential across the sheath has a DC, fundamental and second harmonic components – a linear response to a sinusoidal current would contain only the middle one of these. Similarly, at electrode 'b' there is a rapid change of potential from that in the plasma to zero at the electrode:

$$0 - \phi_p(t) = -V_0 \left[\frac{3}{8} + \frac{1}{2}\cos \omega t + \frac{1}{8}\cos 2\omega t\right]. \qquad (5.14)$$

These voltages are plotted in Figure 5.4. Although each sheath oscillates, the sum of the two sheath thicknesses (and so by subtraction the spatial extent of the plasma region) is independent of time. One source of non-linearity therefore vanishes.

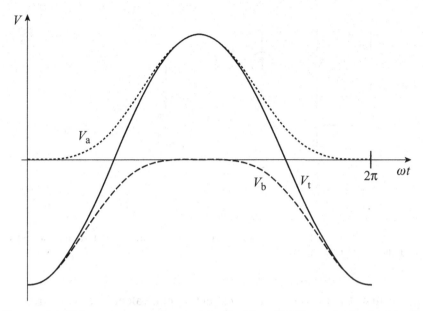

Figure 5.4 Voltages across sheath 'a', V_a (Eq. (5.13)), and sheath 'b', V_b (Eq. (5.14)), along with the sum of the two, $V_t = -V_{ab} = -V_0 \cos \omega t$.

Q Show that at given density n_0, the amplitude of the voltage that appears across the plasma, V_0, has a non-linear dependence on the amplitude of the RF current, I_0.

A Equations (5.3) and (5.12) combine to show

$$V_0 = \frac{2}{n_0 e \varepsilon_0 A^2 \omega^2} I_0^2,$$

so the amplitude response of the system seems non-linear. However, this result may be misleading because in practice the plasma density is a function of V_0 (or equivalently I_0). It will be shown in Section 5.3 that depending on the heating mechanisms, there are regimes in which $n_0 \propto V_0$, so that $I_0 \propto V_0$, as observed experimentally by Godyak [59].

Also plotted in Figure 5.3 is the time-averaged potential (dotted line) at $3V_0/8$; note that both sheaths have an average size of s_0. In this model the ion density is constant everywhere between the electrodes, so the motion of ions, even in the mean field, has been neglected. It is not strictly appropriate therefore to consider the energy of ions at the electrodes, nevertheless it will later be presumed that the ions crossing the sheath gain an energy equivalent to the mean sheath potential. It still remains to consider the electron dynamics that couple energy into the plasma

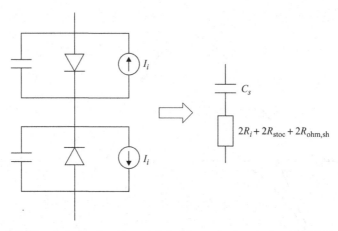

Figure 5.5 The electric circuit model for the two sheaths considered separately (left) and combined (right).

region, but first it is convenient to construct an equivalent electrical circuit for a CCP, according to the constant ion density, high-current-driven model.

5.1.2 Equivalent circuit of a symmetrical CCP

From the outside, a symmetrical CCP can be seen to have a specific relationship between the current that flows in and the voltage that appears across the plates by means of which it absorbs a measurable quantity of power. The aim here is to use standard electronic components to devise an electrical circuit that has equivalent current–voltage characteristic and that dissipates the same power; choosing component values to match the properties of the plasma. If this is done accurately then one can explore parameter space using the circuit model instead of returning each time to the direct solution of Maxwell's equations.

The total sheath region

An equivalent circuit model for an RF sheath was proposed in Section 4.4.1, involving a current source, a diode and a non-linear capacitance. This circuit is used on the left-hand side of Figure 5.5 for the two sheaths in series with a highly conducting plasma region. However, there are three simplifications that can be made for the present situation. First, in the high-current-drive limit the particle currents have already been ignored so the non-linear capacitances predominantly determine the current–voltage relationship. Therefore, so long as the power dissipated in the various components is still included somehow, the diode and current source

5.1 Constant ion density, current-driven model

paths can be omitted. Second, since it has been shown that the combination of the two sheaths becomes linear for any fixed high amplitude of driving current, the combined effects of the sheaths can be modelled by a single capacitance with a single resistance to represent power dissipation in the original model. This much simpler circuit is shown on the right-hand side of Figure 5.5. The value of this capacitance is simply calculated from the following relation between the RF voltage and the RF current:

$$\frac{dV_{ab}}{dt} = -\omega V_0 \sin \omega t \quad \text{(from Eq. (5.11))}$$

$$= -\frac{\omega 2 n_0 e s_0^2}{\varepsilon_0} \sin \omega t \quad \text{(from Eq. (5.12))}$$

$$= -\frac{2s_0}{\varepsilon_0 A} I_0 \sin \omega t \quad \text{(from Eq. (5.3))}.$$

The RF current was defined at the outset to be $I_{RF} = -I_0 \sin \omega t$, so this last equation confirms that together sheaths 'a' and 'b' respond as an effective capacitance:

$$C_s = \frac{\varepsilon_0 A}{2s_0} = \frac{n_0 e \omega \varepsilon_0 A^2}{2I_0}. \quad (5.15)$$

There are three contributions to the power dissipation in the sheath, which can be scaled and lumped together as a single resistance in series with the capacitance. The first two contributions, $2R_{\text{ohm,sh}}$ and $2R_{\text{stoc}}$, are due to electron heating processes within each sheath by collisional (ohmic) and collisionless (stochastic) mechanisms. The third contribution, $2R_i$, is due to power dissipation by ions accelerated across each sheath. Parametric formulas for these resistive terms will be determined in the following sections. The complex impedance for the combination of the two sheaths can be written

$$Z_s = \frac{1}{i\omega C_s} + 2(R_i + R_{\text{stoc}} + R_{\text{ohm,sh}}). \quad (5.16)$$

The total resistance is usually small compared with the impedance of the capacitor.

The plasma region

In solving equations for the distribution of the potential in the previous section, the electric field in the plasma was neglected. The next step is to resolve the detail of the plasma region with a view to tracking the flow of energy. Provided the potential changes across the plasma remain small, it will not be necessary to revisit the sheath model.

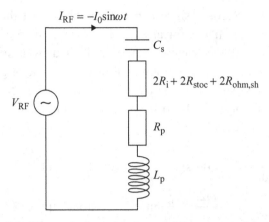

Figure 5.6 The equivalent circuit model for a symmetrical CCP.

It was shown in Chapter 2 that the electrical circuit equivalent to the plasma bulk is composed of a capacitance in parallel with a power-dissipating resistance and an electron inertia inductance in series. The capacitance accounts for the displacement current crossing the plasma, though this is negligible since $\omega \ll \omega_{\mathrm{pe}}$. The equivalent circuit therefore reduces to a resistance and an inductance in series. The potential difference across the plasma is then

$$V_{\mathrm{p}} = R_{\mathrm{p}} I_{\mathrm{RF}} + L_{\mathrm{p}} \frac{\mathrm{d} I_{\mathrm{RF}}}{\mathrm{d} t}. \tag{5.17}$$

In circuit theory, it is often convenient to use complex notations. With the complex plasma impedance $Z_{\mathrm{p}} = R_{\mathrm{p}} + i\omega L_{\mathrm{p}}$ and the complex current amplitude $\widetilde{I}_{\mathrm{RF}}$, the complex amplitude of the potential difference across the plasma is

$$\widetilde{V}_{\mathrm{p}} = Z_{\mathrm{p}} \widetilde{I}_{\mathrm{RF}}. \tag{5.18}$$

The whole CCP

Finally, the total equivalent circuit of the capacitive discharge is a series combination of plasma and total sheath components. The resulting circuit is sketched in Figure 5.6.

RF power is mostly deposited into the plasma electrons, which respond to the RF fluctuations. The power absorbed by the electrons leads to heating of the electron population, by collisional and collisionless mechanisms (Section 5.1.3). Since ions only respond to time-averaged fields, they do not gain energy directly from the RF field; ion heating by the RF field is negligible. However, there is a steady (DC) component of potential across each sheath. The acceleration of ions within

5.1 Constant ion density, current-driven model

this sheath accounts for a significant fraction of the power dissipated within the discharge. The ionic power dissipation is studied in Section 5.1.4.

5.1.3 Power dissipated by electrons

The local, instantaneous ohmic power dissipation was introduced in Chapter 2 via Eq. (2.74). To evaluate the mean ohmic heating power in the plasma, this expression must be integrated over time and space. To achieve this, it is worth utilizing the stratification of the discharge into the plasma and the sheath regions.

The region $s_m \leq x \leq l - s_m$ is always occupied by quasi-neutral plasma, for which the constant ion density model makes the conductivity, σ_m, also constant. The current has been taken to be purely sinusoidal, so Eq. (2.77) takes care of the time average while the integration over space trivially leads to the total power in the plasma volume being

$$A \int_{s_m}^{l-s_m} \frac{I_0^2}{2A^2 \sigma_m} dx = \frac{I_0^2}{2A \sigma_m} \int_{s_m}^{l-s_m} dx$$

$$\equiv \frac{1}{2} R_{ohm,p} I_0^2, \tag{5.19}$$

with

$$R_{ohm,p} = \frac{m \nu_m (l - 2s_m)}{n_0 e^2 A} \approx R_p. \tag{5.20}$$

This is the resistance of the region that is always quasi-neutral.

The situation is more complex in regions explored by the sheaths, i.e., $0 \leq x \leq s_m$ and $l - s_m \leq x \leq l$, because the volume over which the power is deposited depends on time and it is more convenient to integrate first over space and thereafter over time. Adapting Eq. (2.76) to sheath 'a', the instantaneous total power dissipated in the electron front while it is penetrating the region $0 \leq x \leq s_m$ is

$$P_{ohm,sh}(t) = A \int_{s(t)}^{s_m} \frac{J_0^2}{\sigma_m} \left(\frac{1 - \cos 2\omega t}{2} - \frac{\omega}{\nu_m} \sin \omega t \cos \omega t \right) dx. \tag{5.21}$$

where $J_0 = I_0/A$. In deriving Eq. (2.76), the RF current density was chosen to be $J_0 \sin \omega t$, but changing the sign of J_0 to match the situation described above has no effect here so the equation still applies. The fact that the integration domain is not constant in time can be accommodated by using the sheath position to change the variable from space (x) to temporal phase (θ) through the sheath motion $s_0(1 - \cos \theta)$ and so $dx \equiv s_0 \sin \theta d\theta$; in the high-current-drive limit

$s_m = 2s_0$. Transforming the integral from one over space to one over temporal phase gives

$$P_{\text{ohm,sh}}(\theta_1) = \frac{I_0^2 s_m}{2A\,\sigma_m} \int_{\theta_1}^{\pi} \left(\frac{1-\cos 2\theta}{2} - \frac{\omega}{\nu_m}\sin\theta\cos\theta\right)\sin\theta\,d\theta \quad (5.22)$$

$$= \frac{I_0^2 s_m}{2A\,\sigma_{\text{DC}}}\left[\frac{4}{3}(2-\cos\theta_1)\cos^4\left(\frac{\theta_1}{2}\right) + \frac{\omega}{3\nu_m}\sin^3\theta_1\right]. \quad (5.23)$$

The temporal average amounts to an integral over the phase, θ, from zero to 2π which loses the sinusoidal terms, leaving

$$\overline{P}_{\text{ohm,sh}} = \frac{1}{3}\frac{I_0^2 s_m}{A\,\sigma_m}. \quad (5.24)$$

The resistance accounting for ohmic heating in one sheath is therefore

$$R_{\text{ohm,sh}} = \frac{1}{3}\frac{m\nu_m}{n_0 e^2}\frac{s_m}{A}. \quad (5.25)$$

Comparing Eqs (5.20) and (5.25) one can see that ohmic heating in the bulk is larger than ohmic heating in the two sheaths if $l \geq 8s_m/3$. Although the sheath size is usually significantly smaller than the plate separation, this relation shows that for discharges between narrowly separated plates, ohmic heating in the sheath is not negligible compared to ohmic heating in the plasma bulk.

In addition to collisional (ohmic) heating, there is a collisionless mechanism for heating in the sheath, sometimes called stochastic heating. This mechanism results from the interaction of the strongly non-uniform sheath electric field with the plasma electrons. Collisionless (stochastic) heating is strictly zero for a uniform density (homogeneous) sheath, so here $R_{\text{stoc}} = 0$. This mechanism will be discussed in the section devoted to the inhomogeneous model.

5.1.4 Power dissipated by ions

In the high-frequency regime of interest, ions do not respond to the RF field; they do, however, pick up energy from the DC fields and deposit it on the electrodes where the impact energy is absorbed (at significantly high pressure, they also transfer some energy to the neutral gas, leading to neutral gas heating). The DC potential structure within a CCP plasma region is small ($\simeq kT_e/e$) compared with that in the sheaths ($\gg kT_e/e$), so it is the sheath region that dominates. Although the ion conduction current in the sheaths is generally much smaller than the RF displacement current, the power dissipated by the acceleration of ions in the sheath may still be a significant fraction of the power delivered by the generator. It is easily calculated by taking the product of the ion current and the magnitude of the

5.1 Constant ion density, current-driven model

time-averaged voltage across the RF sheath. The former can be assumed to be due to the Bohm flux arriving at the sheath/plasma boundary and the latter has already been calculated (Eq. (5.13) or (5.14)):

$$P_i = \frac{3}{8} e A n_0 u_B V_0$$

$$= \frac{3 u_B}{4\varepsilon_0 A \omega^2} I_0^2. \tag{5.26}$$

As before, an equivalent resistance, R_i, that accounts for this power dissipation can be placed in the equivalent electrical circuit. Placing it in series with the sheath capacitance means that it will pass the entire current I_{RF} so that the mean ion power dissipation would be

$$P_i = \frac{1}{2} R_i I_0^2.$$

That then sets

$$R_i = \frac{3 u_B}{2\varepsilon_0 \omega^2 A}. \tag{5.27}$$

Note that R_i is independent of the plasma density, and that for a fixed RF current flowing in the discharge, it scales with $1/\omega^2$, which indicates that the power dissipation in ions will strongly decrease as the frequency increases. This important point will be revisited later.

5.1.5 Limitations of the uniform ion density model

The uniform ion density model has captured a number of features of a real CCP but the simplification of the constant ion density profile forces a number of compromises, including those in the following list.

- The ions are presumed to enter the sheath at the Bohm speed and to gain energy in freely falling through the sheath, but the ion space charge is taken as constant throughout.
- The ion space charge in the sheath is overestimated – more realistic models will have thicker sheaths.
- The two constant ion density sheaths, though each separately non-linear, combine in such a way that a sinusoidal current in the model CCP would force a sinusoidal voltage across the electrodes – current-driven and voltage-driven CCPs in this model are identical. More realistic models will lead to non-linearities that cause harmonic generation even when the system is driven by a single frequency.

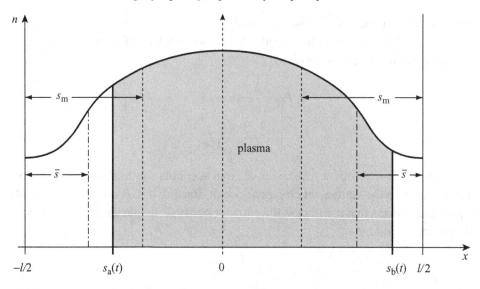

Figure 5.7 Schematic of the inhomogeneous model.

However, it is still the case that the harmonic content of the combination of the two symmetrical sheaths is considerably smaller than that of one single sheath.
- The flat ion profile completely suppresses electron stochastic heating mechanisms that more realistic models exhibit.

At this stage, the various components of the equivalent circuit of a CCP have been specified on the basis of a constant ion density (homogeneous) approximation. From this, one should be able to construct a global model of the whole discharge. Before examining this global model, the circuit parameters are reconsidered in the context of the more realistic situation of non-uniform ('inhomogeneous') ion density between the electrodes.

5.2 A non-uniform ion density, current-driven model

A schematic of the non-uniform capacitive discharge model is shown in Figure 5.7. Note that to be consistent with the results in Chapter 3, the origin has been shifted to the mid-plane. The principles of calculation in this model are similar to those developed for uniform ion density, however, there are some mathematical complications that will not be addressed here in detail. The effects of using the more realistic profile of ion density will be discussed and in particular, the stochastic heating (collisionless dissipation) in the RF sheaths that did not arise in the uniform ion density model will be examined. Stochastic heating is of primary importance in the physics of CCPs. There are two main approaches, but here the emphasis will be

5.2 Non-uniform ion density, current-driven model

on the fluid approach promoted by Turner and co-workers [60, 61]. The alternative approach, based on 'Fermi acceleration' and often called the hard-wall model, is discussed in detail in Lieberman and Lichtenberg [2]. A quantitative comparison of the two models will be given at the end of Section 5.2.3. Finally, the new circuit components that can be used in a global model of CCPs are summarized and compared to those of the uniform ion density model.

5.2.1 Sheath impedance and RF Child law

One of the most important modifications compared to the uniform model is the changed relationship between RF current, voltage and sheath thickness. The law relating the RF current (or voltage), the sheath size and the electron density was discussed at the end of Chapter 4:

$$s_m = \frac{5}{12e(h_1 n_0)^2 \varepsilon_0 k T_e} \left(\frac{I_0}{A\omega}\right)^3 \quad (5.28)$$

for a collisionless sheath [56]; this has the form anticipated in Eq. (4.38). The analysis can also be done for a collisional sheath in the intermediate pressure range [58], which leads to

$$s_m = 0.88 \left(\frac{\lambda_i}{\varepsilon_0 k T_e \omega^3 e h_1^2 A^3}\right)^{1/2} \frac{I_0^{3/2}}{n_0}. \quad (5.29)$$

These relationships can be compared with Eq. (5.3), which holds for homogeneous sheaths. Note that, as illustrated in Figure 5.7, the ion profile is non-uniform, but symmetrical, and in these formulas, n_0 is the electron and ion density at the discharge centre. To take into account the profile of quasi-neutral plasma density, the h_l factor is used in specifying conditions at the boundaries between the quasi-neutral plasma and the sheath region. The h_l factor was discussed in Chapter 3.

The potential across an inhomogeneous sheath in response to an RF current was considered in Section 4.4.3. The sheath voltage has a non-linear dependence on RF current and contains harmonics of the driving frequency up to 4ω. However, unlike in the homogeneous case, analysis reveals that the sum of the voltages across the two out-of-phase sheaths in a CCP modelled with non-uniform ion density is not purely sinusoidal, as the harmonics generated in each sheath do not exactly compensate. This is because the width of the plasma bulk is not constant within the inhomogeneous model. However, it has been shown [39, 56] that the RF voltage across two sheaths in series has no even harmonics and that the third harmonic represents only 4% of the fundamental. Therefore, the voltage is almost sinusoidal and the combination of two sheaths is still reasonably well modelled by a capacitor.

The effective capacitance turns out to be [56]

$$C_\text{s} = K_\text{cap} \frac{\varepsilon_0 A}{s_\text{m}}, \qquad (5.30)$$

where K_cap is a constant given later in Table 5.2 (this constant has different values for collisionless and collisional sheaths). It follows that the voltage amplitude across the discharge is approximately

$$V_0 \approx \frac{s_\text{m} I_0}{K_\text{cap} \omega \varepsilon_0 A} \qquad (5.31)$$

if one again neglects the RF voltage drop across the plasma region.

5.2.2 Ohmic heating and ion power dissipation

The calculation of ohmic heating within the plasma bulk starts by noting that the electron density is now a function of x. Letting $n(x) = n_0 f(x)$, the time and volume-averaged power dissipated due to ohmic heating within the plasma bulk is given by

$$\frac{I_0^2}{2A} \frac{m\nu_\text{m}}{n_0 e^2} \int_{-(l/2-s_\text{m})}^{(l/2-s_\text{m})} \frac{\mathrm{d}x}{f(x)} = \frac{1}{2} R_\text{ohm} I_0^2, \qquad (5.32)$$

where $l - 2s_\text{m}$ is the width of the region that is always occupied by quasi-neutral plasma. Taking a Schottky cosine density profile (valid at rather high pressure), $f(x) = \cos(\pi x/l)$ and Eq. (5.32) integrates to give

$$R_\text{ohm} = \frac{1}{A} \frac{m\nu_\text{m}}{n_0 e^2} \frac{4l}{\pi} \tanh^{-1}\left[\tan\left(\frac{\pi(l-2s_\text{m})}{4l}\right)\right]. \qquad (5.33)$$

What this means is that, compared with the constant ion density model (everywhere n_0), the resistance of the current path through an inhomogeneous plasma is increased since the plasma density only reaches n_0 on the axis. Inserting numbers into the above expression, with a cosine profile and sheaths that together occupy 10% of the space between the electrodes, the plasma resistance is twice what the constant density model would suggest.

When the pressure is lower, the electron density profile is flatter and to a good approximation the resistance remains close to that given by Eq. (5.20).

Ohmic heating in the sheath is determined in the same way as for the homogeneous model. However, because of the complexity of the inhomogeneous sheath model, the calculation is more complicated. The result is given in Table 5.1 [62–64].

5.2 Non-uniform ion density, current-driven model

> **Q** The uniform ion density model found ohmic heating in the sheath and bulk to be comparable when the quasi-neutral plasma occupied only about one-third of the region between the electrodes. How will this be changed by the more realistic ion profile of the inhomogeneous model?
>
> **A** In the inhomogeneous sheath the conductivity is reduced (resistance increases) because the ion profile forces the electron density to be markedly lower than in the constant ion density sheath. This means that sheath heating is more significant in the inhomogeneous sheath and so the equality of bulk and sheath terms occurs for narrower sheaths and when a greater fraction of the gap is filled with quasi-neutral plasma.

The ion power dissipation in one sheath is again given by the product of the ion flux entering the sheath and the mean potential across it:

$$P_i = \overline{V} \cdot \overline{I}_i, \tag{5.34}$$

where the time-averaged voltage across one sheath and the positive ion current are given by

$$\overline{V} = K_s V_0 = K_s \frac{s_m I_0}{K_{cap} \omega \varepsilon_0 A}, \tag{5.35}$$

$$\overline{I}_i = eAh_1 n_0 u_B, \tag{5.36}$$

with K_s a constant given in Table 5.2 (this constant was 3/8 in the homogeneous sheath). The resistance associated with the power dissipated by ions is therefore

$$R_i = \frac{2 K_s e h_1 n_0 u_B s_m}{K_{cap} \omega \varepsilon_0 I_0}. \tag{5.37}$$

5.2.3 Stochastic (collisionless) heating

It has been shown experimentally that ohmic heating is not sufficient to explain the electron power absorption and the high resulting electron density observed in low-pressure capacitively coupled plasmas [65, 66]. It seems clear that electrons gain their energy via collisionless heating, which results from the interaction of the localized sheath electric field with the plasma electrons. An essential feature of this mechanism is that the electron thermal velocity is much larger than the sheath edge velocity. This means that electrons gain their energy in the sheath in a short interval during which the electric field in the sheath is almost constant (independent of time). After interacting with the sheath, electrons then release this energy away from the sheath. A similar type of interaction may exist in the skin depth of a low-pressure inductive discharge, resulting in anomalous field penetration and collisionless inductive heating. There are several approaches to the modelling of this phenomenon, and this subject is still an active area of research. The first approach, proposed initially by Godyak [67] and completed by Lieberman [56] (see also [68,69]), is the so-called 'hard-wall' model. The time-varying sheath edge is treated as a rigid barrier (the hard wall) that specularly reflects electrons coming from the plasma. The argument considers the velocity component that is parallel to the direction of the sheath motion. The reflected velocity of an individual electron is

$$v_r = -v + 2v_s, \tag{5.38}$$

where v is the incident speed and $v_s(t) = u_0 \cos \omega t$ is the sheath edge speed. Electrons interacting with a forward-moving sheath (towards the plasma) gain energy, while those interacting with a retreating sheath lose energy. It is therefore necessary to average over the velocity distribution, and over time, to calculate the net power transferred to the electron population. The number of electrons per unit area interacting with the moving sheath in a time interval dt and in a velocity range v to $v + dv$ is $(v - v_s) f_{es}(v, t) dv\, dt$, where $f_{es}(v, t)$ is the distribution function at the sheath edge for the velocity parallel to the direction of the sheath motion. The infinitesimal power per unit area is then the product of this number times the net energy gain per unit time:

$$dS_{stoc} = \frac{1}{2} m (v_r^2 - v^2)(v - v_s) f_{es}(v, t) dv. \tag{5.39}$$

Integrating this over the appropriate range of velocity is tricky because $f_{es}(v, t)$ evolves as the sheath oscillates. Lieberman [2, 56] assumed a shifted Maxwellian distribution and used the fact that the sheath velocity is much smaller than the thermal velocity, $v_s \ll \bar{v}_e$, to calculate the following time-averaged stochastic heating per unit area:

$$\bar{S}_{stoc,hardwall} = \frac{3\pi}{32} n_s m \bar{v}_e u_0^2 H, \tag{5.40}$$

in which n_s is the density at the ion sheath edge, i.e., where the sheath is fully expanded, and H is the inhomogeneous sheath parameter defined earlier in Eq. (4.39).

The hard-wall model is appealing because it is a kinetic calculation that attempts to represent the interaction between electrons and the sheath electric field. However, it has been shown that this model is not entirely self-consistent, because it violates the conservation of RF current at the instantaneous (moving) sheath edge. An alternative approach has been proposed to repair this deficiency. Turner and

5.2 Non-uniform ion density, current-driven model

co-workers proposed a kinetic fluid (compressional heating) model to describe collisionless heating [60, 61, 70], inspired by earlier work by Surendra and Graves [71] and Surendra and Dalvie [72]. The idea behind the model is that net heating takes place as a consequence of the cyclic compression and expansion of the electron fluid that occurs as the sheaths expand and contract. Within this model, it is clear that heating must vanish in the homogeneous sheath, where there is no compression (no density change). The model is concerned with the quasi-neutral part of the space defined by the maximum sheath expansion; i.e., the model is not strictly speaking dealing with the instantaneous sheath, but rather with the part of the plasma that explores the time-averaged sheath thickness.

The Turner model combines the first three moments of the Vlasov equation (kinetic equation appropriate for collisionless plasmas), which leads to the following equation for the electron fluid:

$$\frac{\partial}{\partial t}\left(\frac{1}{2}nkT\right) + \frac{\partial}{\partial x}\left(\frac{3}{2}nukT + Q\right) - u\frac{\partial}{\partial x}(nkT) = 0, \quad (5.41)$$

where n, u and T are the electron fluid density, speed and temperature, respectively, and Q is the heat flux carried by the electron fluid. There are four variables, so additional physics is required to supply three further equations to solve this problem. First note that the density n is in fact imposed by the time-independent ion density profile, which is already known from the RF sheath model; remember that the aim is to calculate compression of the electron fluid in the *quasi-neutral part* of the time-averaged sheath volume. It is also a reasonable approximation to assume that the electron temperature T is independent of space, because the electron thermal conductivity is very high and the sheath thickness is relatively small. The evolution of the electron temperature is followed by specifying the heat flux Q. To close the set of equations, Turner and co-workers [60, 61, 70] further assumed that:

- the electron heat flux to the electrode is negligible;
- the random fluxes of electrons and heat at the ion sheath edge (at maximum expansion) may be characterized by separate densities and temperatures for electrons entering and leaving the sheath region;
- electrons in the plasma bulk have a fixed temperature T_b.

These assumptions allowed the heat flux at the sheath/plasma interface to be expressed in the following way:

$$Q = \frac{1}{2}n_s \bar{v}_e kT_b \left(\frac{T}{T_b}\right)\left(1 - \frac{T}{T_b}\right) = 0. \quad (5.42)$$

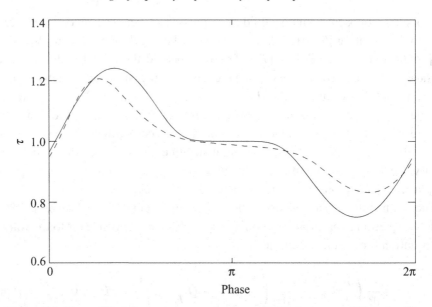

Figure 5.8 Normalized electron temperature $\tau \equiv T/T_b$ during one RF cycle. The dashed line is from a PIC simulation [73] and the solid line is from Eq. (5.43).

With this expression for the heat flux and with the listed assumptions, Eq. (5.41) can be reduced to the following first-order ordinary differential equation for T:

$$\frac{u_0}{\bar{v}_e}\left[(1+\cos\theta)\frac{1}{T_b}\frac{\partial T}{\partial \theta}+2\frac{T}{T_b}\sin\theta \ln\left(\frac{n_{es}}{n_s}\right)\right]+\frac{T}{T_b}\left(\frac{T}{T_b}-1\right)=0, \quad (5.43)$$

where $\theta = \omega t$ is the phase, n_{es} is the density at the instantaneous sheath edge, so that n_{es}/n_s is a function of θ and is given by the sheath model. This equation can be solved numerically to calculate the oscillations of the electron temperature in the sheath and the related heat flux to the plasma. The normalized electron temperature is shown in Figure 5.8 during one RF cycle. During the first part of the cycle the electron temperature in the sheath region is larger than the bulk temperature, leading to a heat flux out of the sheath region into the plasma, while in the second part the opposite occurs. Taking the average over a cycle shows that there is a net heat flux into the plasma. The associated collisionless heating per unit area [61, 70] is found to be

$$\bar{S}_{stoc,fluid}=\frac{\pi}{16}n_s m \bar{v}_e u_0^2\left(\frac{36H}{55+H}\right). \quad (5.44)$$

It is easy to see that for $H \ll 55$, this expression has the same scaling with all parameters as the hard-wall model expression obtained by Lieberman and

5.2 Non-uniform ion density, current-driven model

co-workers. The quantitative difference between the two models (again at moderate H) is

$$\frac{\overline{S}_{\text{stoc,fluid}}}{\overline{S}_{\text{stoc,hardwall}}} = \frac{72}{165} \approx 0.4. \tag{5.45}$$

In this book both models will be used, depending on the original published work, remembering that there is not a large difference between the results of both models. Note that collisionless heating is still an active area of research, and these models may be subject to change in the future.

Since the sheath oscillates with a speed $v(t) = u_0 \cos \omega t$, the amplitude of the current is $I_0 = e n_s u_0 A$. Taking the case of a collisionless sheath, for which $s_m/s_0 = 5\pi H/12$, Eq. (5.40) becomes

$$\overline{S}_{\text{stoc}} = \frac{9}{40} m \overline{v}_e \left(\frac{\omega s_m I_0}{eA} \right). \tag{5.46}$$

The resistance to be inserted in the equivalent circuit model of the capacitive discharge is therefore

$$R_{\text{stoc}} = \frac{2\overline{S}_{\text{stoc}} A}{I_0^2} = \frac{9}{20} m \overline{v}_e \left(\frac{\omega s_m}{e I_0} \right) = \frac{9}{20} \left(\frac{m \overline{v}_e}{n_s e^2 A} \right) \left(\frac{\omega s_m}{u_0} \right). \tag{5.47}$$

Note that it is also possible to use a collisional sheath law to calculate this resistance, with a slightly different result (see Table 5.2).

5.2.4 Comparison with the homogeneous model

Table 5.1 summarizes the values of the elements of the circuit models for the homogeneous and the inhomogeneous models. As mentioned previously, these elements are functions of the electron density, the sheath thickness and the RF current. However, there is a relation between these three quantities (the equivalent of the Child–Langmuir law in DC sheaths). In the table, the electron density has been systematically substituted using the appropriate sheath laws, namely Eq. (5.3) for the homogeneous sheath and Eqs (5.28) and (5.29) for the inhomogeneous sheaths. This will prove convenient when solving the energy balance, although the expressions look somewhat cumbersome. In particular, this hides the important fact that the main resistances accounting for transfer of energy from the electric fields to the electron population, i.e., R_{stoc} and R_{ohm}, scale with $1/n_0$.

The various coefficients introduced in this section come from various integrations; they are summarized in Table 5.2. Some of them have not been obtained in this section, but come from calculations described in the cited publications.

Table 5.1 *Expressions for the components of the equivalent circuit in the different models*

Component	Homogeneous model	Inhomogeneous model
C_s	$\dfrac{\varepsilon_0 A}{s_m}$	$K_{cap}\dfrac{\varepsilon_0 A}{s_m}$
R_{stoc}	0	$K_{stoc}(mkT_e)^{1/2}\left(\dfrac{\omega s_m}{eI_0}\right)$
$R_{ohm,sh}$	$\tfrac{2}{3}m\nu_m s_m\left(\dfrac{\omega s_m}{eI_0}\right)$	$K_{ohm,sh}m\nu_m s_m\left(\dfrac{\omega s_m}{eI_0}\right)$
R_i	$\tfrac{3}{2}\left(\dfrac{u_B}{\varepsilon_0 A\omega^2}\right)$	$K_i\left(\dfrac{es_m I_0}{M\varepsilon_0^3 A^3\omega^5}\right)^{1/2}$
R_{ohm}	$m\nu_m(l-2s_m)\left(\dfrac{\omega s_m}{eI_0}\right)$	$K_{ohm}h_l m\nu_m(l-2s_m)\left(\dfrac{\omega}{eI_0}\right)^{3/2}(A\varepsilon_0 s_m kT_e)^{1/2}$
L_p	R_{ohm}/ν_m	$\nu_m R_{ohm}$

Table 5.2 *Constants used in the inhomogeneous model components of the equivalent circuit*

Constants	Collisionless sheath	Collisional sheath
K_{cap}	0.613	0.751
K_{stoc}	0.72	0.8
$K_{ohm,sh}$	0.33	0.155
K_{ohm}	1.55	$1.14\sqrt{s_m/\lambda_i}$
K_s	0.42	0.39
K_i	0.87	$0.9\sqrt{\lambda_i/s_m}$

5.3 Global model

The expression 'global' here refers to the fact that the spatial dependence of the variables related by the model has been integrated out through prior analysis and reasoning. So far the details have been established for a symmetrical, single-frequency CCP. There are four external (control) parameters that are in principle chosen by the user, namely the neutral gas pressure p, the electrode spacing l, the driving frequency ω and the RF current I_{RF} (when required Eq. (5.31) enables a convenient way of switching to the more intuitive voltage drive). Three global variables then characterize the CCP: the electron temperature T_e, the central electron density in the plasma n_0 and a parameter s_m that represents the size of the region swept out by an RF sheath, i.e., the maximum sheath thickness.

5.3 Global model

To solve for the three variables, three equations are required: the particle balance, the power balance and the RF sheath law (or RF Child law), given by Eq. (5.28) for collisionless sheaths and Eq. (5.29) for collisional sheaths.

The simultaneous solution of these equations can be used to evaluate important quantities from a plasma processing point of view, such as the ion flux to the electrodes or the energy of ions bombarding those electrodes. In the next section, the model equations are gathered together and a simple procedure is given to solve these equations. The effect of the external parameters on the plasma variables will then be analysed. Finally, to account for all the power dissipation, the modelling is extended to include the electrical matching circuitry that is usually incorporated into processing reactors.

5.3.1 Model equations

The electrode surface is A and the plasma volume is $A(l - s_m)$ such that, according to Eq. (3.84), the steady-state particle balance can be written

$$n_g \bar{n}_e K_{iz}(l - s_m) = 2h_l n_0 u_B \quad (5.48)$$

where, from Eq. (2.27), $K_{iz} = K_{iz0} \exp(-e\varepsilon_{iz}/kT_e)$ is the ionization rate constant and ε_{iz} is the ionization potential. Note that this uses the simplification that there is no volume recombination, which is a very good approximation in electropositive plasmas because electron–ion recombination is negligible. Note also that multi-step ionization has been neglected.

> **Q** In the remainder of this section the assumption will be made that $\bar{n}_e = n_0$. To what extent is this a valid approximation?
> **A** In Section 3.4.2 it was shown that $2/\pi \leq \bar{n}_e/n_0 \leq 1$, so setting axial and mean densities equal will not seriously affect the density scalings but may limit confidence in absolute values to $\pm 30\%$.

In the limit of small sheath size, $s_m \ll l$, the particle balance is independent of the electron density, and this equation determines the electron temperature, independently from the two other variables. This key result was part of the discussion of discharge plasmas in Chapter 3. However, many industrial capacitive discharges have a rather small electrode gap (2–3 cm) and consequently it is not always appropriate to neglect s_m in Eq. (5.48). One may anticipate that as the RF current increases, the sheath size increases which reduces the volume of the plasma bulk, without changing the loss flux to the boundaries. Consequently, the ionization rate has to increase to sustain the plasma from a smaller production volume: thus the electron temperature has to increase.

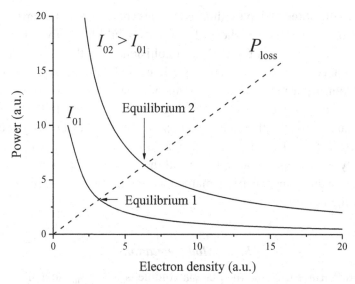

Figure 5.9 Schematic of the power absorbed by electrons as a function of the electron density for two different RF current amplitudes (solid lines) compared with power loss from the electron population (dashed line). The electron temperature is presumed fixed. The crossings are equilibrium points that indicate potentially stable plasmas.

The steady-state electron power balance can be written in terms of the equivalent circuit with Eqs (2.45) and (3.29) as

$$\frac{1}{2}(R_{\text{ohm}} + 2R_{\text{stoc}} + 2R_{\text{ohm,sh}})I_0^2 = 2h_1 n_0 u_B \varepsilon_T(T) A. \quad (5.49)$$

The left-hand side is the absorbed power in the equivalent circuit model. The power dissipated by *ions* crossing the sheaths, represented by R_i in the equivalent circuit, does not feature in the *electron* power balance.

5.3.2 Scaling with external control parameters

A simple graphical analysis of the discharge equilibrium is informative. Figure 5.9 plots schematically the power absorbed by electrons from electric fields, $P_{\text{abs}}(n_0)$, and the power lost from the electron population, $P_{\text{loss}}(n_0)$, for two different values of the amplitude of the RF current, I_0. Since the main resistances in Eq. (5.49) scale with $1/n_0$, at constant current $P_{\text{abs}}(n_0) \propto 1/n_0$ (if one neglects the $R_{\text{ohm,sh}}$ and the weak density dependence of $l - 2s_m$). On the other hand, the rate of energy loss from the electron population increases linearly with n_0, with a slope determined by the electron temperature T_e. The equilibrium must be at the crossing of the two curves $P_{\text{abs}}(n_0)$ and $P_{\text{loss}}(n_0)$, that is when losses exactly balance the power delivered to electrons from the RF generator. In particular, Figure 5.9 shows that when

5.3 Global model

the RF current amplitude is increased, the balance occurs at higher electron density. This graphical representation of the power equilibrium is very useful in understanding the physics of more complex dynamics such as mode transitions and related hysteresis and instabilities – some of these aspects will be treated in later chapters.

For a fully quantitative solution the challenge is to solve simultaneously Eqs (5.48), (5.49) and (5.28) (or (5.29)). A convenient procedure is first to choose a value for s_m, then to solve the particle balance to obtain T_e, and thereafter to solve the power balance together with the appropriate sheath law for n_0 and I_0 [63]. It is possible to obtain analytical solutions in limiting cases and these will be discussed next.

Electron temperature

The electron temperature is essentially set by the particle balance and increases only slowly with the applied voltage as a consequence of the reduction of the plasma volume when the sheath size increases. This effect is more pronounced at lower pressure (<1 Pa) and low density, where the sheaths are largest. Here we should point out that other subtle mechanisms, ignored in this simple modelling, may be responsible for variations in the electron energy distribution, and therefore the effective electron temperature. Among others, multi-step ionization is particularly important.

Plasma density

First consider a situation where (i) the sheath thickness is small compared to the plate separation ($s_m \ll l$), and (ii) stochastic heating dominates in the power absorption term. This is valid at low pressure, typically less than a few Pa. In this case, the electron temperature is determined from Eq. (5.48), which for given pressure and plate separation will fix $\varepsilon_T(T_e)$. Given the expression of R_{stoc} in Table 5.1, and using Eq. (5.31), one easily obtains

$$n_0 = \left[\frac{\varepsilon_0 K_{stoc} K_{cap} (mM)^{1/2}}{4eh_1 \varepsilon_T(T_e)} \right] \omega^2 V_0. \quad (5.50)$$

Under these circumstances the electron density increases linearly with the applied RF voltage, and scales with the square of the applied frequency. The same analysis can be done in the opposite limit, where ohmic heating in the plasma dominates (typically at much higher pressure).

Q Show that $n_0 \propto \omega^2 V_0^{1/2}$ when ohmic heating in the plasma dominates.
A The ohmic power balance sets $R_{ohm} I_0^2/2$ proportional to n_0. The former term contains the product $s_m I_0$, which can be linked to ωV_0, through Eq. (5.31). When this is done and one assumes $l \gg s_m$, what remains contains n_0 scaling with two of the four parameters, as given: ω^2 and $V_0^{1/2}$.

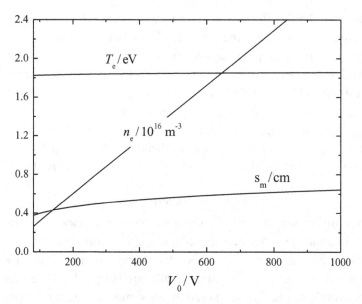

Figure 5.10 Electron density, sheath size and electron temperature as a function of the RF voltage amplitude. The frequency is 13.56 MHz, the argon gas pressure is 19.5 Pa, the electrode radius is 15 cm, and the electrode gap is 3 cm.

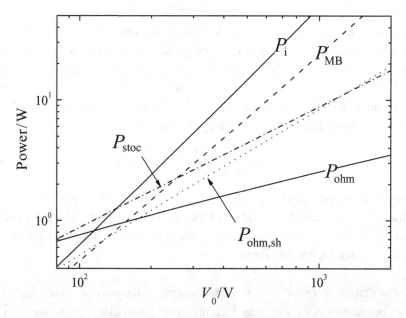

Figure 5.11 Components of the total power dissipation in the inhomogeneous model as a function of the RF voltage amplitude. The frequency is 13.56 MHz, the argon gas pressure is 19.5 Pa, the electrode radius is 15 cm, and the electrode gap is 3 cm. Losses in the match-box P_{MB} will be discussed in Section 5.3.4.

5.3 Global model

Sheath thickness

It is also interesting to examine the scaling of the sheath size with external parameters. Consider first a collisionless sheath. Then, combining Eqs (5.28) and (5.31), one obtains $s_m^4 \propto V_0^3 n_0^{-2} h_1^{-2}$. Since from Eq. (5.50) we have $h_1 \, n_0 \propto \omega^2 V_0$, it follows that

$$s_m \propto \frac{V_0^{1/4}}{\omega}. \tag{5.51}$$

For a given frequency, the sheath size increases moderately with voltage, whereas for fixed voltage, the sheath size decays significantly with frequency. Note that here we considered collisionless sheaths and found that s_m is independent of pressure. However, in the intermediate pressure range, one should use the collisional sheath law Eq. (5.29). Then, following the same procedure,

$$s_m \propto \left(\frac{V_0}{p\,\omega^4}\right)^{1/5}. \tag{5.52}$$

The trends with the RF voltage amplitude and frequency are similar (with slightly different scalings), and we see that the sheath tends to shrink with the neutral gas pressure p (remember that $\lambda_i \propto 1/p$).

The two scaling laws in Eqs (5.51) and (5.52) were obtained on the assumption that stochastic heating dominates, which may be questionable for typical gas pressure used in plasma processing discharges. To assess the validity of these scalings, the exact numerical solution of the three equations is shown in Figure 5.10, where the electron density, sheath size and electron temperature are plotted as a function of the RF voltage amplitude. The frequency has been set at 13.56 MHz and the argon gas pressure is 19.5 Pa; the electrode radius is 15 cm, and the electrode gap is 3 cm. The sheath thickness increases slowly with the voltage (as anticipated by the scaling laws (5.51) and (5.52)), which leads to a very weak increase in the electron temperature. The gas pressure is relatively high ($\lambda_i < s_m < l$), so that ohmic heating should be appreciable. However, the electron density increases almost linearly with the voltage, consistent with the result in (5.50), that is with collisionless heating.

To better understand the above observation and the relative importance of the various heating mechanisms, Figure 5.11 shows the individual components as a function of the RF voltage amplitude for the same conditions. Even for this relatively high pressure, ohmic heating in the bulk plasma never dominates. Stochastic heating at the boundaries of the strongly modulated sheaths is dominant for voltages in the range 200–1000 V, after which ohmic heating in the now relatively thick sheaths takes over. The diagram shows that ohmic heating in the plasma has a weaker dependence on voltage than stochastic heating, and that ohmic heating in the sheaths has the strongest scaling with voltage.

Table 5.3 *Effect of changing the electrode spacing on the plasma parameters for a fixed voltage between the electrodes, $V_0 = 200$ V. The frequency is 13.56 MHz, the electrode radius is $r_0 = 15.0$ cm and the gas pressure is 19.50 Pa*

$l/10^{-2}$ m	3.00	6.00	9.00
kT/e/eV	1.83	1.63	1.53
$s_m/10^{-2}$ m	0.47	0.49	0.51
$n_0/10^{16}$ m^{-3}	0.60	0.80	0.94
I_0/A	1.72	1.62	1.57

Table 5.4 *Effect of changing the gas pressure on the plasma parameters for a fixed voltage between the electrodes, $V_0 = 200$ V. The frequency is 13.56 MHz, the electrode radius is $r_0 = 15.0$ cm and the electrode spacing is $l = 3.00$ cm*

p/Pa	1.33	6.65	19.50	33.25
kT/e/eV	4.00	2.25	1.83	1.69
$s_m/10^{-2}$ m	0.76	0.58	0.47	0.41
$n_0/10^{16}$ m^{-3}	0.10	0.32	0.60	0.85
I_0/A	0.86	1.37	1.72	1.94

Gap width and pressure variations

The electrode spacing and the gas pressure both predominantly affect the electron temperature. Table 5.3 sets out how changes in the electrode separation would affect the plasma while holding $V_0 = 200$ V, starting from reference conditions that correspond with those of Figure 5.10: 13.56 MHz CCP, 19.50 Pa Ar, electrode radius $r_0 = 15$ cm. Increasing the gap at fixed pressure decreases the electron temperature and increases the electron density. Both of these effects are due to reduced losses (h_l decreases) – it was shown in Chapter 2 that pressure and system size determine the electron temperature when volume production is balanced by surface loss. The sheath size increases slightly as the electrode gap is widened. The current associated with the fixed voltage decreases slightly, keeping the product $s_m I_0$ constant, as required by Eq. (5.31).

Exercise 5.1: Global modelling Table 5.4 sets out how changes in the pressure would affect the plasma while holding $V_0 = 200$ V, starting from reference conditions that correspond with those of Figure 5.10. Account for the observation that decreasing the pressure at fixed electrode spacing increases the electron temperature, decreases the electron density and increases the spatial extent of the sheath region.

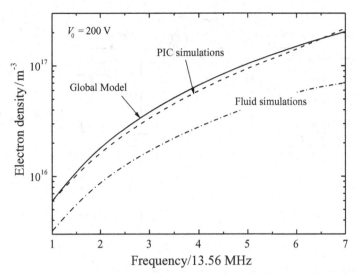

Figure 5.12 Electron density as a function of the driving frequency, as calculated by a PIC simulation, a fluid simulation and the global model. The RF voltage amplitude is fixed at $V_0 = 200$ V, the argon gas pressure is 19.5 Pa and the electrode gap is 3 cm. From [75].

5.3.3 Comparison with numerical models

The scaling with frequency that is predicted by the zero-dimensional global model provides a useful basis for comparing the model with one-dimensional (1-D) fluid and PIC calculations.

> **Q** Identify the parameters that could increase the rate of plasma–surface interactions in a processing step during semiconductor device fabrication.
> **A** Reaction rates are likely to scale with plasma density. According to Eq. (5.50), of the four (p, l, ω, V_0 (or I_{RF})), the frequency and output level of the electrical supply should strongly influence the process times.

Many authors have pointed out the advantages of increasing frequency in order to increase the plasma density [12, 73, 74], though this strategy does not leave everything else unchanged. For instance, if one tries to maintain a fixed plasma density while increasing the frequency, then the voltage across the electrodes will be lower (to maintain constant $\omega^2 V_0$) and therefore there will be a reduction in ion bombardment energy. In a single-frequency CCP the ion flux ($\propto$ plasma density) and ion energy ($\propto$ sheath voltage) cannot be varied independently.

Figure 5.12 compares the predictions of the global model described here with 1-D numerical solutions of the fluid equations [76–78] and a 1-D Monte Carlo, particle-in-cell simulation [79]. All calculations predict a dramatic increase of the

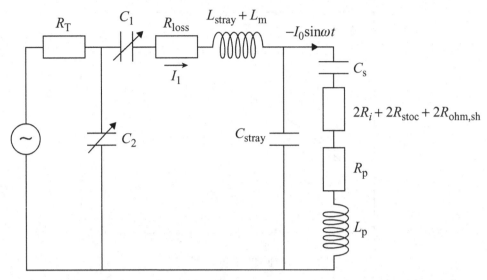

Figure 5.13 Electric circuit model for the capacitively coupled plasma reactor, including match-box and stray elements.

electron density with the driving frequency. There is remarkably good agreement between the global model and the PIC simulation. The fluid code predicts lower densities and a weaker scaling with frequency. This can be attributed to the facts that (i) the fluid approach does not capture the physics of stochastic heating in the sheaths, and (ii) it neglects electron inertia, both of which become important as the frequency increases.

5.3.4 Global model with match-box

The equivalent circuit of a CCP describes the electrical load that such an arrangement presents to an RF power source. To match the load (the reactor with a plasma between the electrodes) to a standard generator with 50 Ω output impedance, the system must include a matching network or 'match-box'. The complete electrical analysis of a CCP should include the match-box and any other elements of the real electrical system. A typical matching circuit comprising two variable capacitors and an inductor is shown in Figure 5.13. The diagram also includes realistic circuit losses, stray series inductance and stray capacitance to ground, respectively R_{loss}, L_{stray} and C_{stray}. The turns of wire that form the inductor are the chief source of circuit resistance, which is significant at radio frequency, so match-boxes usually include a cooling fan to disperse the heat that is generated. In addition to the inductor, there is additional inductance because of the specific path of current to and from the plasma-facing surfaces of the electrodes – there is inductance both in the power feed and in the earth return path. Charge must be supplied to set any conductor at a given potential, and where those potentials are between large areas that are closely separated there is a significant capacitance – the RF-powered electrode in a CCP is usually coaxially sleeved by a few millimetres of insulator and grounded metal.

The match circuit itself is composed of an inductor, L_m, and two variable capacitors, C_1 and C_2, that are adjusted to achieve impedance matching. For RF generators that have an output impedance of 50 Ω (resistive), the matching condition essentially requires that the matched load also amounts to 50 Ω. The practical circuit analysis is simplified if C_2 is considered as part of the output circuit of the generator, and the matching condition is restated as requiring that this arrangement is presented with a total load that amounts to its complex conjugate.

Q Suggest how to determine C_{stray}, L_{stray} and R_{loss} when the match circuit is connected to an empty CCP chamber.

A (i) The simplest quantity to measure is C_{stray}. Using a conventional low-frequency bridge, the capacitance between the input to the match circuit and ground is measured, with the generator and C_2 disconnected. The result is due to C_1 in series with the parallel combination of C_{stray} and the parallel plate capacitance of the inter-electrode gap ($\varepsilon_0 A/l$), which replaces the components of the equivalent circuit that represents the plasma in normal operation.

(ii) Still with the generator and C_2 disconnected, so again no plasma, the total circuit inductance can be found by finding the natural resonance of this inductance with the net capacitance from (i).

(iii) R_{loss} is the hardest to measure directly since it is the net resistance of the circuitry at the main excitation frequency – at RF the skin effect makes resistance a function of frequency. A neat way to measure R_{loss} would be to repeat step (ii) with C_1 or C_{stray} augmented as necessary to make the circuit resonant at the excitation frequency – then, driving the circuit with a rectangular pulse train, the resistance can be deduced from the decay time of the natural resonance initiated by each pulse.

The power dissipated in the match-box is of particular interest. This quantity clearly depends on the reactor design, via the values taken by C_{stray} and L_{stray} for instance, which contribute to determining the current drawn from the generator. For a given reactor design, the power dissipation also depends on the external parameters, pressure, frequency and RF voltage amplitude. The following discussion focuses on the driving frequency.

The resistance R_{loss} included in the circuit accounts for losses in conductors, in the match-box and in parts after the match-box. The RF current flows in the skin

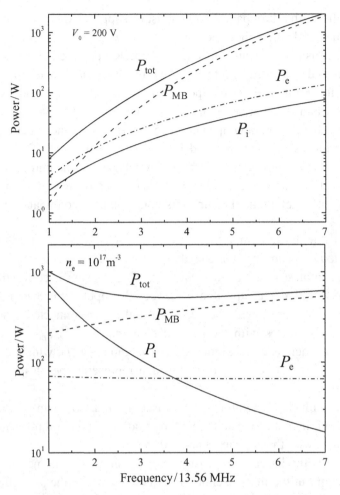

Figure 5.14 Various components of power dissipation as a function of the driving frequency for fixed voltage, $V_0 = 200\,\text{V}$ (upper), or fixed electron density, $n_e = 10^{17}\,\text{m}^{-3}$ (lower). The argon gas pressure is 19.5 Pa and the electrode gap is 3 cm; for illustrative purposes, R_{loss} (13.56 MHz) = 0.5 Ω (in practice it is system-dependent). When the generator is matched to the load, the power dissipated within the output circuit of the generator also equals P_{tot}.

of these conductors, that is in a small area; the (resistive) skin depth at 13.56 MHz in aluminium is approximately 22 μm. In addition, and as shown in Eq. (2.58), the skin depth scales with $\omega^{-1/2}$, which implies that the RF resistance of a conductor increases with the square root of the frequency. One can therefore anticipate that the power dissipated in the match-box should increase with the driving frequency. But this is not quite the end of the story, because the RF currents also tend to be higher at higher frequency. Figure 5.14 compares the various sources of power dissipation in the total circuit of Figure 5.13, as a function of the driving frequency, for a fixed voltage amplitude, $V_0 = 200$ V (upper), and for a fixed electron density, $n_e = 10^{17}$ m^{-3} (lower).

In the upper graph, the voltage is held constant and the frequency is increased, such that the electron density increases with the square of the frequency (this corresponds to the case in Figure 5.12). The power deposited in electrons, P_e, increases accordingly. The ion power also increases because at constant V_0 the time-averaged sheath voltage remains constant while the ion current increases linearly with the electron density. Finally, notice that the power dissipation in the match-box increases more rapidly than the other components. This is because the loss resistance increases with the square root of the frequency while the RF current increases; by contrast, the plasma resistance decreases with frequency because of the very fast increase in the electron density. At very high frequency, most of the power delivered by the RF generator is dissipated in the match-box and circuitry external to the plasma.

In the lower graph of Figure 5.14, the electron density is held constant. This time, the electron power and the RF current remain fairly constant (since n_e is constant). However, the power dissipated by the ion bombardment of electrodes decreases dramatically. This is due to the decrease in the sheath voltage. Indeed, since the sheath impedance is capacitive, the voltage drops when the frequency increases at fixed current. This shows the advantage of increasing the frequency to increase the coupling to the plasma, that is to increase the amount of power dissipated by electrons against the power lost by accelerating ions in the sheath. Since the RF current is fairly constant, the losses in the external circuit increase only through the skin effect. Again, at high frequency, most of the power is dissipated outside the plasma.

Q Explain the result shown in Figure 5.11, which indicates that the loss in the match-box increases with the applied voltage faster than the power dissipated by electron heating.

A In the equivalent circuit, the RF current flows through R_p, $R_{\mathrm{ohm,sh}}$ and R_{loss}; the last component also passes the current in the stray capacitance. Whereas R_{loss} is independent of the applied voltage, the resistances accounting for electron heating decay with $1/n_0$, and so decrease with increasing voltage. Taken together, these factors account for the observed trends.

5.4 Other regimes and configurations

So far CCPs have been considered only in perfectly symmetrical arrangements and at rather low pressure. Often the earthed electrode is electrically connected to the walls of a vacuum vessel and therefore the area of grounded surface is much larger than that of the RF-powered electrode; the implication of electrode asymmetry is discussed in the next section. Another interesting issue is related to higher-pressure operation, in which secondary electrons emitted from electrodes may play a role in the ionization processes and in the electron power balance. The final topic in this section concerns the series resonance that occurs between the sheath capacitance and the plasma inductance.

5.4.1 Asymmetric discharges

Consider an asymmetrical CCP, that is one between electrodes of different area: A_a is the area of electrode 'a' and A_b is the area of electrode 'b'. Since the areas are different, the size and potential associated with sheaths 'a' and 'b' must be different, otherwise the sheaths could not pass the same RF current – they must do so since current flowing through the system is conserved.

For the constant ion density model, according to Eq. (5.3), the continuity of RF current would require that

$$s_{0,a} A_a = s_{0,b} A_b.$$

Since in this model $s_m = 2s_0$, it follows that

$$s_{m,a} A_a = s_{m,b} A_b,$$

so the time-averaged sheath, that is the region from which electrons are swept out once during each RF cycle, is smaller in front of the electrode of larger surface. The voltage ratio is determined by using Eq. (5.12), and the DC component of (5.13) to replace the sheath thicknesses. Rearranging the result:

$$\frac{\overline{V}_a}{\overline{V}_b} = \left(\frac{A_b}{A_a}\right)^2. \tag{5.53}$$

Thus, the larger DC (and RF) voltage is adjacent to the smaller electrode. In fact, the different sheath voltages mean that an asymmetric CCP develops a DC bias between the electrodes – a symmetrical one does not. The DC bias is

$$\overline{V}_b - \overline{V}_a = \overline{V}_b \left(1 - \left(\frac{A_b}{A_a}\right)^2\right).$$

In principle, at large asymmetry, this means that almost all the RF potential is rectified by the larger sheath, through which ions reach the surface with a bombardment energy almost equal to the RF amplitude. Figure 5.15 shows a sketch of potentials in an asymmetric CCP.

5.4 Other regimes and configurations

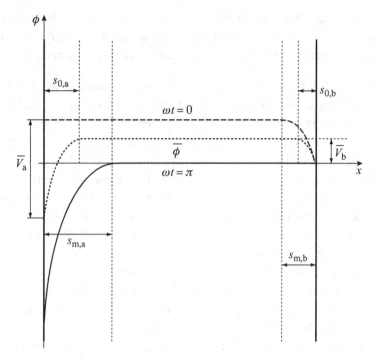

Figure 5.15 Potential $V(x, t)$ between the electrodes for two different times during the RF cycle, along with the time-averaged value, in the case of an asymmetric capacitive discharge. $\overline{V}_b - \overline{V}_a$ is the self-bias voltage.

Q The scaling between voltage and area from the constant ion density model is unlikely to be perfect because of the unrealistic representation of ion space charge. What other assumption(s) of the homogeneous CCP model might affect its usefulness here?

A In setting up the current-driven CCP it was convenient to ignore the self-consistent, self-bias voltage that properly balances the particle currents – instead, the instantaneous sheath is supposed to vanish for an instant. In the asymmetrical CCP, although this works well at the small area electrode where there is a large RF current density and large voltage, it is not a good assumption for the large area side where the current density is relatively small. The lack of spatial structure in the plasma is another compromise – the two electrodes may not be in exactly the same plasma.

Using the more realistic inhomogeneous sheath model, Eq. (5.28), the continuity of RF current requires that

$$\frac{A_b}{A_a} = \left(\frac{s_{m,a}}{s_{m,b}}\right)^{1/3}. \tag{5.54}$$

Together with Eq. (5.51) this implies a fourth-power scaling between area ratios and voltage ratios; however, experimental measurements generally reveal much weaker scalings. Various important factors can reconcile the disparity.

- First, the neglect of particle currents is a serious omission, at least as far as the larger area electrode is concerned. The RF-enhanced floating potential discussed in Chapter 4 determines the DC voltage between the plasma and ground, though to invoke the basic formula in terms of a modified Bessel function, one needs to assume a sinusoidal RF plasma potential. At least one voltage-to-area scaling calculation has been based entirely on particle currents [80], but the analysis is flawed by an assertion that ion and electron loss rates are instantaneously in balance, whereas the true balance is established over an RF cycle.
- Second, the inhomogeneous CCP features a symmetrical plasma volume with central density n_0 and densities at the boundaries given by $h_1 n_0$, but this cannot be so for an asymmetric CCP. The easiest asymmetric CCP to think about would be one between a pair of concentric cylinders – analysis shows that the densities at the inner and outer sheath/plasma boundaries are not equal and one should employ different h_1 terms for the two electrodes [2].
- Third, in practice the effective area of the grounded surface of a CCP in a metallic vacuum chamber is difficult to quantify because of the presence of ports and re-entrant, shielded-electrode structures.
- Fourth, there is only one proper earth point, and so-called grounded surfaces are connected to it via distributed, inductive current paths. Indeed, in some circumstances there may be some RF current that circulates through grounded surfaces but without intercepting the earth point.

A universal scaling between the area ratio and the DC bias of an asymmetric CCP is not available. Nevertheless, it is clear that ions will have higher energies at the smaller electrode in strongly driven, highly asymmetric CCPs. This is the archetype for reactive ion eching (RIE).

5.4.2 Higher-pressure regimes

The foregoing discussions have ignored aspects that may become important when varying the neutral gas pressure from very low values (about a Pa) to fairly high values (hundreds of Pa).

Electron energy distribution functions

To illustrate this, we start by examining the electron energy distribution function changes, as measured by Godyak and Piejak [66] in a single-frequency capacitive discharge in argon. Figure 5.16 shows how the semi-log plot of the so-called

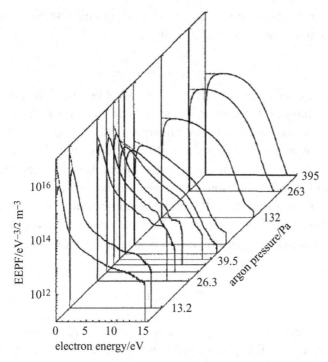

Figure 5.16 Evolution of the electron energy probability function (EEPF) as a function of pressure in a single-frequency capacitive discharge. From Godyak and Piejak [66].

electron energy probability function (EEPF) changes as pressure is varied from 9.3 to 400 Pa (0.07–3 torr). The EEPF is the energy distribution function divided by the square root of the energy ($f(\varepsilon)/\varepsilon^{1/2}$) and for a Maxwellian this gives a single straight line on a semi-log plot (see Chapter 10). The first striking observation is that the distribution is never close to a Maxwellian distribution. In previous discussions the electron population has always been presumed to have a Maxwellian energy distribution – though convenient, that is clearly not strictly appropriate, so care must be exercised in expressing absolute confidence in numerical results. A second observation is that there is a clear transition in the middle of the range. The distribution at low pressure has a high-energy tail superimposed on a cooler bulk. At higher pressure the distribution becomes dome-shaped, with a distinct reduction in the numbers of higher-energy electrons.

Godyak and Piejak attributed this effect on the distribution function to a transition from a discharge dominated by stochastic heating at low pressure, to a discharge dominated by ohmic heating at higher pressure [66]. This transition occurs at around 65 Pa in their experiment, which is close to where the global model in

Section 5.3 anticipates a change-over between stochastic and ohmic mechanisms for a discharge in a symmetrical CCP chamber.

Secondary electrons

In DC discharges, secondary electrons are essential because ions produced in the plasma cannot carry all the current in the cathode sheath. RF discharges, on the other hand, do not in general require secondary emission to take place. Nevertheless, Godyak and co-workers have shown that the electron energy distribution function may be modified by the emission of secondary electrons at the electrode [81]. Even when the sheath is collisional, ions reach the electrode with sufficient kinetic plus internal energy to enable around 10% of the recombination of ions and electrons to liberate bound electrons from the surface. These secondary electrons are accelerated back into the plasma by the sheath electric field. The flux of secondary electrons is written

$$\Gamma_{se} = \gamma_{se}\Gamma_i, \quad (5.55)$$

where γ_{se} is the electron secondary emission coefficient and Γ_i is the ion flux. Note that at high frequency, although the ion flux is time-independent, the acceleration of secondary electrons varies through the RF cycle such that their energy and density fluctuates at the frequency of the generator.

The secondary electrons may participate in ionization and they may be responsible for a significant fraction of the power dissipation [39, 82, 83]. This effect may be included in global models of CCPs similar to the one described in this book (see for instance [62, 82, 83]). In order for secondary electrons to be important in the discharge power balance, two conditions must be fulfilled. First of all, these electrons need to get significant energy in the sheath; the effect is going to be larger at large sheath voltage, that is at large power. Second, they need to be able to ionize (i.e., experience inelastic ionizing collisions) before escaping the discharge or being thermalized; the effect is not significant at low pressure. To illustrate this, Figure 5.17 shows the evolution of the plasma density as a function of the RF voltage amplitude in a 3.2 MHz capacitive discharge in helium. The pressure was 400 Pa (3 torr) and the electrode spacing was 7.8 cm. The dashed line is an experimental result obtained by Godyak and Khanneh [82], while the three other curves are fluid calculations by Belenguer and Boeuf [84] with various values of γ_{se}. There is a transition at around 400 V, corresponding to a threshold voltage above which the discharge becomes sustained by secondary electrons. This is sometimes called the γ-*mode* transition.

The above work was carried out at rather low frequency, for which voltages tend to be larger, and at rather high pressure. In typical etching plasmas, with

5.4 Other regimes and configurations

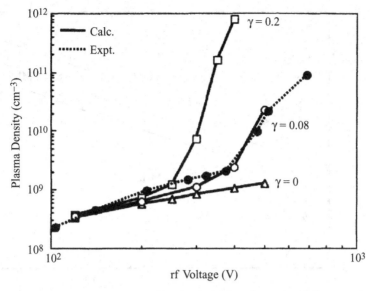

Figure 5.17 Plasma density as a function of RF voltage for different values of $\gamma (\equiv \gamma_{SE})$ showing the transition to the γ-mode. From Belenguer and Boeuf [84].

pressures around a few Pa and frequencies above 13.56 MHz, secondary electrons are usually not dominant, unless they are deliberately enhanced [83]. However, the next chapter explores reactors that operate with more than one frequency, with the lower frequency sometimes significantly below the usual 13.56 MHz. In such systems, secondary electrons may play a role.

5.4.3 Series resonance

It is clear from the analysis of the last two chapters that although one may choose to excite the plasma at one particular frequency, the non-linearities of the sheaths in particular give rise to harmonics that introduce a wide range of frequencies into the system – under these circumstances it is important to check whether any of the 'natural modes' are likely to become excited. This final section looks briefly at the consequences.

As widely discussed in this chapter, the equivalent circuit of a capacitive discharge is composed of resistances, accounting for various dissipations, in series with a capacitance, C_s, modelling the combination of the two sheaths (where the electrostatic field energy is high) and an inductance, L_p, due to plasma electron inertia. This equivalent circuit has a series resonance at a frequency $\omega_{res} = (L_p C_s)^{-1/2}$; the resonance couples the electrostatic field energy to the kinetic energy stored in the RF oscillation of the electron population. From Eqs (2.82) and (5.30), it

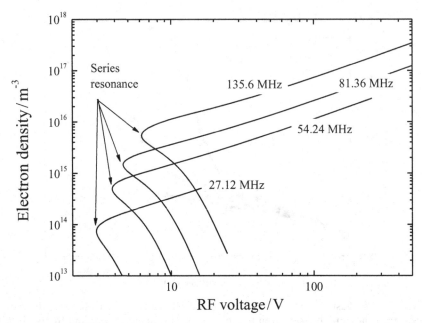

Figure 5.18 Electron density as a function of the RF voltage when crossing the series resonance.

follows that

$$\omega_{\text{res}} = \left(\frac{dK_{\text{cap}}}{\omega_{\text{pe}}^2 s_{\text{m}}}\right)^{-1/2} \approx \omega_{\text{pe}} \sqrt{\frac{s_{\text{m}}}{d}}. \quad (5.56)$$

For a typical electron density of $n_e = 10^{16}$ m^{-3}, and a symmetrical parallel plate CCP, the series resonance is likely to be in the range $0.1 - 0.5$ GHz and therefore much larger than the driving frequency.

Although conventional plasma processing reactors do not operate at a fundamental driving frequency near the series resonance, it has been suggested that enhanced power absorption might be achieved through deliberately addressing the series resonance [85]. It is possible to meet the resonance conditions by running at very low density, that is by decreasing the plasma frequency, but the series resonance in the normal operating regime of a 13.56 MHz capacitive discharge is expected to lie beyond 100 MHz. To reach the resonance at higher density one must therefore increase the driving frequency, noting also the fact that at constant electron density, s_{m} decreases with increasing frequency, based on the scaling $s_{\text{m}} \propto \omega^{-1}$ from Eq. (5.52).

In Figure 5.18, the electron density is plotted as a function of the RF voltage when crossing the series resonance, as calculated by the global model. At low

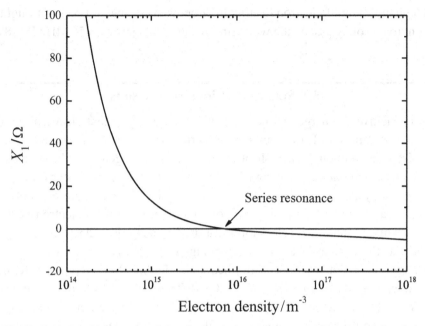

Figure 5.19 Imaginary part of the discharge impedance as a function of the electron density for a 135.6 MHz discharge. The resonance is around $n_e = 10^{16}\,\text{cm}^{-3}$.

voltages, there are two solutions for the electron density at a given voltage. In other words, when scanning the electron density by increasing the power out of the generator, one passes a point of minimum voltage, which corresponds to the series resonance. This was demonstrated experimentally by Godyak and Popov [86]. As expected, the electron density at which the resonance occurs increases with the driving frequency. At 10 times the traditional RF excitation frequency, that is at 135.6 MHz, the resonance occurs around $n_e = 10^{16}\,\text{cm}^{-3}$, which is a typical density for plasma processing. An interesting feature to note is that the reactive part of the discharge impedance, X_1, switches from positive to negative values on crossing the resonance by increasing density, as shown in Figure 5.19. This means that at low density, the discharge impedance is mostly inductive (the electron inertia inductance dominates), while at high density it becomes capacitive (the sheath capacitance dominates).

It is difficult to prevent energy from building up in the natural modes of a system, so it is perhaps not surprising to discover that in practice the harmonics generated by the non-linearities of the sheaths will almost inevitably excite the series resonance to some degree [87], and this is the basis of a diagnostic method described in Chapter 10. This self-excitation in turn has been found to contribute to an enhancement of electron heating in circumstances that at first sight might be thought to be solely associated with currents and voltages at 13.56 MHz [88, 89].

5.5 Summary of important results

- A capacitive discharge can be modelled with an equivalent electrical circuit that is composed of a capacitance in series with an inductance and several resistances. Although each sheath is non-linear (i.e., generates harmonics of the driving frequency), the combination of the two is nearly linear for a symmetrical system, this is why they can be modelled by a single capacitor. A significant fraction of the power is dissipated in sheaths by electron heating mechanisms and ion acceleration towards the electrodes. Electron heating in the sheaths may exceed electron heating in the plasma in narrow gap discharges. Collisionless (stochastic) heating is dominant in the pressure regime of interest in etching plasmas (0.1–20 Pa). These dissipative effects are modelled with resistances. The plasma does not generate harmonics and can be modelled by the series combination of a resistance (due to electron–neutral elastic collisions, responsible for ohmic heating) and an inductance (due to electron inertia).
- At 13.56 MHz the impedance of the sheaths is much larger than the plasma impedance and almost all the voltage drop takes place across the sheaths.
- The equivalent circuit description allows the match-box to be included in the global model. The fraction of the power dissipated in the match-box is significant (it may easily exceed that delivered to the plasma), and strongly increases with the applied voltage and the driving frequency.
- Increasing the output power of the RF generator increases both the plasma density, which governs the ion flux to the electrodes, and the sheath voltage, which governs the bombarding ion energy. Ion flux and ion energy cannot be varied independently, which is a great limitation of single-frequency CCPs. This issue is discussed further in the next chapter.
- Increasing the driving frequency allows the sheath voltage to be reduced, at given plasma density, or the plasma density to be increased, at given sheath voltage.
- Increasing the gas pressure or the electrode gap size increases the electron density at fixed voltage and decreases the electron temperature.
- The sheath size increases with the voltage, significantly decreases with the driving frequency, and only moderately decreases with the neutral gas pressure; the typical scaling is $s_m \propto (V_0/p\omega^4)^{1/5}$.

- Asymmetric discharges, that is discharges with unequal electrode areas, have a larger voltage sheath in front of the smaller electrode and a smaller voltage sheath in front of the larger electrode.
- Secondary electrons play an important rôle in higher-pressure, high-voltage CCPs.
- A CCP stores energy in electrostatic fields of the sheath and in the inertia of electrons in the plasma bulk – resonances can occur as energy is transferred between these energy reservoirs.

6
Multi-frequency capacitively coupled plasmas

In the previous chapter it was shown that single-frequency capacitive discharges do not allow ion flux and ion energy to be varied independently. To overcome this limitation, inductive discharges may be used, in which the plasma is produced by an RF current in an external coil while the wafer-holder is biased by an independent power supply. These discharges are studied in the next chapter.

It should also be possible to achieve a reasonable level of control of the ion flux independently of the ion energy, by using dual-frequency CCP. Figure 6.1 shows the inspiration for this assertion [90]: the ion energy is plotted as a function of the ion flux at the grounded electrode of a symmetrical CCP for three different single-frequency discharges. The symbols are measurements from a planar probe and from a retarding field analyser inserted in the grounded electrode [75, 90, 91] (see Chapter 10 for background on these measurements). The lines in the figure are from a global model similar to that developed in the previous chapter. It appears as expected that the trajectory in flux–energy space is a single line for each driving frequency. At 13.56 MHz, there is a clear trend towards high ion energies and small ion fluxes, while at 81.36 MHz the opposite arises. Etching often requires ions to have energy in excess of 100 eV to enhance chemical reactions, but less than about 500 eV to avoid physical damage to the surface being etched, or to the photoresist mask. It is clearly not easy to explore this process window with a single-frequency CCP. Hence, it is worth investigating the idea that a dual-frequency excitation might offer control of the ion energy with the lower frequency while controlling the ion flux with the higher frequency. This was first proposed by Goto *et al.* [92], but later it was shown that the ideal separation is not always achieved [93]. In this chapter, it is shown that the two frequencies are indeed coupled via electron heating mechanisms in the dual-frequency sheaths.

The outline of this chapter is as follows. In Section 6.1, dual-frequency capacitive discharges are investigated with the same approach as in the previous chapter. The aim is to establish a global model that can be used to predict the plasma parameters

6.1 Dual-frequency CCP in electrostatic approximation

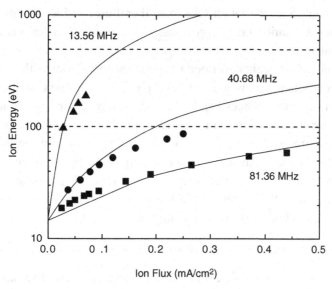

Figure 6.1 The flux–energy diagram for ions bombarding the electrodes in a symmetrical single-frequency capacitive discharge in argon, from [90]. The gas pressure was 2 Pa (15 mtorr). Symbols are experimental results, while solid lines are results from a global model similar to that developed in Chapter 5.

for any given combination of input parameters. To shorten the analysis it is restricted to the electrostatic regime and a full circuit model is not developed. The use of three or more frequencies of excitation is also briefly considered. Section 6.2 addresses the case of excitation by a *single* very high frequency in the electromagnetic regime, that is when the wavelength of the RF excitation is comparable with, or less than, the size of the electrodes. The electromagnetic analysis of dual (or multiple)-frequency CCPs remains to be done.

> **Q** When should one be concerned about electromagnetic effects in a system with electrodes of diameter 30 cm?
> **A** The free space wavelength that corresponds with 1 GHz is 30 cm – for frequencies around and above 100 MHz it is not reasonable to suppose that RF potential is uniformly distributed across a conductor of this size.
> *Comment: For a CCP at high frequency one must also consider effects of skin depth.*

6.1 Dual-frequency CCP in the electrostatic approximation

The global model for a dual-frequency capacitive discharge is based, as usual, on a combination of particle balance and energy balance. The nature of the power source enters only into the latter. One must therefore consider the effect of using dual-frequency excitation on the processes that transfer electrical energy into the electron population.

The plasma bulk is again described by a resistance in series with an inductance, i.e., linear components – as a result, the passage of current arising from two independent frequency sources can be treated by simple superposition. It is more complicated for the combination of the two sheaths, which, unlike in the single-frequency case, do not behave like a simple capacitor because the inherent non-linearity of the sheaths leads to a strong coupling of the two frequencies.

Experience from earlier chapters suggests that one should go straight to a current-driven scenario, with the current expressed as the sum of a basic sinusoidal component (at ω) and a higher sinusoidal component (at ω_h):

$$I_{RF}(t) = -I_0 \sin \omega t - I_h \sin \omega_h t = -I_0 (\sin \omega t + \beta \sin \alpha \omega t), \qquad (6.1)$$

where $\alpha \equiv \omega_h/\omega$ and $\beta \equiv I_h/I_0$. If the higher frequency is an exact harmonic, $\alpha = 2, 3, 4, \ldots$ the relative phase between the components of the current is important but in general it is not; one particular case will be mentioned in Section 6.1.4.

The following definitions that were introduced earlier will again be used to characterize the strength of the current and the range of the RF motion of electrons:

$$H = \frac{1}{\pi \varepsilon_0 k T_e h_1 n_0} \left(\frac{I_0}{A\omega}\right)^2, \qquad (6.2)$$

$$s_0 = \frac{1}{e h_1 n_0} \left(\frac{I_0}{A\omega}\right). \qquad (6.3)$$

6.1.1 Electron heating in the plasma bulk

The plasma bulk is again defined as that part of the space between the electrodes that is always quasi-neutral; its spatial extent is $l - 2s_m$. In that region the total current transfers electrical energy into the electron population at a mean rate determined by the resistance of the plasma bulk. The time-averaged power per unit volume is

$$\overline{S}_{ohmDF} = \frac{1}{2} \frac{m v_m (l - 2s_m)}{e^2 A^2 n_0} I_0^2 (1 + \beta^2), \qquad (6.4)$$

where the mean plasma density has been approximated by n_0. There is no coupling of the two frequencies in the plasma region – the ohmic heating term is averaged over time so the product ($\sin \omega t \times \sin \alpha \omega t$) averages to zero and neither frequency appears in the final expression. As found in the single-frequency case, in conditions that are typical of plasma etching, ohmic heating in the plasma bulk turns out to be small compared with heating in the sheaths.

6.1.2 Electron heating in the dual-frequency sheath

A model of a dual-frequency sheath is required so that the size of the plasma $(l - 2s_m)$ and sheaths $(s(t))$ can be properly specified. The situation has been studied by Robiche et al. [94] and by Franklin [43] in the limit when $\beta/\alpha \ll 1$, which corresponds with there being either a small fraction of current at a slightly higher frequency or perhaps comparable current at a much higher frequency. In this limit the effective RF Child–Langmuir law is found to be

$$\frac{s_m}{s_0} = 2\left(1 + \frac{\beta}{\alpha}\right) + \frac{5\pi H}{12}\left(1 + \frac{9\beta}{5\alpha}\right)$$

which, in the limit of high current drive ($H \gg 24/5\pi$), reduces to

$$\frac{s_m}{s_0} \approx \frac{5\pi H}{12}\left(1 + \frac{9\beta}{5\alpha}\right). \tag{6.5}$$

Q Look back at the two single-frequency cases of the previous chapter and then identify the physical process that underlies the factor in brackets on the RHS of Eq. (6.5).

A The process in question is the passage of current through the sheath. As with the single-frequency analyses, the total current is related to the motion of electrons at the plasma/sheath boundary so that

$$I_{RF}(t) = n_i(s)e\frac{ds}{dt}.$$

Given the form of Eq. (6.1), treating the boundary density as constant, a simple integration of this would lead to a factor $(1 + \beta/\alpha)$ in s_m. *Comment: The added complexity of the changing density at the instantaneous boundary between quasi-neutral plasma and sheath introduces additional numerical factors which can only be found through numerical analysis.*

It has been shown also that a good approximation of the time-averaged potential drop in the dual-frequency sheath is

$$\frac{e\overline{V}}{kT_e} = \frac{1}{2}\left[1 + \pi H\left(\frac{3}{4} + \frac{\beta}{\alpha}\right)\right]^2 - \frac{1}{2}.$$

In the high-current-driven case this reduces to

$$\frac{e\overline{V}}{kT_e} = \frac{\pi^2}{2}H^2\left(\frac{3}{4} + \frac{\beta}{\alpha}\right)^2. \tag{6.6}$$

Q Check that the single-frequency results of the previous chapter can be recovered from Eqs (6.1) and (6.6).

A Setting $\beta = 0$ in each and substituting using Eqs (6.2) and (6.3) gives

$$s_m = \frac{5\pi}{12} \frac{1}{\pi\varepsilon_0 k T_e(h_1 n_0)} \left(\frac{I_0}{A\omega}\right)^2 \frac{1}{e(h_1 n_0)} \left(\frac{I_0}{A\omega}\right)$$

and

$$\frac{e\overline{V}}{kT_e} = \frac{\pi^2}{2} \left(\frac{3}{4\pi\varepsilon_0 k T_e(h_1 n_0)}\right)^2 \left(\frac{I_0}{A\omega}\right)^4.$$

Tidying up and inserting numerical factors, the results are readily shown to be equivalent to Eqs (5.28) and (5.35).

If the lower frequency is significantly above the ion plasma frequency, then one can neglect transit-time effects which lead to a complex ion–energy distribution function. The mean ion energy at the electrode is then simply $w_i = e\overline{V}$.

Electron heating mechanisms in the dual-frequency sheath have been investigated by Turner and Chabert [64, 95, 96]. Both collisionless (stochastic) heating and collisional (ohmic) heating were calculated following the same procedure as that described in the previous chapter. The calculations are somewhat complicated and will not be detailed here. The result for collisionless heating is [95]

$$\overline{S}_{\text{stocDF,fluid}} = \frac{\pi}{16} n_s m \overline{v}_e u_0^2 \left(\frac{36H}{55+H}\right)(1 + 1.1\beta^2). \tag{6.7}$$

This expression is the dual-frequency extension of the fluid model described in the single-frequency case; it reduces to Eq. (5.44) for $\beta = 0$ as it should. Similarly, Kawamura et al. [69] have extended the hard-wall model to the dual-frequency case and found that in the parameter range of interest,

$$\frac{\overline{S}_{\text{stocDF,fluid}}}{\overline{S}_{\text{stocDF,hardwall}}} \approx 0.3, \tag{6.8}$$

which is close to the ratio of 0.4 obtained in the single-frequency case. Again, the scalings with the RF current (or voltage) were essentially identical for both models. Therefore, the scaling of the important qualitative results presented below does not depend upon the model used for collisionless heating. Examination of Eq. (5.44) reveals that a dual-frequency sheath is not simply the sum of two separate terms for high frequency and low frequency. The following question explores this point.

6.1 Dual-frequency CCP in electrostatic approximation

Q Use Eq. (5.44) to write down the equivalent stochastic terms at frequencies ω and $\alpha\omega$ if applied separately, with the latter component β times the amplitude of the former. Then for $H \gg 1$ compare the sum of these terms with Eq. (6.7).

A The low-frequency term, $\overline{S}_{\text{stoc},l}$, is simply given by Eq. (5.44) (or Eq. (6.7) with $\beta = 0$), while the higher-frequency term is

$$\overline{S}_{\text{stoc},h} = \frac{\pi}{16} n_s m \overline{v}_e \beta^2 u_0^2 \left(\frac{36 H (\beta/\alpha)^2}{55 + H (\beta/\alpha)^2} \right). \tag{6.9}$$

For sufficiently large H:

$$\frac{\overline{S}_{\text{stocDF,fluid}}}{\overline{S}_{\text{stoc},h} + \overline{S}_{\text{stoc},l}} \approx \frac{1 + 1.1\beta^2}{1 + \beta^4/\alpha^2}. \tag{6.10}$$

This number is typically greater than one, which shows that collisionless electron heating within the sheath is enhanced by dual-frequency excitation.

For a given value of the ratio of frequencies, α, the maximum enhancement predicted by Eq. (6.10) is approximately $(1 + \alpha)/2$, and occurs when $\beta \approx \sqrt{\alpha}$. Taking the example of $\alpha = 11$ (for instance, 13.56 MHz combined with 149.16 MHz) leads to a maximum enhancement of about 6 times. The effect is thus very pronounced.

Ohmic heating in the dual-frequency sheath has been calculated in [64], resulting in

$$\overline{S}_{\text{ohmDF,sh}} = n_s m u_0^3 \left(\frac{\nu_m}{\omega} \right) F_0(\alpha, \beta, H), \tag{6.11}$$

where $F_0(\alpha, \beta, H)$ is a complicated function of the three parameters. Perhaps not surprisingly, the non-linearity of the sheath again makes cross terms dominate, so that ohmic heating is also enhanced by the combination of two frequencies.

Phase-resolved optical emission spectroscopy (PROES) experiments have been used to demonstrate the coupling of the two frequencies and their combined effect on electron heating [97,98], leading to the following qualitative insight into the heating enhancement in dual-frequency sheaths. The higher frequency alone produces a significant heating through the faster sheath motion, but this heating is limited to a small ion density range because the sheath size of a single high-frequency discharge is small. In contrast, the lower frequency acting alone explores a larger range of ion density because of the larger sheath extension (voltage) but does not produce a significant heating because of the slow sheath motion. The dual-frequency excitation provides a synergy between the larger sheath caused by the lower frequency and the higher heating rates caused by the higher frequency. However, this coupling between the two frequencies reduces the capability of dual-frequency discharges to

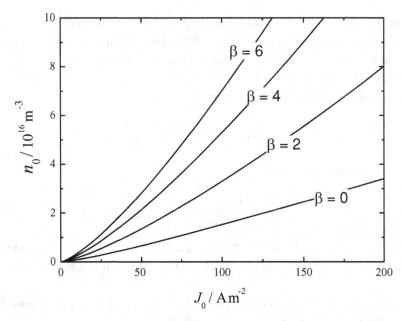

Figure 6.2 Electron density as a function of the low-frequency current density amplitude $J_0 = I_0/A$, for different values of the ratio β. The argon gas pressure was 1.33 Pa, the electrode spacing was 5 cm, the low frequency was 13.56 MHz and the high frequency was 149.16 MHz, i.e., $\alpha = 11$.

provide independent control of the ion flux and the ion energy. This is investigated in the next section.

6.1.3 Global model of a dual-frequency CCP

All the necessary ingredients of a global model of the dual-frequency capacitive discharge are now ready. The particle balance remains similar to that detailed in the single-frequency case:

$$n_g K_{iz}(l - s_m) = 2h_l u_B, \tag{6.12}$$

where the sheath size is given by the dual-frequency Child–Langmuir law, Eq. (6.5). The power balance is

$$\overline{S}_{\text{ohmDF}} + 2\overline{S}_{\text{stocDF}} + 2\overline{S}_{\text{ohmDF,sh}} = 2h_l n_0 u_B \varepsilon_T(T_e). \tag{6.13}$$

The three equations, Eqs (6.5), (6.12) and (6.13), have been solved by Levif [99] in order to calculate all the plasma parameters (n_0, T_e and s_m). Given the relative complexity of the heating power expressions, the calculation is not straightforward. The strategy is similar to that detailed in the previous chapter, i.e., first choose a sheath size, s_m, then calculate T_e using Eq. (6.12), and finally solve for H and n_0

6.1 Dual-frequency CCP in electrostatic approximation

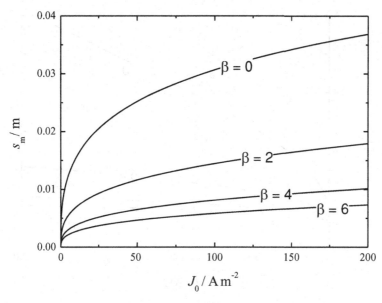

Figure 6.3 Sheath size as a function of the low-frequency current density amplitude $J_0 = I_0/A$, for the same conditions as in Figure 6.2.

using Eqs (6.5) and (6.13). In Figure 6.2 the result of the calculation for the electron density as a function of the low-frequency current density amplitude $J_0 = I_0/A$ is plotted, for different values of the ratio of higher to lower frequency current β. The argon gas pressure is 1.33 Pa, the electrode spacing is 5 cm, the low frequency is 13.56 MHz and the high frequency is 149.16 MHz, i.e., $\alpha = 11$. The effect of adding a high-frequency component to the RF current drive is clearly seen: at a fixed value of j_0, the electron density increases dramatically with β. The ion flux to the electrode will consequently increase as well. Note that PIC simulations, including secondary electron emission at the electrodes, have found that the plasma density increases as the low-frequency current [100] or voltage [101] increases.

For the same conditions, the sheath size s_m decreases with β, as shown in Figure 6.3. This is because the electron density increases at fixed J_0, which in turn decreases the parameter H. However, the exact scaling is not so simple and will be obtained later in the low-pressure limit.

Q Figure 6.4 shows that under the same conditions, the electron temperature also decreases as β increases. Account for this observation.

A As a consequence of the decrease in the sheath size, the plasma bulk volume, where ionization takes place, increases; so a smaller electron temperature can sustain the plasma.

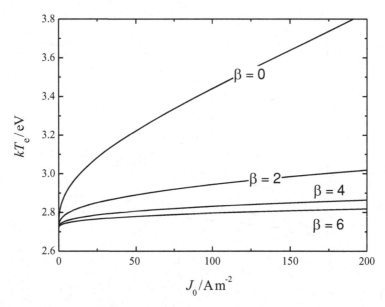

Figure 6.4 Electron temperature as a function of the low-frequency current amplitude $J_0 = I_0/A$, for the same conditions as in Figure 6.2.

These results have practical implications for plasma processing. Firstly, the electron temperature is a very important quantity in determining the degree of dissociation of molecular gases. Secondly, changing the sheath size changes the ratio s_m/λ_i (where λ_i is the ion–neutral mean free path), that is the sheath collisionality, which is also of great importance in determining the ion energy distribution at the electrode.

So far, the model has been analysed through input parameters $J_0 = I_0/A$ and β. In practice, RF currents are not the control parameters at the experimentalist's disposal. Experimentalists usually use the RF power delivered by the generator as the control parameter. Unfortunately, as seen in the previous chapter, the power supplied by the generator is not all delivered to the electrons. These considerations make comparisons between theory and experiments quite complicated.

A useful way of looking at the model results is to plot the flux/energy diagram, as done at the beginning of this chapter in the single-frequency case. This is done in Figure 6.5 for the same conditions as in Figure 6.2. The curve for $\beta = 0$ corresponds to the 13.56 MHz single-frequency case, and can be compared to the result obtained by Perret *et al.* in Figure 6.1. The slight difference observed between the two plots has several origins: (i) the pressures are slightly different, (ii) the hard-wall model for collisionless heating was used in Figure 6.1 while the collisionless fluid model is used in Figure 6.5, (iii) the ion energy was calculated including the DC floating potential limit at low RF voltage in Figure 6.1 while it was not included

6.1 Dual-frequency CCP in electrostatic approximation

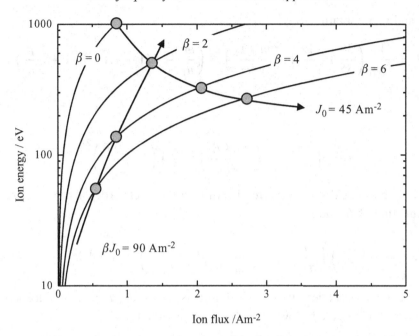

Figure 6.5 Mean ion energy as a function of the ion flux to the electrode, for the same conditions as in Figure 6.2. The arrow pointing to the right represents the trajectory for constant low-frequency current density $J_0 = 45\,\text{A m}^{-2}$, while the arrow pointing up represents the trajectory for constant high-frequency current density $\beta J_0 = 90\,\text{A m}^{-2}$; $\alpha = 11$.

in Figure 6.5. Nevertheless, the general trend is similar: 13.56 MHz alone does not produce high ion fluxes at moderate ion energies.

Starting from a point determined by j_0 on the single-frequency (13.56 MHz, $\beta = 0$) reference curve and then increasing the RF current provided by the 149.16 MHz power supply, it is possible to explore most of the space in the diagram. However, it is not possible to move horizontally or vertically in this space by changing only one component of the RF current. This is shown in the figure by the trajectories corresponding to a fixed low-frequency current of $J_0 = 45\,\text{A m}^{-2}$ and by a fixed higher-frequency component $\beta J_0 = 90\,\text{A m}^{-2}$. Both flux and energy change under the influence of each control parameter. There is no clear decoupling of the effects of the two frequencies in a dual-frequency capacitive discharge; if there were, one trajectory would be vertical and the other horizontal.

To quantify this very important result, it is useful to analyse the scalings of a simplified solution in which stochastic heating dominates, and the parameters are such that $24/(5\pi) \ll H \ll 55$. This solution is typically valid at low pressure (around 1 Pa) and at rather high power. In this limit the Child–Langmuir law and the collisionless heating expressions can be manipulated, along with the definition of H and s_0, to obtain

$$H = \frac{1}{2\sqrt{8}} \left(\frac{36}{55}\right) \left(\frac{12}{5}\right)^2 \left(\frac{m_e k T_e}{\pi^2 e^2}\right)^{1/2} \left(\frac{\omega^2 s_m^2}{u_B \varepsilon_T}\right) (1 + 1.1\beta^2) \left(1 + \frac{9\beta}{5\alpha}\right)^{-2} \quad (6.14)$$

and

$$n_e = \left(\frac{5\pi H}{12}\right)^2 \left(\frac{\pi \varepsilon_0 k T_e}{e^2 h_1}\right) \left(\frac{H}{s_m^2}\right) \left(1 + \frac{9\beta}{5\alpha}\right)^2. \quad (6.15)$$

Finally, combining these two expressions allows the electron density to be expressed as a function of I_0 and β:

$$n_e = \left[\frac{1}{2\sqrt{8}} \left(\frac{36}{55}\right) \left(\frac{m}{\pi k T_e}\right)^{1/2} \left(\frac{I_0^4}{\varepsilon_0 A^4 \omega^2 e^3 h_1^3 u_B \varepsilon_T}\right) (1 + 1.1\beta^2)\right]^{1/3}. \quad (6.16)$$

Since the ion current density to the electrode is $J_i = e h_1 n_e u_B$ and the ion energy is given by Eq. (6.6), the following scalings are obtained:

$$J_i \propto I_0^{4/3} \omega^{-2/3} (1 + 1.1\beta^2)^{1/3}, \quad (6.17)$$

$$E_i \propto I_0^{4/3} \omega^{-8/3} \left(\frac{3}{4} + \frac{\beta}{\alpha}\right)^2 (1 + 1.1\beta^2)^{-2/3}, \quad (6.18)$$

$$s_m \propto I_0^{1/3} \omega^{-5/3} \left(1 + \frac{9\beta}{5\alpha}\right) (1 + 1.1\beta^2)^{-2/3}. \quad (6.19)$$

It appears that the three quantities are functions of I_0 and β. The first expression shows that the ion current (or flux) to the electrode depends on I_0 and not only on the high-frequency RF current βI_0. Similarly, Eq. (6.18) shows that the ion energy depends on β. Finally, Eq. (6.19) shows that the sheath size decreases with β at fixed I_0. These scalings are very well illustrated in Figures 6.5 and 6.3.

6.1.4 Further control of ion energy

One means of controlling the ion energy arriving at a surface is through direct control of the voltage waveform between the plasma and the electrode. This is most easily done using tailored voltage waveforms applied between the surface and the plasma with an independent plasma source (such as an ICP or a helicon). Having separated plasma production from the voltage across the sheath in this way, one is free to choose waveforms that set the ion energy distribution. For instance, it has been shown that a combination of a relatively long period (2 μs) of steady ion-accelerating bias is interspersed with short positive pulses of 200 ns

[50]. The latter pulse is too rapid for the ions to respond, while it gathers electrons to neutralize the steady flux of ions. This achieves a near mono-energetic IEDF.

Although ion energy and ion flux are not entirely decoupled by the use of dual-frequency excitation, the additional variables (frequency ratio and amplitude ratio) provide a useful means of addressing the parameter space. When the two frequencies are harmonically related, then one also has the opportunity to use the phase difference between the waveforms to obtain a further degree of control. A deceptively simple effect occurs when the higher frequency is an even harmonic of the lower frequency [102]. In that case when a temporally symmetric, multi-frequency voltage waveform containing one or more even harmonics is applied to a symmetric, parallel plate CCP, the two sheaths on the two electrodes are necessarily asymmetric. To balance the charged particle fluxes, a self-bias voltage has to develop (in much the same way as it does for an asymmetric electrode configuration excited by a single frequency); the magnitude and sign of this self-bias depend on the phase difference between the component waveforms. In the simplest case using voltage waveforms at ω and 2ω the self-bias, and therefore the ion energy, is almost a linear function of the phase angle between them. By simply varying the phase between zero and 2π, the apparent asymmetry and bias can be shifted from one electrode to the other. This remarkably simple result is by no means obvious, but it emerges from extensive analytical modelling of a dual-frequency, symmetrical CCP and it has been confirmed by experiments and simulations [103].

6.2 Electromagnetic regime at high frequency

Warning: First, note that in the previous section J_0 represented current density whereas in this section $J_0(x)$ and $J_1(x)$ represent Bessel functions. Second, in contrast to earlier chapters where the prime mark was used to indicate the derivative of a quantity, in this section it will signify a quantity 'per unit length', so Z', etc. implies the impedance per unit length of a transmission line. As in Chapter 2, when studying waves, k_B is Boltzmann's constant but k and k_z are wavenumbers.

The electrostatic model cannot be used for a CCP at an arbitrarily high excitation frequency in pursuit of higher electron density since nonuniformities arise when the excitation wavelength λ becomes comparable to the electrode radius, and the plasma skin depth δ becomes comparable to the electrode spacing. These conditions define the change-over from an electrostatic to an electromagnetic regime. This section reviews electromagnetic effects [21, 63, 90, 91, 104–119].

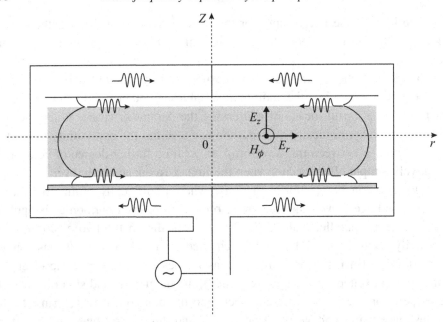

Figure 6.6 Schematic of wave propagation in a capacitive discharge operated in the electromagnetic regime.

A schematic of a parallel plate discharge operated in the electromagnetic regime is shown in Figure 6.6. In order to understand what follows, one should think of the reactor as a waveguide (or cavity) loaded by a plasma, rather than two parallel electrodes across which an oscillating voltage is applied. It is also very important to realize that in the electromagnetic regime, the potential difference between the electrodes is not constant in space, even though the electrodes are made of conducting material. The arrows in the figure symbolize the propagation of waves in the system. In vacuum and dielectric regions, the waves are mostly transverse and propagate at the speed of light. For $\omega \ll \omega_{pe}$ (which is usually the case of interest), waves will not propagate into the plasma, instead they propagate along the surface between the sheath (a dielectric) and the plasma (a conductor), with a characteristic decay length into the plasma which is typically the skin depth derived in Chapter 2. Owing to the symmetry of the system, the waves propagate radially inwards on the plasma-facing surfaces of the electrodes, establishing a standing wave. It will be seen that when the electromagnetic fields do not fully penetrate the plasma, i.e., when the skin depth is not infinite, the RF current does not flow perpendicularly to the electrodes and therefore the electric field has a component parallel to the electrode, E_r. From the various phenomena described briefly above, electromagnetic effects have been divided into three categories by Lieberman *et al.* [108]: (i) standing wave effect, (ii) skin effect and (iii) edge effect.

6.2 Electromagnetic regime at high frequency

The origin of these will be explained in the following. The standing wave effect is the most important from the processing point of view, as it appears in typical operating conditions and it leads to severe non-uniformities.

The first step is to examine a parallel plate vacuum capacitor at high frequency. This will be done initially using electromagnetic fields directly and afterwards by constructing a circuit model. At high frequency one needs to allow for the time it takes for the electromagnetic fields (or equivalently the voltages and currents) to propagate through the system – for an electrode of radius comparable with the wavelength of electromagnetic waves at a given frequency, the fields at the centre lag behind those at the edge. That means that the circuit model here is more properly called a distributed transmission line model, whereas in the previous chapter the model was interpreted in terms of local circuit components. Introducing a low-density plasma between the electrodes slows the electromagnetic waves, leading to a standing wave becoming established on the scale of typical laboratory CCPs. Finally, the more general case of high-density plasmas is treated, in which the skin effect becomes important.

It should be noted that electromagnetic effects have also been studied in the context of capacitive discharges for CO_2-lasers [120, 121]. Although the basic phenomenon is similar to those considered here, CO_2-lasers are operated at relatively high pressure (of the order of 10^4 Pa), which results in different mechanisms of power dissipation. Also, the discharges for lasers are essentially shaped like a long rectangular prism, whereas plasma processing reactors are either cylindrical or cuboidal. Finally, unlike in discharge-lasers, the electron density in plasma processing reactors is such that skin effects may be important.

6.2.1 The capacitor at high frequency

Figure 6.7 shows a sketch of a cylindrical capacitor with a vacuum between the plates. It is supposed that the plates' diameter is much larger than the separation. When an RF current transfers charge from one plate to the other through an external circuit, an axial electric field, amplitude E_z, is established in the space between them. Edge effects will be neglected. According to Maxwell's equations there must also be an azimuthal magnetic field, amplitude B_θ, such that the Faraday and Ampère laws are satisfied in the vacuum:

$$\operatorname{curl} \mathbf{E} = -\frac{\partial \mathbf{B}}{\partial t},$$

$$\operatorname{curl} \mathbf{B} = \frac{1}{c^2}\frac{\partial \mathbf{E}}{\partial t}.$$

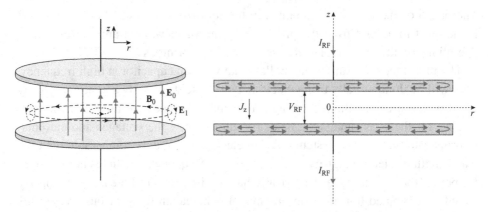

Figure 6.7 Electric and magnetic field between the plates (left), current in the electrodes and voltage between the electrodes (right).

In the present situation, the cylindrical symmetry rules out azimuthal variations and the short axial range means that the predominant variation is in the radial direction, so these equations can be combined and simplified to give an equation for the axial electric field:

$$\frac{c^2}{r}\frac{\partial}{\partial r}\left(r\frac{\partial E_z}{\partial r}\right) = \frac{\partial^2 E_z}{\partial t^2}. \tag{6.20}$$

This is in fact an equation for radially propagating cylindrical waves. The solution can include inward and outward travelling waves that interfere to create a standing wave structure in the electric field.

Using exponential notation for the temporal variation of the fields, i.e.,

$$E_z = \mathrm{Re}\left[\widetilde{E}_z \exp\mathrm{i}\omega t\right],$$

Eq. (6.20) simplifies to

$$\frac{\partial^2 \widetilde{E}_z}{\partial r^2} + \frac{1}{r}\frac{\partial \widetilde{E}_z}{\partial r} + k_0^2 \widetilde{E}_z = 0, \tag{6.21}$$

which is a standard equation that has as its solution the zero-order Bessel function of the first kind, $J_0(k_0 r)$, where here $k_0 = \omega/c$ is the wavenumber in free space; see Figure 6.8. So the electric field is

$$\widetilde{E}_z = E_0 J_0(k_0 r). \tag{6.22}$$

A simple two-term expansion of the Bessel function gives

$$\widetilde{E}_z \approx E_0 \left(1 - \frac{k_0^2 r^2}{4}\right). \tag{6.23}$$

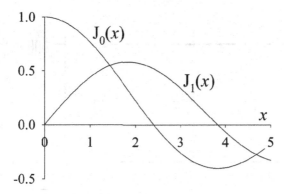

Figure 6.8 Zero-order and first-order Bessel functions.

The structure of the magnetic field can be found using either of the above Maxwell equations, giving

$$\tilde{B}_\theta = -B_0 J_1(k_0 r), \qquad (6.24)$$

where $B_0 = E_0/c$ and $J_1(k_0 r)$ is the first-order Bessel function (Figure 6.8). The leading term in the expansion of the first-order Bessel function gives

$$\tilde{B}_\theta = -\left(\frac{\omega r}{2c^2}\right) E_0. \qquad (6.25)$$

The Bessel function that describes the amplitude of the axial electric field passes through zero (a node) when $k_0 r = 2.405$; the two-term expansion in Eq. (6.23) crosses zero slightly earlier at $k_0 r = 2$. By this stage the parallel plate arrangement no longer has a pure (electrostatic) capacitance. Indeed, it ceases to be purely capacitive as soon as there is any significant non-uniformity of the electric field. To get a sense of the frequencies at which non-uniformities become problematic, one can identify the condition for a 10% edge-to-centre non-uniformity of the electric field.

Q For a parallel plate gap, in vacuum, between plates of diameter $2r_0 = 30$ cm, estimate the frequency at which the electric field at the edge has fallen by 10% from the value on the axis.

A According to Eq. (6.23), the field falls to 90% of its axial value when

$$(\pi r_0/\lambda)^2 = 0.1. \qquad (6.26)$$

For plates with 15 cm radius, non-uniformities greater than 10% are expected at frequencies above $f_0 = c/(10 \times r_0) \approx 200$ MHz.
Comment: This result is independent of the plate separation.

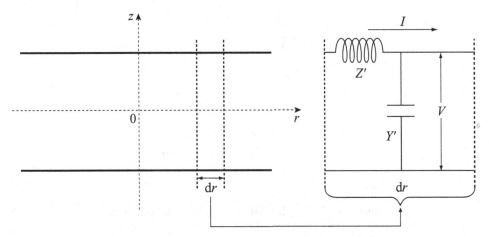

Figure 6.9 Schematic of the transmission line model of a parallel plate capacitor for VHF, showing the inductance and capacitance contributed by a radial element between r and $r + dr$.

For 1 m electrode radius, the frequency limit goes down to about 30 MHz. It is important to remember that the above calculation considers a vacuum between the electrodes. It will be shown in the next section that when a plasma is present between the electrodes, the wavelength is considerably shorter and hence the limiting frequency is markedly lower.

The non-uniformity is described as the standing wave effect because in reality, propagating waves enter the 'capacitor' from the sides and therefore establish a standing wave, with an antinode (maximum magnitude) of the electric field in the centre, and a node of the magnetic field in the centre. Equivalently, the RF voltage between the electrodes exhibits an antinode in the centre, while the RF current flowing in the electrodes has a node in the centre. Note that the RF current circulates in a very thin layer, as shown on the RHS of Figure 6.7.

Circuit models use voltage and current in preference to electric and magnetic fields. At VHF the circuit elements become distributed as inductance per unit length and capacitance per unit length. The radial distribution of RF voltage and RF current can then be viewed in terms of a transmission line, as described in Figure 6.9 (a good description of transmission line theory can be found in [122]). A short element dr is represented by an impedance Z' and admittance Y', where the prime indicates 'per unit length'. The RF voltage drops across this section by $IZ'dr$ owing to the inductance and the current in the line is similarly reduced by $VY'dr$, which passes through the capacitance. The following propagation equations

6.2 Electromagnetic regime at high frequency

therefore encapsulate the effect of a section of line of length dr:

$$\frac{dV}{dr} = -Z'I, \quad (6.27)$$

$$\frac{dI}{dr} = -Y'V. \quad (6.28)$$

The impedance per unit length Z' is due to the inductance encountered by the current flowing (in opposite directions) in the two plates: electromagnetic analysis gives

$$Z' = i\omega\mu_0 \frac{l}{2\pi r}. \quad (6.29)$$

Similarly, Y' is due to the capacitance per unit length:

$$Y' = i\omega\varepsilon_0 \frac{2\pi r}{l}. \quad (6.30)$$

Note that this transmission line has no dissipation and that $|Y'Z'| = k_0^2$. Taking the first derivative of Eq. (6.27) with respect to r, and inserting the result in Eq. (6.28), gives the following (Bessel) equation for the voltage:

$$\frac{d^2V}{dr^2} + \frac{1}{r}\frac{dV}{dr} + k_0^2 V = 0, \quad (6.31)$$

which has the following solution:

$$V(r) = V_0 J_0(k_0 r), \quad (6.32)$$

which satisfies the symmetry requirement that $dV/dr = 0$ at $r = 0$. This solution is equivalent to that obtained with Maxwell's equations with $V_0 = E_0 l$.

On the basis of the above discussion, the modification induced by having a plasma between the electrodes can now be explored. It seems likely that the voltage non-uniformities due to the standing wave effect will still exist and that they will affect the uniformity of the plasma parameters. However, radial non-uniformities of the voltage are transferred to the plasma differently by the non-local power deposition of the low-pressure regime compared with the local power deposition that prevails at higher pressure.

6.2.2 The low-density CCP at VHF

The first case to consider is that of a plasma at a pressure sufficiently high that the power deposition is local, i.e., in terms of the transmission line model the power balance is achieved in each plasma/sheath slab of width dr, and sheaths are collisional. The electron density will also be presumed to be low enough for the skin depth to be much larger than the electrode spacing. In this case, the magnetic field is not perturbed by the presence of the plasma. Consequently, the electric field and the RF current between the electrodes remain mostly perpendicular to the electrodes (along the z-axis).

The series impedance per unit length of the transmission line, Z', remains that given by Eq. (6.29), since it is due to the passage of current in the electrodes and the magnetic field is not changed by the presence of the plasma. However, the parallel branch of the transmission line is profoundly modified. Each slab must be described by an equivalent circuit based on that derived in the previous chapter, shown in Figure 5.6. The simple capacitance of the vacuum case is now replaced by the sheath capacitance, in series with resistances and the plasma inductance. The plasma dissipates power, which means that the model becomes a lossy transmission line. The plasma inductance is not of great importance, unless the system is driven at extremely high frequency.

Although the resistances are essential to model the power dissipation and therefore to calculate the electron density, they do not play a significant role in determining the wavelength of the standing wave. Indeed, as in the electrostatic approximation described in the previous chapter, the parallel admittance is essentially that of the sheath capacitance. Using Eq. (5.30),

$$Y' = i\omega\varepsilon_0 K_{cap} \frac{2\pi r}{s_m}. \tag{6.33}$$

For K_{cap} see Table 5.2. The wavenumber in the presence of plasma therefore becomes $k^2 = |Y'Z'| = k_0^2 K_{cap} l / s_m$, so

$$\lambda = \lambda_0 \left(\frac{s_m}{K_{cap} l} \right)^{1/2}. \tag{6.34}$$

The wavelength in the presence of plasma is therefore significantly shorter than the wavelength in vacuum, because the sheath size is generally much smaller than the electrode spacing, $s_m \ll l$. Typically, the wavelength is 3 to 5 times smaller in the presence of a plasma.

Equation (6.34) captures the essential physics responsible for the wavelength reduction – within the dependence on sheath size s_m there is an implicit dependence on the magnitude of the voltage across the plates, the frequency, electron temperature and gas pressure. To express the wavelength as a function of external parameters, one needs to couple the transmission line, Eqs (6.27) and (6.28), to the particle and power balance, and to the Child–Langmuir law [63, 111]. The parallel

6.2 Electromagnetic regime at high frequency

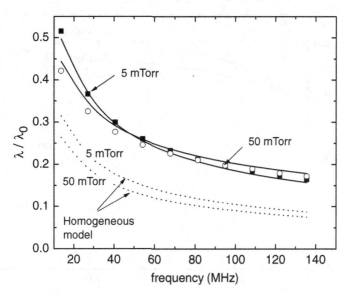

Figure 6.10 Wavelength in the presence of plasma as a function of the driving frequency. The solid line refers to this model. The dashed line is for a homogeneous model; from [63].

admittance per unit length may then be expressed as

$$Y' = i\omega\varepsilon_0 \frac{2\pi r}{l} \alpha V_0^{-1/5}, \tag{6.35}$$

$$\alpha = \left[\frac{K_{cap}^2 K_{stoc}^2 m T_e^2 \omega^4 l^5}{12.32 \, \lambda_i e u_B^2 \varepsilon_T^2} \right]^{1/5}, \tag{6.36}$$

if one considers collisional sheaths. Note that Y' is a function of V_0 so that Eqs (6.27) and (6.28) become two non-linear coupled differential equations. It is not possible to obtain analytical expressions for $V(r)$ and $I(r)$ in this case (unlike in the vacuum case). Nevertheless, using the appropriate values of all the coefficients in α, one obtains the following practical formula for the scale of the standing wave ($\equiv 2\pi/k$) in the presence of plasma:

$$\frac{\lambda}{\lambda_0} \approx 40 V_0^{1/10} f^{-2/5} l^{-1/2}, \tag{6.37}$$

where λ_0 is the vacuum wavelength (in metres), V_0 is the RF voltage at the electrode centre (in volts), f is the driving frequency (in hertz) and l is the electrode spacing (in metres). As anticipated, the wavelength in the presence of plasma is shorter than the wavelength in vacuum ($\lambda/\lambda_0 < 1$), becoming more so as the frequency increases, as can be seen in Figure 6.10. Equation (6.37) also indicates that the

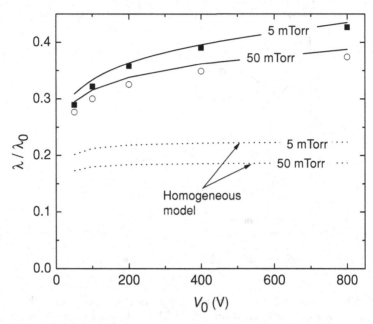

Figure 6.11 Wavelength in the presence of plasma as a function of the RF voltage at the discharge centre; from [63]

effect of wavelength shortening is enhanced as the electrode spacing increases. Interestingly, increasing the voltage produces the opposite effect, as seen in Figure 6.11, since the sheath size increases with the voltage, tending to decrease the extent of plasma in the gap.

Figure 6.12 shows results from a numerical solution of the transmission line equations coupled to the power and particle balance and the Child–Langmuir model for the RF sheath. The electron density, the sheath size and the electron temperature are plotted as a function of the radius for a capacitive discharge in argon, with an electrode spacing of 4 cm. The electrode radius is 20 cm, the driving frequency is 80 MHz and the argon gas pressure is 20 Pa. The standing wave effect is very pronounced, since the density and sheath size pass through a minimum (also the voltage passes through zero and the current through a maximum) within the radius of the electrodes. In this case, the power balance is completely local, and the radial diffusion of particles and energy has been ignored.

> **Q** Since the model considers local power balance and ignores radial diffusion of particles, does the model overestimate or underestimate the radial non-uniformities of the plasma density?
> **A** The model overestimates the non-uniformities because energy and particle radial transport would tend to level off the gradients.

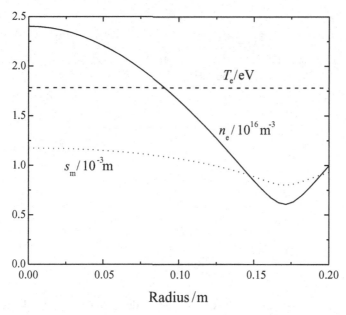

Figure 6.12 Electron density, electron temperature and sheath size as a function of radius for a capacitive discharge in argon, with an electrode spacing of 4 cm. The electrode radius is 20 cm, the driving frequency is 80 MHz and the argon gas pressure is 20 Pa.

Comparison with experiments

The standing wave effect can be visualized experimentally by mapping the ion flux across the electrodes [91]. The discharge was produced in argon gas at 20 Pa between two square plates (40 cm × 40 cm) separated by a distance of 4.5 cm, confined laterally by a 4 cm thick teflon barrier. The discharge was therefore virtually symmetrical. The lower electrode was powered by three different RF generators, operating at 13.56, 60 and 81.36 MHz. The ion flux variation across the electrodes was measured by a matrix of 64 planar electrostatic probes inserted in the grounded upper electrode. Figure 6.13 shows maps of the ion flux at low RF power (50 W) corresponding to the three excitation frequencies. At this low power the skin depth is large compared to the electrode spacing, which is assumed in the model above. The ion flux is fairly uniform at 13.56 MHz (although it is slightly higher near the edges), but for 60 MHz and 81.36 MHz the flux is maximal at the centre with a dome-like distribution. The experiment is compared to the model in Figure 6.14, which displays the radial profile of the plasma density. The symbols are measurements, the dashed line is the transmission line model results, and the solid line is the result when the calculation is done using the wavelength in vacuum for the radial distribution of the voltage. The model overestimates the non-uniformity of the electron density, as anticipated. The scale length of the standing wave structure

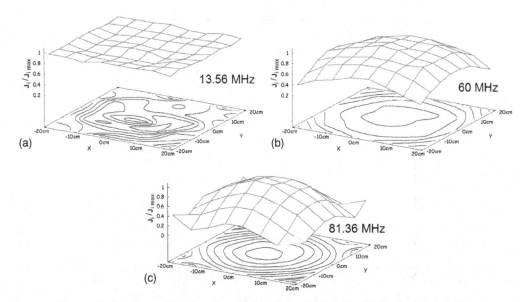

Figure 6.13 Maps of the ion flux to the electrode at 28 Pa/50 W for 13.56 MHz, 60 MHz and 81.36 MHz. The electrode separation was 4.5 cm, the electrode dimensions were 40 cm × 40 cm.

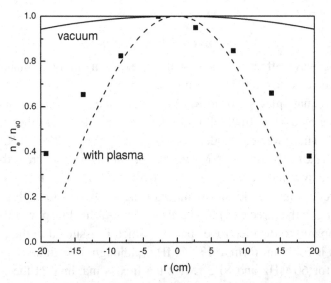

Figure 6.14 Comparison between the transmission line model and experiment for the non-uniformity of the electron density due to the standing wave effect at 81.36 MHz. The symbols are measurements (obtained from a cross-section of Figure 6.13(c)), the dashed line is the transmission line model result, and the solid line is the result when the calculation is done using the wavelength in vacuum for the voltage radial distribution.

6.2 Electromagnetic regime at high frequency

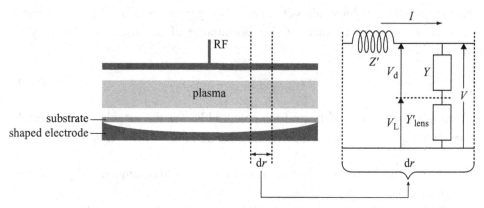

Figure 6.15 The lens concept employs a shaped electrode behind the substrate to suppress the standing wave effect (left) and the equivalent transmission line model (right).

in the presence of plasma is clearly much shorter than that based on the vacuum wavelength, as predicted by the model.

Restoring uniformity with a dielectric lens

The standing wave effect is a severe problem for large-area, industrial plasma processing at very high frequency owing to the inherent lack of spatial uniformity. An ingenious solution to this challenge has been proposed involving a specially shaped electrode to form a dielectric lens [109], as shown in Figure 6.15. The idea behind this is to compensate for the radial fall-off of the electrode voltage by matching it to a reduction in the effective gap so that the field remains constant. The transmission line theory described above can be used to calculate the required shape of the electrode [111]. The modification of the transmission line model is shown on the RHS of Figure 6.15.

The dielectric lens adds admittance Y'_{lens} in series with that of the plasma/sheath slab (or vacuum), Y'. In cylindrical geometry, and in the limit of no skin effect ($\delta \gg l$), the lens admittance and series impedance of the electrodes are given by

$$Y'_{\text{lens}}(r) = i\omega\varepsilon_0\varepsilon_r \frac{2\pi r}{x(r)}, \qquad (6.38)$$

$$Z'(r) = i\omega\mu_0 \frac{\xi(r)}{2\pi r}, \qquad (6.39)$$

where $x(r)$ and ε_r are respectively the thickness and relative permittivity of the lens, and $\xi(r) = l + x(r)$ is the electrode separation. Note that we have considered $d\xi/dr \ll 1$ (i.e., the shaped-electrode curvature is weak). If $V_L(r)$ is the RF voltage across the lens and $V(r)$ remains the voltage across the electrodes:

$$V(r) = V_g(r) + V_L(r), \qquad (6.40)$$

where $V_g(r)$ is the voltage across the gap l; this gap being loaded by the discharge (plasma and sheaths) or vacuum. The admittance of the discharge slab may be written

$$Y'(r) = i\omega\varepsilon_0 \frac{2\pi r}{l} f(V_g), \tag{6.41}$$

where f is any function of the uniform potential V_g. For the vacuum case we simply have $f(V_g) \equiv 1$, whereas for the plasma case we have $f(V_g) = \alpha V_g^{-1/5}$.

Q What is the appropriate condition to use in order to obtain a radially uniform discharge?
A The condition is a uniform radial discharge voltage V_g, which implies $dV_g/dr = 0$.

Inserting the expressions of the various admittances into the transmission line equations and using the condition of constant discharge voltage, $dV_g/dr = 0$, leads, after some algebra [111], to the following differential equation for the electrode separation ξ:

$$\frac{d^2\xi}{dr^2} = \left[\frac{1}{\xi}\frac{d\xi}{dr} - \frac{1}{r}\right]\frac{d\xi}{dr} - \varepsilon_r k_0^2 \xi, \tag{6.42}$$

which can be integrated to obtain

$$\xi(r) = l + x(r) = [l + x(0)] \exp\left(-\frac{\varepsilon_r k_0^2 r^2}{4}\right). \tag{6.43}$$

Hence, the dielectric lens should have a Gaussian shape in order to obtain a uniform voltage across the discharge, and thus suppress the standing wave effect. A comparable calculation using Maxwell's equation was performed in [115]. The lens profile and the voltages across the lens and across the electrodes are both Gaussian, with or without plasma. In addition, the voltage across the lens is significantly increased in the presence of a plasma. This is because the plasma impedance is much lower than that of vacuum. In fact, $V_L(r)$ increases by a factor of $f(V_g) = \alpha V_g^{-1/5}$, a number much larger than one, in the presence of a plasma. Hence, the voltage non-linearity of the sheath does not affect the electrode shape but it changes the voltage amplitude across the lens. This rather high voltage could be a problem if vacuum (or low-pressure gas) is used in the lens instead of a dielectric material, because a parasitic plasma may be struck behind the substrate.

Figure 6.16 reproduces results that demonstrate the effectiveness of the dielectric lens [110]. The standing wave effect makes a clear imprint on the ion flux when

6.2 Electromagnetic regime at high frequency

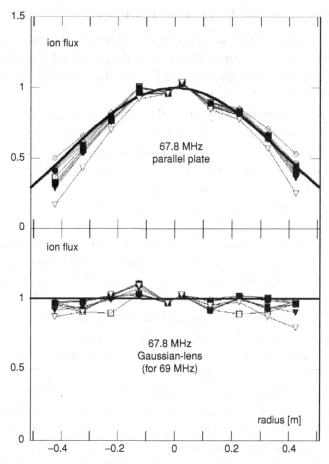

Figure 6.16 Experimental evidence of the suppression of the standing wave effect using the shaped electrode and lens concept. From [110].

using parallel electrodes, whereas the ion flux is uniform when using an electrode that incorporates a dielectric lens.

> **Q** The lens concept seems very promising for suppressing non-uniformities in the context of moderate electron densities, e.g., in plasma-enhanced chemical vapour deposition. Explain why it is less effective for high-density and multiple-frequency sources.
>
> **A** The lens concept will not be effective when the skin effects become significant, that is at higher plasma densities, since then the electric field has a component parallel to the electrode. The shape of the lens is determined by the standing wave structure which is frequency-dependent – multiple-frequency sources cannot be matched in the same way.

6.2.3 The general CCP at VHF

In the previous section the plasma skin depth was large compared to the electrode gap and the pressure was sufficiently high for the power balance to be localized. These limitations are now relaxed to establish a more general model of the capacitive discharge at very high frequency.

The main effect introduced by finite skin depth is that the electric field is no longer perpendicular to the electrodes; it has radial and axial components. When the electron density is small, the skin depth is large and the radial component is negligible, but it becomes significant when the plasma density is high enough for the skin depth to shrink below the size of the gap. Whereas the axial field is electrostatic, at least at low frequency, the radial electric field is in essence an electromagnetic induction field, associated with currents in the electrode; the power transferred to the plasma electrons by this field can therefore be associated with an *inductive* heating mechanism. As in inductive discharges, when the power deposited by the inductively coupled current is larger than the power deposited by current that is driven by the electrostatic field, the discharge can be said to be in the H-mode. In the other limit, the discharge is in the E-mode. In this section it will be shown that CCPs at VHF can undergo E to H transitions.

The effect of non-local power deposition at low pressure will also be introduced into the model. When the pressure is low, there is a regime in which the electron temperature and, consequently, the ionization rate are independent of space, even though the electric field is strongly non-uniform in the radial direction. Such an approximation has already been invoked in presuming that the electron temperature is independent of the shorter dimension, z.

The principle of the modelling that follows is first to examine the form of the electromagnetic fields in Maxwell's equations, for a given, uniform electron density and constant sheath size. Then the transmission line equations are adapted to the more general regime and a dispersion relation for electromagnetic waves is found. Finally, the power balance, the particle balance and the RF Child–Langmuir law are used to obtain self-consistent solutions for the RF voltage, the RF current and the plasma parameters.

Electromagnetic fields and the dispersion relation

The situation again involves two parallel circular electrodes of radius r_0 separated by a distance l, as shown schematically in Figure 6.17. The transmission line model, to be discussed later, is shown on the RHS. The plasma (width d) is separated from the electrodes by sheaths and is locally modelled as a uniform dielectric stationary in time having a relative permittivity given by the complex relative permittivity (Eq. (2.52)) $\varepsilon_p = 1 - \omega_{pe}^2/(\omega(\omega - i\nu_m))$.

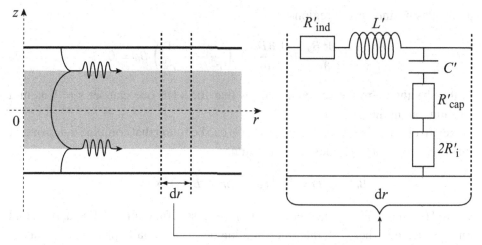

Figure 6.17 Schematic of the transmission line model constructed from the electromagnetic model.

Q Is it appropriate to ignore the sheath motion, which is at the same frequency as the electromagnetic waves – are the sheaths and plasma stationary in time?
A Although each sheath oscillates in width, the oscillations are 180° out-of-phase. Therefore the total sheath width, and thus the sheath capacitance, is roughly constant in time. One might suppose that in the interests of tractability the volume occupied by the sheaths can be treated as being steady.
Comment: This model ignores the temporally non-linear sheath behaviour that leads to harmonic generation.

Assuming an azimuthal magnetic field B_θ, the corresponding components of the electric field, E_z and E_r, are determined from Maxwell's equations for harmonic waves (which means that temporal variation is included as $\exp i\omega t$ as before), in a medium with relative permittivity ε:

$$-\frac{\partial \tilde{B}_\theta}{\partial z} = \frac{i\omega}{c^2/\varepsilon} \tilde{E}_r, \qquad (6.44)$$

$$\frac{1}{r}\frac{\partial (r\tilde{B}_\theta)}{\partial r} = \frac{i\omega}{c^2/\varepsilon} \tilde{E}_z, \qquad (6.45)$$

$$\frac{\partial \tilde{E}_r}{\partial z} - \frac{\partial \tilde{E}_z}{\partial r} = -i\omega \tilde{B}_\theta. \qquad (6.46)$$

Here $\tilde{E}_z$ is the capacitive electric field (perpendicular to the electrodes) and $\tilde{E}_r$ is the inductive field (parallel to the electrodes). Substituting for $\tilde{E}_r$ and $\tilde{E}_z$ from Eqs (6.44) and (6.45) into (6.46) yields a (2-D cylindrical) wave equation for $\tilde{B}_\theta$

in the plasma and sheath regions:

$$\frac{\partial^2 \tilde{B}_\theta}{\partial z^2} + \frac{\partial^2 \tilde{B}_\theta}{\partial r^2} + \frac{1}{r}\frac{\partial \tilde{B}_\theta}{\partial r} + \left(\frac{\omega^2}{c^2/\varepsilon} - \frac{1}{r^2}\right)\tilde{B}_\theta = 0. \qquad (6.47)$$

In the sheaths there are so few electrons that for VHF one can set $\varepsilon = 1$ as in a vacuum, but in the plasma $\varepsilon = \varepsilon_p$.

Relatively simple solutions for Eq. (6.47) can be found that combine independent functions of r, z and t. These can be written

$$B_\theta(r, z, t) = \text{Re}\left[H(\sqrt{\varepsilon}\omega r/c)\tilde{B}_\theta(z)\exp i\omega t\right],$$

where $H(\sqrt{\varepsilon}\omega r/c)$ is a linear combination of Bessel functions of first and second kinds which together with the time variation describe radially propagating waves, much as $\exp i(\omega t \pm kz)$ does in planar geometry. In Section 6.2.1, where B_θ was not a function of z, this is what gave rise to the standing wave profiles featuring J_0 and J_1 for E_z and B_θ, respectively. A standing wave is indeed established and the electric fields are not uniform along r (the E_z field is maximum in the centre while the E_r field is maximum away from the centre). The standing wave and the radial non-uniformity will be introduced in the transmission line description. Edge effects will be considered at the end of this chapter.

One needs to build solutions for the z-dependence in the plasma and in the sheath that match at the plasma/sheath interface, using the boundary conditions that $\tilde{E}_r = 0$ at $z = 0$ and $z = \pm l/2$, and $\tilde{E}_r$, $\tilde{B}_\theta$ and $\varepsilon \tilde{E}_z$ are all continuous at $z = \pm d/2$. The derivation is rather complicated, so it is simpler here to give the results [21] – such solutions may be checked by differentiation. The fields in the plasma are

$$\tilde{E}_r(z) = -\frac{A\alpha_p \cosh \alpha_0 s}{i\omega\varepsilon_0\varepsilon_p} \sinh \alpha_p z, \qquad (6.48)$$

$$\tilde{B}_\theta(z) = \mu_0 A \cosh \alpha_0 s \cosh \alpha_p z, \qquad (6.49)$$

$$\tilde{E}_z(z) = \frac{Ak \cosh \alpha_0 s}{i\omega\varepsilon_0\varepsilon_p} \cosh \alpha_p z. \qquad (6.50)$$

The fields in the sheath regions are

$$\tilde{E}_r = \frac{A\alpha_0 \cosh \alpha_p d/2}{i\omega\varepsilon_0} \sinh \alpha_0(l/2 - z), \qquad (6.51)$$

$$\tilde{B}_\theta = \mu_0 A \cosh \alpha_p d/2 \cosh \alpha_0(l/2 - z), \qquad (6.52)$$

$$\tilde{E}_z = \frac{Ak \cosh \alpha_p d/2}{i\omega\varepsilon_0} \cosh \alpha_0(l/2 - z), \qquad (6.53)$$

6.2 Electromagnetic regime at high frequency

where A is an arbitrary amplitude. It is necessary to introduce k as the wavenumber along r, and α_p and α_0 as the wavenumbers along z in the plasma and in the sheath, respectively, with $k^2 - \alpha_p^2 = k_0^2 \varepsilon_p$ and $k^2 - \alpha_0^2 = k_0^2$. These functions effectively describe evanescent field structures in the z-direction. The continuity of $\tilde{B}_\theta$ and $\varepsilon \tilde{E}_z$ at the sheath/plasma boundary is already assured by these equations, but the continuity of $\tilde{E}_r$ is still to be used as a constraint. Imposing it now yields a relationship between wavenumber and frequency that amounts to a dispersion relation for the surface waves that are guided along the interface between the plasma and the sheath and that are responsible for the standing wave effect:

$$\alpha_0 \varepsilon_p \sinh \alpha_0 s \cosh \alpha_p d/2 + \alpha_p \cosh \alpha_0 s \sinh \alpha_p d/2 = 0. \tag{6.54}$$

In the regime of interest, $|\alpha_0 s| \ll 1$, $|kc/\omega_p| \ll 1$ and $\omega \ll \omega_p$, whereupon the dispersion relation of the surface wave can then be simplified to

$$\frac{k^2}{k_0^2} = 1 + \frac{\delta}{s}\left(1 - i\frac{\nu_m}{\omega}\right)^{1/2} \tanh\left[\frac{d}{2\delta}\frac{1}{(1 - i\nu_m/\omega)^{1/2}}\right], \tag{6.55}$$

where $\delta \equiv c/\omega_p$, the inertial plasma skin depth. Equation (6.55) is valid if $(\omega_p/\omega)^2 \gg \max(1 + d/2s, s/\delta)$, a condition satisfied from small to large skin depth. In the low-pressure limit, $\nu_m \ll \omega$, and the dispersion relation reduces to

$$k^2 \approx k_0^2 \left[1 + \frac{\delta}{s} \tanh \frac{d}{2\delta}\right]. \tag{6.56}$$

From this relation, one sees that in the limit of very large skin depth, the wavelength becomes

$$\lambda \approx \lambda_0 \left[1 + \frac{d}{2s}\right]^{-1/2} \equiv \lambda_0 \left[\frac{S_m}{l}\right]^{1/2}. \tag{6.57}$$

This expression is similar to Eq. (6.34), though K_{cap} is missing from Eq. (6.57), because here the sheaths have been modelled simply as vacuum regions with no periodic penetration by electrons which would be part of a more realistic model. In the opposite limit of small skin depth, the dispersion relation is

$$k^2 \approx k_0^2 \left[1 + \frac{\delta}{s}\right]. \tag{6.58}$$

Q When $\delta \to 0$, it seems that the wavelength comes back to the wavelength of vacuum. Explain this phenomenon.

A When the skin depth is infinitely small, the plasma behaves like a perfectly conducting metal. In this case, the wave propagates in each sheath (electron-free), as in a classical waveguide made of metal. The wavelength is that of vacuum and is independent of the gap (sheath) size.

Transmission line model

The transmission line model is derived from the fields (see the RHS of Figure 6.17). Here $Z' = R'_{\text{ind}} + i\omega L'$ and $Y'^{-1} = R'_{\text{cap}} + R'_i + (i\omega C')^{-1}$, where L' is the series inductance per unit length and C' is the parallel capacitance per unit length, and R'_{ind}, R'_{cap} and R'_i are dissipative terms, which are small and shall be treated as perturbations in the following.

It is more usual to express losses in the capacitive branch of Figure 6.17 in terms of conductance per unit length. This means that whereas R'_{ind} is in ohms per metre, R'_{cap} and R'_i are in units of (ohms^{-1} per metre)$^{-1}$, that is ohm metres. Be careful to consider this when checking dimensions.

For transmission line modes, the voltage and current in the transmission line are generally calculated from the power flow of the travelling wave [122]. However, in the regime of interest here, from Eqs (6.48)–(6.53) the electric field in the sheaths is very nearly 'transverse', that is to say perpendicular to the direction of propagation, and is much greater than that in the plasma. In that case, the voltage amplitude for a single radially propagating wave is approximately

$$\tilde{V} = -2 \int_0^{l/2} \tilde{E}_z(z)\,dz,$$

the current amplitude is

$$\tilde{I} = 2\pi r \mu_0 \tilde{B}_\theta(z = l/2)$$

and the characteristic impedance of the line is

$$\tilde{V}/\tilde{I} = \sqrt{L'/C'}.$$

In addition, $k = \omega\sqrt{L'C'}$, so that one can solve for L' and C' using Eq. (6.56), to obtain

$$L' = \mu_0 \frac{s}{\pi r}\left(1 + \frac{\delta}{s}\tanh\frac{d}{2\delta}\right)\left(1 - \frac{\omega^2}{\omega_p^2}\frac{\delta}{s}\tanh\frac{d}{2\delta}\right) \quad (6.59)$$

and

$$C' = \frac{\varepsilon_0 \pi r}{s}\left(1 - \frac{\omega^2}{\omega_p^2}\frac{\delta}{s}\tanh\frac{d}{2\delta}\right)^{-1}. \quad (6.60)$$

The second terms in the last brackets of Eqs (6.59) and (6.60) will be neglected in the following, since $\omega \ll \omega_p$. When δ is large, Eq. (6.59) reduces to its value in vacuum, as expected.

6.2 Electromagnetic regime at high frequency

Q Examining Eq. (6.60), it is evident that the value of C' may become ∞ at a particular combination of frequency and plasma density. Account for this phenomenon.

A The parallel branch of the transmission line was constructed in such a way that the inductance due to electron inertia was not explicitly calculated, but was buried in the expression for L' and C'. In other words, C' includes both the sheath capacitance and the electron inertia inductance. When $1/\omega C' = 0$, the discharge is driven at the series resonance described in Chapter 5.

The resistances in the transmission line are calculated from the power dissipation. The power loss per unit length due to the inductive $\tilde{E}_r(z)$ field is

$$\text{Re}\left[\int_0^{d/2} \tilde{J}_r(z) \cdot \tilde{E}_r^*(z)\,dz\right] 2\pi r = \frac{1}{2}|\tilde{I}|^2 R'_{\text{ind}},$$

(* implies complex conjugate) which yields the series transmission line resistance per unit length:

$$R'_{\text{ind}} = \frac{1}{2\pi r \sigma_m \delta}\left[\frac{\sinh(d/\delta) - (d/\delta)}{1 + \cosh(d/\delta)}\right], \quad (6.61)$$

where $\sigma_m = e^2 n_e / m\nu_m$ is the plasma conductivity. When the skin depth is large ($n_e \to 0$), R'_{ind} increases linearly with n_e, whereas when $n_e \to \infty$, R'_{ind} decreases as $1/n_e^{1/2}$. This is similar to the plasma resistance of a classical inductive discharge driven by a coil through a dielectric window (see Chapter 7).

The parallel resistance due to ohmic heating by the capacitive field $\tilde{E}_z(z)$ is found similarly:

$$R'_{\text{ohm}} = \frac{\delta}{2\pi r \sigma_m}\left[\frac{\sinh(2d/\delta) + 2d/\delta}{1 + \cosh(2d/\delta)}\right]. \quad (6.62)$$

Unlike the inductive resistance, R'_{ohm} always decreases with n_e. At low pressure, ohmic heating is dominated by stochastic heating and ohmic heating in the sheath, which are both independent of skin effects since they occur within the sheath region. From expressions obtained in the previous chapter (see also [21] for details), we introduce the following resistances to account for these power dissipations (for the collisionless sheath case):

$$R'_{\text{stoc}} = \frac{4 K_{\text{stoc}} (m k_B T_e)^{1/2} s^2}{e^{1/2} \varepsilon_0 \pi r |\tilde{V}|}, \quad (6.63)$$

$$R'_{\text{ohm,sh}} = \frac{2 K_{\text{cap}} K_{\text{ohm,sh}} m \nu_m s^3}{e \varepsilon_0 \pi r |\tilde{V}|}, \quad (6.64)$$

where the constants are given in Table 5.2 (note that the maximum sheath expansion is defined here as $s_m = 2s K_{cap}$). Here the stochastic resistance was calculated using the hard-wall model. Finally, $R'_{cap} = R'_{ohm} + R'_{stoc} + R'_{ohm,sh}$.

The resistance due to power dissipation by ions flowing in the sheaths is also included [63]:

$$R'_i = \frac{4 K_v e h_1 n_e u_B s^2}{\omega^2 \varepsilon_0^2 \pi r |\widetilde{V}|}. \tag{6.65}$$

So far, L', C', R'_{ind}, R'_{cap} and R'_i are functions of n_e, s and $|\widetilde{V}|$. In order to find a self-consistent solution for the plasma parameters and the voltage, the transmission line equations must be coupled to the particle and power balance equations, and to the Child–Langmuir law for the RF sheath. Two limiting cases can be treated: (i) non-local power deposition, when the electron energy relaxation length is greater than the discharge radius (low pressure); (ii) local power deposition in the opposite limit (high pressure).

Non-local power deposition: global E–H transitions

In the non-local case the radial profile of the electron density is not determined by the power deposition profile, but rather by solving the transport equations with constant ionization rate (cf. Chapter 3). In the low-pressure regime under consideration here, Chabert et al. [21] suggested using the following radial electron density profile:

$$n_e(r) = n_{e0} \left[1 - (1 - h_{r0}^2) \frac{r^2}{r_0^2} \right]^{1/2}, \tag{6.66}$$

which is a good approximation of the solution proposed by Godyak [39] in the cylindrical geometry. Here, $h_{r0} = 0.80(4 + r_0/\lambda_i)^{-1/2}$ is the radial edge-to-centre density ratio for a plasma with cold ions, from Eq. (3.83).

The density at the reactor centre, n_{e0}, must then be determined from the global power balance, $P_e = P_{loss}$, where

$$P_e = \frac{1}{2} \int_0^{r_0} R'_{cap} \left| \frac{d\widetilde{I}}{dr} \right|^2 dr + \frac{1}{2} \int_0^{r_0} R'_{ind} |\widetilde{I}|^2 dr \tag{6.67}$$

is the absorbed power, which includes capacitive heating (first term), that is heating provided by the E_z field, and inductive heating (second term), that is heating provided by the E_r field. The loss power has its usual form, extended for a two-dimensional cylindrical geometry:

$$P_{loss} = 2 n_{e0} u_B \left(\pi r_0^2 h_1 + 2\pi r_0 d h_{r0} \right) \varepsilon_T(T_e). \tag{6.68}$$

6.2 Electromagnetic regime at high frequency

The power loss is a function of T_e, which is determined by the global particle balance:

$$n_g K_{iz} \pi r_0^2 d = u_B \left(\pi r_0^2 h_l + 2\pi r_0 d h_{r0} \right). \tag{6.69}$$

For a symmetrically driven system, the boundary conditions are $V(r = 0) = V_0$ and $I(r = 0) = 0$ (a standing wave with no edge effects at $r = r_0$). In practice, the model is solved as follows: for a given V_0 and a set of n_{e0} values, (i) calculate T_e from the particle balance equation, (ii) insert $n_e(r)$ into the Child–Langmuir law, Eq. (5.28), to obtain $s(r, V)$, (iii) solve the transmission line equations to obtain $V(r)$ and $I(r)$, (iv) calculate P_e and P_{loss}, and plot both against n_{e0}, the intersection of the two curves being the equilibrium.

Typical power against n_{e0} graphs are shown in Figure 6.18 at 4 Pa, 200 MHz, with electrodes of radius $r_0 = 15$ cm spaced by $l = 4$ cm, for (a) $V_0 = 60$ V and (b) $V_0 = 800$ V. This simple graphic way of analysing the discharge equilibrium was introduced in Figure 5.9. This graph is the equivalent when inductive heating is taken into account. At low voltage, capacitive heating dominates (E-mode) whereas at high voltage the inductive heating takes over (H-mode), such that the discharge experiences an E–H transition as the voltage is raised. Unlike in inductive discharges, the transition is smooth and is not clearly defined. For the sake of simplicity, one can define the E–H transition as the condition $P_{ind} = P_{cap}$.

The E–H transition does not occur at a specific electron density, but also depends on the frequency. To analyse the role of the driving frequency, one has to remember that the voltage and the current are not radially uniform because of the standing wave effect. The voltage is maximum in the centre, where the current is zero, and decreases with radius. The radial position where the voltage reaches its minimum (and the current its maximum) will be denoted $r = r_1$ in the following. The standing wave effect is weak if $r_1 \gg r_0$, and strong if $r_1 \leq r_0$.

The ratio of inductive-to-capacitive power, P_{ind}/P_{cap}, is plotted in Figure 6.19 as a function of n_{e0}, at equilibrium (n_{e0} was varied by varying V_0), for various frequencies. Consider first the case $r_0 = 0.15$ m, for which the standing wave effect is moderate ($r_1 > r_0$). The inductive heating is barely seen at 27 MHz but it increases significantly with frequency; E to H transitions are obtained for frequencies above 170 MHz. The frequency dependence can be understood from a simplified analytical solution obtained for $r_1 \gg r_0$, considering that the electron density and the voltage are independent of r; $n_e(r) = n_{e0}$ and $V(r) = V_0$. Assuming that $R'_{cap} \approx R'_{stoc}$, the ratio at equilibrium is given by

$$\frac{P_{ind}}{P_{cap}} \propto v_m h_1^{1/2} r_0^2 \omega \left[\frac{\sinh(d/\delta) - d/\delta}{1 + \cosh(d/\delta)} \right] \tag{6.70}$$

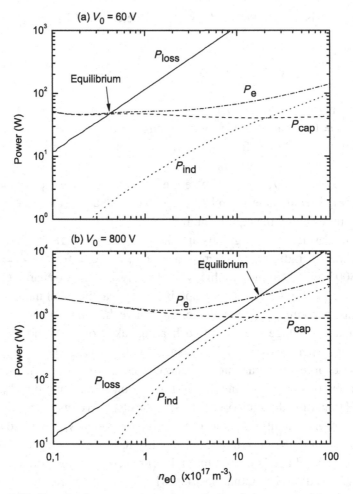

Figure 6.18 Contributions to power absorption and the power loss term as a function of central electron density, n_{e0}, for a 4 Pa (30 mtorr), 200 MHz capacitive discharge with electrodes of radius $r_0 = 15$ cm separated by $l = 4$ cm, for (a) $V_0 = 60$ V and (b) $V_0 = 800$ V. From [21].

in the limit of small inductive heating, $P_{ind} \ll P_{cap}$ [21]. This approximation explains most of the variations shown in Figure 6.19(a). At given density, the inductive heating indeed increases with the frequency and, if the frequency is fixed, the ratio increases with n_{e0} at low density and saturates at high density. Equation (6.70) also predicts that inductive heating will increase with the electrode radius. This is well verified as long as the standing wave effect remains weak (compare 60 MHz at $r_0 = 0.15$ and at $r_0 = 0.25$).

The situation becomes more complicated when the standing wave effect is strong, $r_1 \leq r_0$. For instance, in Figure 6.19(b) note that P_{ind}/P_{cap} is smaller at 250 MHz than at 150 MHz, in contradiction with the above discussion. The regime

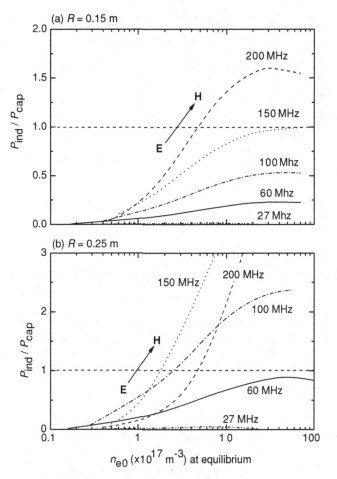

Figure 6.19 Inductive-to-capacitive power ratio, P_{ind}/P_{cap}, vs. n_{e0} at equilibrium, for several frequencies and for (a) $r_0 = 15$ cm and (b) $r_0 = 25$ cm. The other conditions are similar to that in Figure 6.18. From [21], where the frequency unit should be MHz and R is equivalent to r_0 in this text.

of a strong standing wave can be understood by plotting in Figure 6.20 the ratio P_{ind}/P_{cap} against frequency for a fixed equilibrium density $n_{e0} = 5 \times 10^{17}$ m^{-3}. The ratio increases with frequency while $r_1 \geq r_0$, reaches its maximum when $r_1 \approx r_0$ (in fact a little after) and decays for $r_1 \leq r_0$. This is because when $r_1 \leq r_0$, the voltage has a node and increases again for $r \geq r_1$, as does the capacitive heating, and the current decreases for $r \geq r_1$, as does the inductive heating.

Local power deposition: spatial E–H transitions

When the pressure is higher, the energy relaxation length becomes smaller than the discharge radius and the power deposition is local rather than global. In this situation, the electron density profile is determined by the local voltage and current.

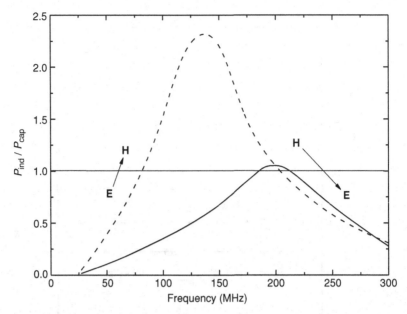

Figure 6.20 Inductive-to-capacitive power ratio, P_{ind}/P_{cap}, vs. frequency for $n_{e0} = 5 \times 10^{17}$ m^{-3}. The dashed line is for $r_0 = 0.25$ m and the solid line is for $r_0 = 0.15$ m. The other conditions are similar to those in Figure 6.18. From [21], where R_1 is equivalent to r_1 in this text.

The electron temperature also becomes a function of the radius, since the sheath width varies. The local power balance is $P'_e = P'_{loss}$, with

$$P'_e = \frac{1}{2} R'_{cap} \left| \frac{d\tilde{I}}{dr} \right|^2 + \frac{1}{2} R'_{ind} |\tilde{I}|^2 \qquad (6.71)$$

and

$$P'_{loss} = 4\pi r h_1 n_e u_B \varepsilon_T(T_e), \qquad (6.72)$$

where the electron temperature is determined from the local particle balance equation

$$n_g K_{iz} d = 2h_1 u_B. \qquad (6.73)$$

Note that d, and therefore T_e, varies with r since the RF voltage varies with r.

Figure 6.21 shows the electron density profile, normalized to the central density, for a 200 MHz discharge at 20 Pa (the electrode radius is 15 cm). The profile moves from being dominated by the standing wave effect at moderate density ($n_{e0} = 1.5 \times 10^{17}$ m^{-3}) to being dominated by the skin effect at higher density

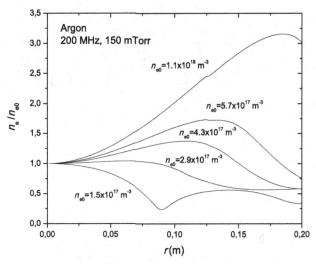

Figure 6.21 Electron density normalized to its central value n_{e0} vs. radius for various RF voltages (and thus various n_{e0}). The frequency is 200 MHz, the argon pressure is 20 Pa and the electrode radius is $r_0 = 0.15$ m. From [21].

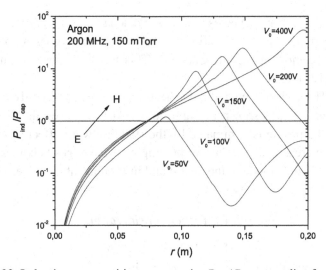

Figure 6.22 Inductive-to-capacitive power ratio, P_{ind}/P_{cap}, vs. radius for the same conditions as in Figure 6.21. The spatial E–H transitions are clearly seen at high voltage. From [21].

($n_{e0} \approx 10^{18}$ m^{-3}). This is more clearly seen in Figure 6.22, which shows the ratio of inductive to capacitive power as a function of the radius for the same conditions. At low voltage (50 V corresponding to the $n_{e0} = 1.5 \times 10^{17}$ m^{-3} case), the discharge operates in the capacitive E-mode at almost all radial positions. At higher voltage,

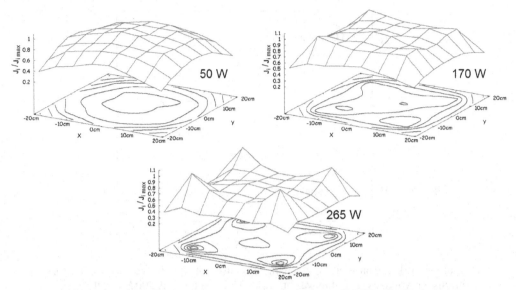

Figure 6.23 Ion flux mapping over the electrode for a 60 MHz discharge driven at different powers. Inductive heating is observed near the edges. From [91].

inductive heating (H-mode) takes over at the discharge periphery while the centre remains in E-mode. In this situation, the discharge undergoes a spatial E to H transition when moving radially outward. This has been observed in numerical simulations [123].

Spatial E–H transitions and inductive heating have been observed experimentally by Perret *et al.* [91], as shown in Figure 6.23. At low power, the discharge is in E-mode and the profile is dominated by the standing wave effect (dome-like). As the power is increased, the inductive heating near the edges becomes significant. The discharge is in E-mode at the centre and in H-mode at the edge.

6.2.4 Further considerations

The issue of ion energy uniformity

It has been shown that the flux of ions leaving a CCP is strongly non-uniform when the frequency is high enough for the standing wave effect to occur. However, under the same conditions, the ion energy may remain uniform according to Perret *et al.* [90]. They measured the ion velocity distribution functions (IVDFs) using retarding field analysers (the IVDF is equivalent to dN/dv – see Chapter 10). The excitation frequency was 81 MHz and the gas was argon at 2 Pa, resulting in an ion flux of about 3 A m^{-2} ($n_s \sim 10^{16}$ m^{-3}). Data from identical analysers at the centre, the side and the corner of the electrodes of a square CCP were compared.

6.2 Electromagnetic regime at high frequency

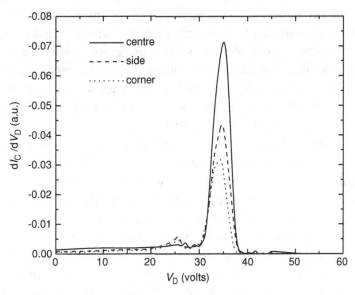

Figure 6.24 Ion velocity distribution function at the centre, the side and the corner of the grounded electrode of a capacitive discharge at 81.36 MHz. The horizontal scale is effectively kinetic energy in eV, which is $Mv^2/2e$. From [90].

Q IVDFs from Perret *et al.* [90] are shown in Figure 6.24. What can be inferred from these data about the following?
(i) The collisionality of the sheath.
(ii) The time-averaged plasma potential.
(iii) The standing wave effect.
(iv) The variation of ion bombardment energy across the electrode.

A (i) The absence of low-energy ions implies a collisionless sheath.
Comment: The ion–neutral mean free path ($\lambda_i \approx 2\,mm$) is larger than the sheath size (the calculated sheath size is $s = 0.9\,mm$).
(ii) The highest energy at which ions are detected corresponds with the time-averaged plasma potential one mean free path or so from the sheath/plasma boundary and this potential is the same at all positions.
(iii) There is some evidence of density structure in the plasma – the area under the IVDF in the centre is largest, that for the corner is smallest, suggesting that there are more ions created in and leaving the central region, probably because of the standing wave effect on the RF voltage across the electrodes.
(iv) Since the position of the IVDF peak (around 35 V) is independent of the analyser location, it must be that ions gain more or less the same energy in crossing the sheath, right across the surface, even though the RF voltage and the ion flux are strongly non-uniform.

The ion energy uniformity may at first sight be surprising since one might perhaps anticipate a smaller time-averaged (DC) plasma potential near the edges where the RF voltage is considerably smaller owing to the standing wave effect. However, the plasma DC conductivity is much too high to maintain a large potential difference and instead a significant radial current must flow in the plasma to short it out. Perret *et al.* [90] estimated that the DC potential difference in the plasma, $\Delta \overline{V}$, was ≤ 0.1 V at 2 Pa and ≤ 1 V at 20 Pa.

Having established that the time-averaged plasma potential must be uniform to within a volt, the next question must be about what the absolute value should be. There is still a requirement that in a steady-state CCP during one RF cycle the net current to an electrode must be zero. Note first that the RF self-bias Eq. (4.15) does not depend on the density of the plasma at the plasma/sheath boundary, so the standing wave effect does not affect the boundary fluxes through its effect on the local plasma density. However, the tendency of the standing wave is also to set the amplitude of the RF to be maximum at the centre of an electrode, so one expects that this would make the self-bias voltage largest here. To even out the mean potential there must be a slight lessening of bias in the centre. A slight local decrease in the magnitude of the self-bias would not fully suppress the electron current, leading to a net local electron current – given the exponential dependence of electron flux, small potential adjustments lead to large changes in current. Elsewhere, a slight decrease in RF potential across the sheath would allow a net positive ion current. Clearly the criterion that sets the total potential across the electrode must be that the net current to the electrode is zero. This requires a slight modification of the standing wave effect that arises when current circulates from near the axis, radially across the plasma edge and back at larger radii and from there radially inwards in the electrode surface. Of course if the standing wave brings a potential node within the radius of the electrodes the current flow will be more complicated. Howling *et al.* [114] have corroborated this, showing that a DC current does indeed flow parallel to the surface in conducting electrodes.

> **Q** Suggest what would happen to the ion flux variation across an electrode of a VHF CCP on which there was a dielectric substrate.
> **A** A dielectric substrate would not allow the potential levelling DC current to flow in the electrode, therefore the ion energy will not be uniform and will be lower in the region of the nodes of the standing wave. See [114].

Edge effects and asymmetric discharges

In addition to standing wave and skin effects, one should also consider the effects of finite geometry. The edge effects are difficult to study in a general way since they depend crucially on the strategy used to confine the plasma at the electrode edge. Lieberman *et al.* [108] have incorporated edge effects in their electromagnetic model by introducing evanescent waves at the radial interface between the plasma and a dielectric (or vacuum). The role of these evanescent waves is to bridge the propagating wave solution in vacuum and in the plasma. The assumption is that all the RF current flows across the plasma, which in turn constrains the magnetic field to be independent of z at the plasma edge, $r = r_0$. As a result, the inductive field E_r is zero at $r = r_0$, and the axial (capacitive) field exhibits a spike at this position. Therefore, edge effects lead to a strong plasma production near the edge which competes with the skin effect.

Howling *et al.* have examined the effect of edge asymmetry in a large area reactor [112, 113]. They showed that the redistribution of RF current to maintain current continuity near asymmetric side-walls causes a perturbation in RF plasma potential to propagate radially inwards. This effect was termed the 'telegraph effect', since the transmission line (telegraph) equations were used to calculate the typical damping length of the perturbation. Asymmetry of the electrodes requires an asymmetric field solution to be added to that for purely symmetric electrodes, Eqs (6.48)–(6.53). This supplementary solution gives rise to a mode associated with the telegraph effect [118]. This study was done in the context of somewhat higher-pressure plasmas used for deposition, in which resistive effects are important. Indeed, the plasma resistance may be large enough for the wave to be absorbed when propagating inward. In the limit of very high resistivity, the wave would not reach the centre of the discharge and therefore the standing wave would not be established; the power deposition then falls from the edge to the centre [118].

Multiple-frequency excitation and non-linear effects

The electromagnetic regime of the capacitive discharge has been explored so far for the single-frequency case. The generalization of this model to multiple-frequency excitation has not been done so far. Earlier in this chapter it was shown that low frequencies generate larger sheaths. Since the scale of the standing wave varies with $(s_m/l)^{1/2}$ for any given electrode size, adding a frequency would tend to suppress some of the non-uniformity of a VHF CCP.

Miller *et al.* [119] have measured the RF magnetic field in a high-frequency capacitive discharge. They found that the magnetic field does indeed start to decrease from the edges to the centre, in line with the elementary analysis presented earlier. However, they also found that the waveform was far from a pure sinusoid, as has been assumed here. Therefore, it seems that non-linearities of the sheaths generate higher harmonics that should be included in more sophisticated electromagnetic models.

6.3 Summary of important results

- The independent control of ion energy and ion flux is not possible in a single-frequency CCP because both quantities depend on the absorbed power. Although ion flux and ion energy cannot easily be decoupled by using a single frequency, the use of two well-separated frequencies provides additional degrees of freedom through which the energy–flux parameter space can be explored, even though perfect decoupling is not achieved.
- The use of VHF sources and/or large diameter CCPs requires one to consider wave phenomena in describing how RF power enters the plasma-filled space between the electrodes. Treating the system as a centre-fed radial transmission line enables a circuit approach to the electromagnetic problem. This shows that standing wave effects occur when the CCP exceeds a few percent of the vacuum wavelength of electromagnetic waves at the source frequency, leading to strong radial non-uniformity. The non-uniformity can be reduced markedly by introducing a dielectric with radial structure that compensates for the standing wave effect.
- The non-uniformity can also be understood in terms of spatial structure in the RF fields that change the nature of the power coupling from being dominated by the axial field (so-called E-mode) close to the axis to being dominated by radial electric fields (so-called H-mode) at larger radii.
- Electromagnetic solutions can be constructed and combined with the fluid equations to produce a global model of VHF discharges.

7
Inductively coupled plasmas

Capacitively coupled plasma reactors have some natural limitations. Although very high-frequency CCPs achieve high plasma density (typically $n_e \approx 10^{17}$ m^{-3}), this is accompanied by major uniformity problems. Moreover, the ion flux and the ion energy cannot be varied totally independently, even when multiple-frequency excitation is employed. Inductively coupled discharges overcome these limitations to some extent. They are used in plasma processing and for plasma light sources.

Inductive discharges have been known since the end of the nineteenth century. The principle is to induce an RF current in a plasma by driving an RF current in an adjacent coil. From an electromagnetic point of view, the changing magnetic field associated with the coil current induces an electromagnetic field similar to the H-mode studied in the previous chapter. However, the coil is much more efficient than a pair of parallel plates in exciting an H-mode. Interestingly, the coil also couples to the plasma electrostatically, which means that an inductive discharge may also operate in an E-mode and therefore it can experience transitions between E and H-modes. These transitions are usually sharper than in VHF capacitive discharges, with strong hysteresis effects [18] and instabilities when electronegative gases are used [20, 124–126].

The energy of ions incident on a substrate electrode immersed in an inductively coupled plasma can be adjusted independently of the ion flux using a separate power supply for biasing. This is easily done using the self-bias effect described in Chapter 4, to set the voltage between the electrode and the plasma with the substrate holder capacitively coupled. The amount of power transferred to the plasma electrons by the bias supply is usually such that the RF power from the substrate-holder contributes only marginally to the plasma density (and thus to the ion flux). It is the power supplied to the ICP coil that controls ion flux.

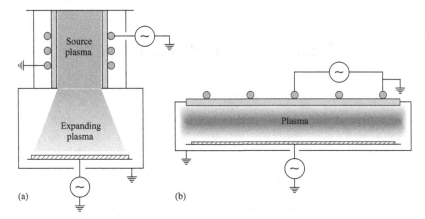

Figure 7.1 Inductively coupled plasma reactors: (a) cylindrical source tube with an expanding chamber, (b) planar coil geometry.

Inductively coupled plasma (ICP) reactors for plasma processing can be divided into two main geometric designs, shown in Figure 7.1. On the LHS the plasma is generated by a coil that surrounds the cylindrical dielectric wall of the source region. Plasma expands from here into the processing chamber where the substrate is placed. This configuration is similar to that used in helicon plasma processing reactors described in the next chapter. On the RHS of Figure 7.1 is an alternative arrangement that is much used in plasma etching in the microelectronics industry – a flat spiral coil, separated from the plasma by a flat dielectric window. The distance between this window and the substrate-holder is significantly smaller than the chamber radius.

Warning:
- The short-cylinder geometry cannot be modelled using a single dimension which impedes insightful analysis. In this chapter, therefore, the long-cylinder arrangement will be treated since the analysis can be done in 1-D and consequently important scalings can be revealed. Most of the general principles are equally valid for the flatter geometry.
- Since this chapter deals with electromagnetic waves, Boltzmann's constant is written k_B so that k can be used as a wavenumber.

The RF current generated in the plasma, or equivalently the induced electromagnetic field, flows in a skin depth δ, which is classically defined by Eq. (2.57) in a collisionless plasma (when $\nu_m \ll \omega$) and Eq. (2.58) in a collisional plasma (when $\nu_m \gg \omega$). It will be shown in this chapter that the skin depth is sometimes different due to non-local effects, and that the electric field is also non-uniform owing to geometric effects.

> **Q** In the absence of significant capacitive coupling, the sheath that forms at the dielectric window develops a voltage defined by the floating potential, $eV_s = e(V_p - V_f) \approx 5k_B T_e$, and is consequently only a few Debye lengths. Compare the typical width of a floating sheath with the collisionless skin depth.
>
> **A** It was shown in Chapter 3 that $\lambda_{De}/\delta = v_e/c$, so the skin depth is much larger than the Debye length because the electron thermal speed is much smaller than the speed of light. In addition.

In ICPs the boundary sheath size is necessarily much smaller than the skin depth and, unlike in capacitive discharges, the physics occurring in the sheath is of minor importance when the inductive discharge operates in its usual H-mode. However, the systems are sometimes driven in a low-current (low-power) regime in which electrostatic coupling between the coil and the plasma may dominate. In such a regime, the sheath physics does play a role.

Inductive discharges are commonly modelled using a transformer analogy; indeed, they are sometimes called TCPs (for transformer-coupled plasmas). The transformer model will be analysed in Section 7.3. The analysis begins with an electromagnetic description of the inductive discharge, following the early work of Thompson [241]. By solving Maxwell's equation in an idealized geometry, the fields and the RF currents can be calculated in the system composed of the coil, the dielectric tube and the plasma. Then the power absorbed in the plasma is calculated and finally the total impedance of the system is determined.

Note: Section 7.1 requires an appreciation of Bessel functions of complex arguments. Some of their useful properties are summarized here. If x is a real number, then

$$J_0(ix) = I_0(x),$$

$$J_1(ix) = iI_1(x),$$

where $I_0(x)$ and $I_1(x)$ are modified Bessel functions, which increase exponentially when x is large. In addition, when $x \to \infty$

$$\frac{I_1(x)}{I_0(x)} \to 1.$$

In the other limit when $x \to 0$, the modified Bessel functions may be approximated by

$$I_0(x) \approx 1,$$

$$I_1(x) \approx \frac{x}{2}.$$

Unfortunately, when the argument has both real and imaginary parts, such simplifications do not arise.

The transformer model is a decomposition of the electromagnetic model presented in the first section. The plasma RF current loop is seen as the secondary of a transformer, the primary of which is the driving coil. This decomposition is not unique and the electromagnetic model will serve as a guide to propose the most appropriate choices. In Section 7.2, we study low-power operation, in which capacitive coupling (i.e., the E-mode) dominates, and introduce E–H transitions. The global model of the inductive discharge is then established by joining the inductive electromagnetic model and the capacitive coupling model to the particle and power balance. The most important results are summarized in Section 7.7.

At the end of the chapter, other designs for the suppression of capacitive coupling and for better power coupling efficiency are discussed. Other regimes are also considered in which the penetration of the fields is anomalous and the power absorption is collisionless (stochastic inductive heating). Finally, non-linear effects are introduced, resulting in harmonic generation and ponderomotive effects.

7.1 Electromagnetic model

The plasma is represented by a uniform, complex permittivity ε_p (i.e., a uniform electron density), generated and contained in a dielectric tube of inner radius r_0, outer radius r_c and length $l \gg r_0$. The tube is surrounded by a coil having N turns uniformly distributed, in which flows an RF sinusoidal current

$$I_{RF}(t) = \text{Re}\left[\tilde{I}_{RF} e^{i\omega t}\right],$$

where $\tilde{I}_{RF}$ is the complex amplitude. The schematic of this model is shown in Figure 7.2. It is convenient to use H instead of $B (= \mu_0 H)$.

Given the long cylindrical geometry, the magnetic field is along z and the electric field is azimuthal, i.e., around θ. The fields must obey Maxwell's equations, which are

$$-\frac{\partial \tilde{H}_z}{\partial r} = i\omega \varepsilon_0 \varepsilon \tilde{E}_\theta, \tag{7.1}$$

$$\frac{1}{r}\frac{\partial (r\tilde{E}_\theta)}{\partial r} = -i\omega \mu_0 \tilde{H}_z, \tag{7.2}$$

where $\varepsilon = \varepsilon_p$ in the plasma and $\varepsilon = \varepsilon_t$ in the dielectric tube. Combining these two equations one obtains the Bessel equation for $\tilde{H}_z$,

$$\frac{\partial^2 \tilde{H}_z}{\partial r^2} + \frac{1}{r}\frac{\partial \tilde{H}_z}{\partial r} + k_0^2 \varepsilon \tilde{H}_z = 0. \tag{7.3}$$

7.1 Electromagnetic model

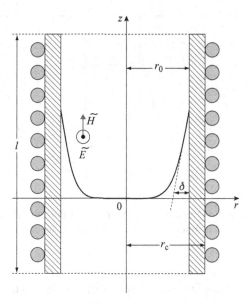

Figure 7.2 Schematic of the inductive discharge. The discharge is contained in a dielectric tube of inner radius r_0, outer radius r_c and length $l \gg r_0$. The induced electric field is azimuthal, while the induced magnetic field is axial. At high plasma density, both fields decay within the skin depth δ.

7.1.1 Fields in the plasma

The following solution is obtained for the plasma:

$$\tilde{H}_z = H_{z0} \frac{J_0(kr)}{J_0(kr_0)}, \qquad (7.4)$$

$$\tilde{E}_\theta = -\frac{ikH_{z0}}{\omega\varepsilon_0\varepsilon_p} \frac{J_1(kr)}{J_0(kr_0)}, \qquad (7.5)$$

where $H_{z0} \equiv \tilde{H}_z(r = r_0)$, and $k \equiv k_0\sqrt{\varepsilon_p}$ is the complex wavenumber in the plasma while $k_0 \equiv \omega/c$ is the wavenumber in free space. Taking H_{z0} to be a real number defines the reference for the phase to be that of the magnetic field at the edge of the plasma. The arguments in the Bessel functions J_0 and J_1 are complex numbers. The modulus of the electromagnetic fields of Eqs (7.4) and (7.5) are plotted in Figure 7.3 for various electron densities. At low electron density, the plasma skin depth is large, $\delta \gg r_0$, and the magnetic field H_z is nearly constant across the radius. This is close to the solution in free space. Note, however, that the electric field E_θ is not uniform, falling linearly with r from the edge to the centre. At high electron density, $\delta \ll r_0$, both fields decay nearly exponentially within the plasma skin depth.

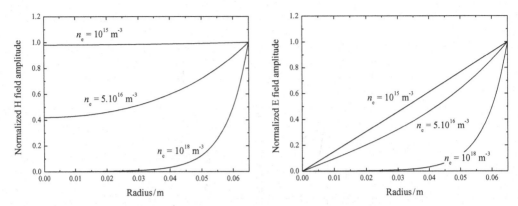

Figure 7.3 Calculation of the electromagnetic field magnitude normalized to the edge value at $r = r_0$, as a function of the cylinder radius and for various electron densities.

Q Explain why the electric field decays from the edge to the centre in the very low-density plasma regime (or in absence of plasma), where the plasma skin depth is infinite. Then propose a characteristic 'geometric' decay length for the electric field.

A The symmetry of the problem requires the electric field to be zero at the centre and finite at the edge. The characteristic 'geometric' decay length is therefore the cylinder radius r_0. It is interesting to note that the electric field profile for the $n_e = 5 \times 10^{16}$ m^{-3} case is not profoundly different from the linear decay of the free space limit.

7.1.2 Fields in the dielectric tube

The fields in the dielectric tube are also solutions of the Bessel equation for $\tilde{H}_z$ (Eq. (7.3)), with wavenumber $k_1 = k_0\sqrt{\varepsilon_t}$ (real). The expressions for the fields are quite cumbersome, but may be simplified drastically by noting that $k_1 r_0 \ll 1$ for the typical frequencies used in inductive discharges.

Q Calculate the value of $k_1 r_0$ for 13.56 MHz with $r_0 = 6.5$ cm and $\varepsilon_t = 4.5$ and then describe the form of the RF magnetic field in the dielectric tube.

A First note that

$$k_1 r_0 = \frac{2\pi \, 13.56 \times 10^6}{3 \times 10^8} \times \sqrt{4.5} \times 0.065 = 0.04.$$

Since the wavenumber k_1 is real, one expects the solution to be of the form $\tilde{H}_z \sim J_0(k_1 r)$. In fact, it is a little more complicated because the field must be

7.1 Electromagnetic model

continuous at the plasma/dielectric interface. However, as long as $k_1 r_0 \ll 1$, the Bessel function $J_0(k_1 r_0) \approx 1$ and the magnetic field is nearly independent of the radius.

The magnetic field is therefore nearly constant in the dielectric tube when $k_1 r_0 \ll 1$, and so

$$\tilde{H}_z \approx H_{z0} \quad r_0 < r < r_c. \tag{7.6}$$

Note that it is not exactly constant because some displacement current flows in this tube. The RF currents flowing in the coil, in the dielectric tube and in the plasma are evaluated in the next section. If the magnetic field is constant in the tube, then using the integral form of Faraday's law $\oint \tilde{E}_\theta dl \equiv \partial/\partial t \iint \tilde{B}_z dS$, it follows that

$$\tilde{E}_\theta(r_c) = \tilde{E}_\theta(r_0)\frac{r_0}{r_c} - i\omega\mu_0 H_{z0}\left(\frac{r_c^2 - r_0^2}{2r_c}\right). \tag{7.7}$$

7.1.3 RF currents

The total current flowing in the plasma may be calculated from the integration of the RF current density between the centre and the edge of the plasma at $r = r_0$,

$$\tilde{I}_p = l\int_0^{r_0} \tilde{J}_\theta \, dr. \tag{7.8}$$

The RF current density is related in the usual way to the electric field, $\tilde{J}_\theta = i\omega\varepsilon_0\varepsilon_p \tilde{E}_\theta$, so with Eq. (7.5) the integral simplifies to

$$\tilde{I}_p = lH_{z0}\frac{1}{J_0(kr_0)}\int_0^{kr_0} J_1(kr)\,d(kr) = lH_{z0}\left[\frac{1}{J_0(kr_0)} - 1\right]. \tag{7.9}$$

Note again here that k is a complex wavenumber and is a function of the electron density. In the plasma density range of interest, the real and imaginary parts of $\tilde{I}_p$ are both negative, and the real part is much larger than the imaginary part. The plasma current may also be calculated from Ampère's theorem, following contour 1 in Figure 7.4:

$$\tilde{I}_p = l\tilde{H}_z(0) - lH_{z0} = lH_{z0}\left[\frac{1}{J_0(kr_0)} - 1\right]. \tag{7.10}$$

Similarly, the (displacement) current flowing in the dielectric tube, $\tilde{I}_t$, is obtained using contour 2, while the current flowing in the coil, $\tilde{I}_{RF}$, is obtained using

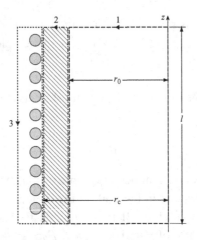

Figure 7.4 Ampère contours for the current in the plasma (1), in the dielectric tube (2) and in the coil (3).

contour 3:

$$\tilde{I}_t = l H_{z0} - l \tilde{H}_z(r_c), \qquad (7.11)$$

$$N \tilde{I}_{RF} = l \tilde{H}_z(r_c). \qquad (7.12)$$

Doing the sum of those three currents leads to

$$\tilde{I}_p + \tilde{I}_t + N \tilde{I}_{RF} = \frac{l H_{z0}}{J_0(k r_0)}. \qquad (7.13)$$

Note here that both $\tilde{I}_t$ and $\tilde{I}_{RF}$ are positive (for $\tilde{I}_t$ it comes from the fact that $\tilde{H}_z(r = r_c) > H_{z0}$). Since, as discussed above, the magnetic field is nearly uniform in the dielectric tube, the RF current flowing in this tube is negligible. From now on, we will set $\tilde{I}_t = 0$. Combining Eqs (7.11) and (7.12) with $\tilde{I}_t = 0$ allows the magnetic field at the plasma edge, H_{z0}, to be related to the current in the coil, $\tilde{I}_{RF}$:

$$H_{z0} = \frac{N \tilde{I}_{RF}}{l}.$$

Note that having neglected the current in the dielectric tube, $\tilde{I}_{RF}$ has become a real number (due to the choice of phase reference), equivalent to I_{coil}, which is the magnitude of the current in the coil. From now on in this chapter, therefore,

$$H_{z0} = \frac{N I_{coil}}{l}. \qquad (7.14)$$

7.1 Electromagnetic model

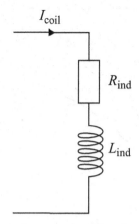

Figure 7.5 The equivalent circuit of the plasma load on the coil.

When the plasma density is high, $kr_0 \ll 1$, the Bessel function of complex argument grows exponentially and consequently,

$$\tilde{I}_p + NI_{coil} \approx 0. \tag{7.15}$$

The current induced in the plasma skin depth flows in the opposite direction to the current in the coil and cancels the magnetic field produced by the coil in the bulk of the plasma.

7.1.4 Poynting theorem for harmonic fields

The impedance of the system, composed of reactive and resistive components, may be derived from the fields using Poynting's theorem for harmonic time variations of the fields [127]. The complex power input, which is the sum of the power dissipated and the power stored by the electromagnetic fields in the cylinder of radius r_c under consideration here, is given by the quantity

$$\tilde{P} = -\frac{1}{2}\tilde{E}_\theta(r_c)\tilde{H}_z(r_c)2\pi r_c l = \frac{1}{2}Z_{ind}I_{coil}^2, \tag{7.16}$$

where Z_{ind} is the total complex impedance of the system. It follows immediately that one can define dissipative and reactive components (see Figure 7.5):

$$R_{ind} = \frac{2\text{Re}\left[\tilde{P}\right]}{I_{coil}^2}, \tag{7.17}$$

$$X_{ind} = \frac{2\text{Im}\left[\tilde{P}\right]}{I_{coil}^2}. \tag{7.18}$$

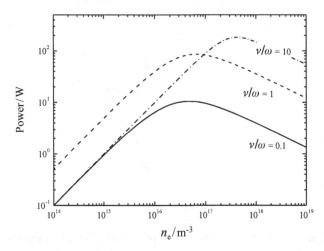

Figure 7.6 Absorbed power as a function of the electron density for fixed coil current and for three values of the ratio ν_m/ω; $I_{coil} = 3$ A, $r_0 = 0.065$ m and $l = 0.3$ m.

Ignoring the displacement current in the dielectric tube and considering that $\tilde{H}_z(r = r_c) = H_{z0}$, then the electric field at $r = r_c$ is given by Eq. (7.7). Using also Eq. (7.14) it can be shown that

$$\tilde{P} = i\frac{\pi N^2 I_{coil}^2}{l}\left[\frac{kr_0 J_1(kr_0)}{\omega \varepsilon_0 \varepsilon_p J_0(kr_0)} + \frac{1}{2}\omega\mu_0\left(r_c^2 - r_0^2\right)\right]. \tag{7.19}$$

7.1.5 Power dissipation: resistance

Electromagnetic calculation

The time-averaged power dissipated in the system is the real part of Eq. (7.19):

$$P_{abs} = \mathrm{Re}\left[\tilde{P}\right] = \frac{\pi N^2}{l\omega\varepsilon_0}\mathrm{Re}\left[\frac{ikr_0 J_1(kr_0)}{\varepsilon_p J_0(kr_0)}\right]I_{coil}^2. \tag{7.20}$$

The absorbed power P_{abs} for a coil current of 3 A is plotted as a function of the electron density in Figure 7.6 for various values of the ratio ν_m/ω, which is proportional to the gas pressure. The low-pressure limit $\nu_m/\omega = 0.1$ corresponds approximately to 0.27 Pa in argon. The absorbed power increases linearly with n_e at low densities, goes through a maximum and then decays with $n_e^{-1/2}$ at high densities. This behaviour results from a transition from the very low-density regime, for which the electromagnetic fields are not modified by the presence of the plasma, to the high-conductivity regime, in which the electromagnetic fields are absorbed by the plasma, in the skin depth, as n_e increases.

7.1 Electromagnetic model

> **Q** Compare the radius of the ICP with the collisionless skin depth for the electron density that corresponds with the peak in the absorbed power ($v_m/\omega = 0.1$) in Figure 7.6.
> **A** The collisionless skin depth is given by Eq. (2.57):
>
> $$\frac{r_0}{\delta} = r_0 \omega_{pe}/c = \frac{6.5}{2.6} = 2.5.$$
>
> *Comment: It is perhaps not too surprising that the peak power absorption should occur when the density is still low enough for most of the plasma volume to be penetrated by the fields.*

The scaling at high electron density will be discussed later, in Section 7.2.1. In the opposite range at low electron density, note that the absorbed power increases linearly with the collision frequency v_m (this corresponds to a translation upwards in the log scale of the figure) when $v_m/\omega \ll 1$, but then decays with v_m in the opposite limit of $v_m/\omega \gg 1$. This scaling with v_m (or equivalently the gas pressure) at low electron density is not immediately obvious from the somewhat complicated expression in Eq. (7.20). However, as seen in the next section, it may be understood by applying the analysis of a simple solenoid loaded by a lossy medium. Before doing this analysis, the resistance is derived from the definition of the absorbed power as

$$R_{ind} = \frac{2P_{abs}}{I_{coil}^2} = \frac{2\pi N^2}{l\omega \varepsilon_0} \text{Re}\left[\frac{ikr_0 J_1(kr_0)}{\varepsilon_p J_0(kr_0)}\right]. \tag{7.21}$$

Approximations in the low-density limit

At low electron density, the fields are similar to the case of a solenoid in free space. From Faraday's law, the circulation of the electric field along a circular loop of radius r is equal to the time derivative of the magnetic flux within this loop:

$$-\frac{d\Phi}{dt} = \oint E \cdot dl = 2\pi r E_\theta. \tag{7.22}$$

Using complex notation, the magnetic flux is $\tilde{\Phi} = \mu_0 \tilde{H}_z \pi r^2$ and the time derivative is $i\omega \tilde{\Phi}$. This gives the azimuthal electric field at any radius in the low-density limit as

$$\tilde{E}_\theta = -\frac{\mu_0 r}{2l} i\omega N I_{coil}. \tag{7.23}$$

The current density in the plasma is related to the electric field by $\tilde{J}_\theta = i\omega \varepsilon_0 \varepsilon_p \tilde{E}_\theta$. In the low-pressure regime, where $v_m/\omega \ll 1$, we have $\varepsilon_p \approx -\omega_p^2/\omega^2$ and consequently

$$\tilde{J}_\theta = -\frac{n_e e^2}{m}\frac{\mu_0 r}{2l} N I_{\text{coil}}. \tag{7.24}$$

The dissipated power is then

$$P_{\text{abs}} = \int_0^{2\pi}\int_0^{r_0}\int_0^l \frac{|\tilde{J}_\theta|^2}{2\sigma_{\text{dc}}} r\,d\phi\,dr\,dz = \frac{n_e e^2 \nu_m \mu_0^2 \pi r_0^4}{4ml} N^2 I_{\text{coil}}^2, \tag{7.25}$$

such that

$$R_{\text{ind}} = \frac{n_e e^2 \nu_m \mu_0^2 \pi r_0^4 N^2}{2ml}. \tag{7.26}$$

Thus the absorbed power increases linearly with n_e at low density. This scaling with the electron density is disguised in the more rigorous electromagnetic calculation inside ε_p. Note also that the absorbed power is proportional to ν_m, that is to pressure, as observed from $\nu_m/\omega = 0.1$ to $\nu_m/\omega = 1$.

In the high-pressure limit, $\varepsilon_p \approx -\omega_p^2/(i\omega\nu_m)$. In this case

$$P_{\text{abs}} = \frac{n_e e^2 \omega^2 \mu_0^2 \pi r_0^4 N^2}{4ml\nu_m} I_{\text{coil}}^2, \tag{7.27}$$

$$R_{\text{ind}} = \frac{n_e e^2 \omega^2 \mu_0^2 \pi r_0^4 N^2}{2ml\nu_m}. \tag{7.28}$$

At high pressure the absorbed power is seen to be inversely proportional to the collision frequency ν_m, that is to the gas pressure. This is consistent with the results of the exact calculation shown in Figure 7.6, where the power falls off as the collisionality increases from $\nu_m/\omega = 1$ to $\nu_m/\omega = 10$.

7.1.6 Stored power: inductance

The imaginary part of the complex power allows the reactance of the system to be determined. Within our approximation that the displacement current in the dielectric tube is neglected, the reactance is essentially an inductance, L_{ind}, such that

$$X_{\text{ind}} = L_{\text{ind}}\omega = \frac{2\text{Im}[\tilde{P}]}{I_{\text{coil}}^2} = \frac{\pi N^2 \omega \mu_0}{l}(r_c^2 - r_0^2) + \frac{2\pi N^2}{l\omega\varepsilon_0}\text{Im}\left[\frac{ikr_0 J_1(kr_0)}{\varepsilon_p J_0(kr_0)}\right]. \tag{7.29}$$

7.1 Electromagnetic model

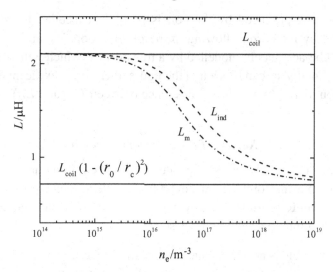

Figure 7.7 Total inductance L_{ind} and magnetic storage inductance L_m as a function of the electron density for $v_m/\omega = 1$. We have chosen $N = 5$, $r_0 = 0.065$ m, $r_c = 0.08$ m and $l = 0.3$ m.

Q What is the inductance of the coil itself?
A The coil inductance is defined as the magnetic flux per unit current in the turns: $L_{coil} = \mu_0 N \pi r_c^2 H_{z0}/I_{coil}$. Since $H_{z0} = NI_{coil}/l$, it follows that

$$L_{coil} = \frac{\mu_0 \pi r_c^2 N^2}{l}. \tag{7.30}$$

Using the expression of the coil inductance, the inductance of a cylindrical ICP becomes

$$L_{ind} = L_{coil}\left(1 - \frac{r_0^2}{r_c^2}\right) + \frac{2\pi N^2}{l\omega^2 \varepsilon_0} \text{Im}\left[\frac{ikr_0 J_1(kr_0)}{\varepsilon_p J_0(kr_0)}\right]. \tag{7.31}$$

This inductance combines the magnetic energy storage inductance, embedded in both terms (not only in the first term), and a contribution due to electron inertia (part of the second term). The inductance due to electron inertia is R_{ind}/v_m, such that the magnetic energy storage inductance is

$$L_m = L_{ind} - \frac{R_{ind}}{v_m}. \tag{7.32}$$

It is interesting to analyse the behaviour of L_{ind} at low and high electron density. In Figure 7.7, the inductance is plotted as a function of the electron density for $v_m/\omega = 1$. In the low electron density limit, $L_{ind} \approx L_{coil}$, i.e., the plasma plays no role. In the high electron density limit, $L_{ind} = L_{coil}\left(1 - r_0^2/r_c^2\right)$. In this limit, the magnetic flux produced by the coil is partially cancelled by the magnetic flux produced by the current flowing in the plasma loop. We anticipate that this situation will be adequately modelled by a transformer, which will be done in the next section. Finally, as can be seen in the figure, there is a significant contribution of the electron inertia (which is the difference between L_{ind} and L_{m}).

7.1.7 Review of the electromagnetic model

The electromagnetic fields can be calculated from Maxwell's equations in a system composed of a plasma contained in a cylindrical dielectric tube (of infinite length) and excited by an RF current I_{coil} flowing in an N-turn coil surrounding the tube. The following summarizes the findings so far.

- The magnetic field is nearly constant within the dielectric tube, and the displacement current flowing in the tube is negligible. Consequently,

$$H_{z0} = \frac{NI_{\text{coil}}}{l}, \tag{7.33}$$

where H_{z0} is the magnetic field amplitude at the plasma edge.
- At low electron density, the magnetic field is nearly constant in the plasma while the electric field decays linearly from the edge to the centre, where it is null.
- At high electron density, both fields decay nearly exponentially from the edge, within a characteristic length: the skin depth δ.
- The power dissipated in the plasma increases linearly with the electron density at low electron density, goes through a maximum and decays with the square root of the electron density at high electron density. The maximum of the absorbed power curve occurs when $\delta \lesssim r_0$.
- The equivalent lumped circuit of the system is constructed from the electromagnetic fields using the complex Poynting theorem. It is composed of a resistance and an inductance:

$$R_{\text{ind}} = \frac{2\pi N^2}{l\omega\varepsilon_0} \text{Re}\left[\frac{ikr_0 J_1(kr_0)}{\varepsilon_p J_0(kr_0)}\right],$$

$$L_{\text{ind}} = L_{\text{coil}}\left(1 - \frac{r_0^2}{r_c^2}\right) + \frac{2\pi N^2}{l\omega^2\varepsilon_0}\text{Im}\left[\frac{ikr_0 J_1(kr_0)}{\varepsilon_p J_0(kr_0)}\right],$$

which both depend on the electron density via k and ε_p. Simplified expressions of R_{ind} have been obtained in the low electron density limit. Simplifications are also available in the high electron density limit, as will be seen in the next section. [Note that to account for power dissipation in the coil itself an extra resistance, R_{coil}, should be added to R_{ind}. This will be done later in this chapter.]

7.2 Impedance of the plasma alone

At this point, all the required information is available to model the inductive mode. One could proceed directly to Section 7.4. However, it is instructive first to study the usual transformer analogy, detailed in several articles and in other textbooks. This is the purpose of the next two sections.

7.2 Impedance of the plasma alone

It will be useful to have an expression for the impedance of the plasma alone, with the contribution of the coil taken out of Z_{ind}. This is the first step towards the decomposition of the transformer model of the next section. This is done by assuming that the current flowing in the plasma loop, denoted $\tilde{I}_p$ in Section 7.1.3, exists on its own (a situation that cannot exist in reality). The resistance and inductance of the plasma are determined respectively from the absorbed power and from the magnetic flux produced by this current alone.

7.2.1 Plasma resistance

The plasma resistance is obtained from the plasma current (I_p) and the absorbed power in the plasma (P_{abs}) so that

$$P_{\text{abs}} = \frac{1}{2} R_p |\tilde{I}_p|^2.$$

Then, using Eqs (7.9), (7.14) and (7.20):

$$R_p = \frac{2 P_{\text{abs}}}{|\tilde{I}_p|^2} = \frac{2\pi}{l \omega \varepsilon_0} \text{Re}\left[\frac{i k r_0 J_1(k r_0)}{\varepsilon_p J_0(k r_0)}\right] \left|\frac{1}{J_0(k r_0)} - 1\right|^{-2}. \quad (7.34)$$

Figure 7.8 shows the plasma resistance and the modulus of the plasma current as a function of the electron density for $v_m/\omega = 0.1$, as before with $N = 5$, $I_{\text{coil}} = 3$ A, $r_0 = 0.065$ m and $l = 0.3$ m. It appears that, unlike R_{ind}, towards low electron density R_p increases strongly. In terms of the mathematics, this comes about because at low density the complex wavenumber $k \to 0$ and so $J_0(k r_0) \to 1$; physically it can be understood because the conductivity goes to zero when the electron density goes to zero. In the regime of high plasma density, for which $J_0(k r_0) \to \infty$, the skin depth is small, $\delta \ll r_0$, and the fields decay exponentially near the edge. The plasma resistance decreases because the conductivity increases faster than the skin depth shrinks. More meaningful expressions for the resistance at high plasma density will be obtained below, in the low and high-pressure limits corresponding respectively to $v_m \ll \omega$ and $v_m \gg \omega$.

The plasma current is very small and initially increases with the electron density at low electron density. When the density becomes sufficiently high, the current localizes in the skin depth and consequently saturates to a value that may be

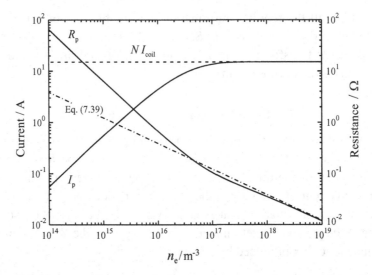

Figure 7.8 Plasma resistance and plasma current as a function of the electron density for $\nu_m/\omega = 0.1$. We have chosen $N = 5$, $I_{coil} = 3\,\text{A}$, $r_0 = 0.065\,\text{m}$ and $l = 0.3\,\text{m}$.

evaluated using Ampère's theorem: $H_{z0} = I_p/l = NI_{coil}/l$, such that $I_p = NI_{coil}$. In this regime, note that the plasma resistance becomes proportional to the total resistance defined previously:

$$R_p = \frac{R_{ind}}{N^2}, \quad (7.35)$$

which now looks like a relationship from the equivalent circuit of a transformer. Before proceeding, note that Piejak et al. [128] have measured the plasma current and the plasma resistance as a function of the discharge power, i.e., equivalent to the trend with electron density, and found the same behaviour as that described above: at low and moderate electron density, the resistance decreases while the current increases with the electron density.

The plasma resistance at high electron density may be approximated by taking the appropriate limits of the Bessel functions in the two limiting cases of low gas pressure, $\nu_m \ll \omega$, and high gas pressure, $\nu_m \gg \omega$.

The low-pressure regime

At low pressure, $\nu_m \ll \omega$, and the square root of the plasma permittivity may be approximated by

$$\sqrt{\varepsilon_p} \approx \pm \frac{\omega_{pe}}{\omega}\left(\frac{\nu_m}{2\omega} + i\right). \quad (7.36)$$

Using the properties of Bessel functions with pure imaginary arguments and their limits at large arguments, i.e., for $kr_0 \to \infty$ (see the introduction to this

7.2 Impedance of the plasma alone

chapter):

$$R_p \approx \frac{2\pi k_0 r_0}{l\omega\varepsilon_0} \text{Re}\left[\frac{-1}{\sqrt{\varepsilon_p}}\right]. \tag{7.37}$$

From Eq. (7.36):

$$\text{Re}\left[\frac{-1}{\sqrt{\varepsilon_p}}\right] \approx \frac{\nu_m}{2\omega_{pe}}. \tag{7.38}$$

In terms of the collisionless skin depth $\delta = c/\omega_{pe}$ and the definition of the plasma conductivity, Eq. (2.54), the resistance may be written in the following way:

$$R_p = \frac{\pi r_0}{\sigma_m l \delta}. \tag{7.39}$$

Q Noting that the RF current flows in a one-turn loop of cross-section $l\delta$ and length $2\pi r_0$, one expects a plasma resistance given by $R_p = 2\pi r_0/\sigma_m l \delta$. Why is the actual expression of Eq. (7.39) smaller by a factor of two?
A The power dissipation profile depends on the square of the current and therefore scales with $\exp -2x/\delta$, halving the effective cross-sectional area.

The approximation of Eq. (7.39) is plotted (dash–dot line) in Figure 7.8. It compares well, at high electron densities, with the exact calculation of Eq. (7.34).

The high-pressure regime

In the opposite high-pressure regime, $\nu_m \gg \omega$,

$$\sqrt{\varepsilon_p} \approx \frac{\omega_{pe}}{\sqrt{2\nu_m \omega}}(1+i) = X(1+i). \tag{7.40}$$

At high density X is also large. The behaviour of the Bessel functions with large and complex arguments in this case allows some simplification, though it is not easily shown algebraically. This then leads to a high-pressure form for the plasma resistance:

$$R_p = \frac{\pi r_0 \omega_{pe}}{\sigma_m l c}\left(\frac{2\omega}{\nu_m}\right)^{1/2}. \tag{7.41}$$

Q Use Eq. (2.58) to show that this time the plasma resistance is given by $R_p = 2\pi r_0/\sigma_m l \delta_{coll}$.
A Starting with $\delta_{coll} = \sqrt{2/\mu_0 \sigma_m \omega}$, rewrite this as

$$\delta_{coll} = \sqrt{\frac{2}{\mu_0} \frac{m\varepsilon_0}{ne^2} \frac{\nu_m}{\omega}}$$

so $\delta_{coll} = \delta(2\nu_m/\omega)^{1/2}$, such that the substitution gives the required result.

7.2.2 Plasma inductance

The current path in the plasma has both resistance and inductance. Some inductance can be attributed to the electron inertia, $L_p = R_p/\nu_m$, as seen previously (Eqs (2.82) and (2.83)). The loop of current flowing in the plasma also generates a magnetic flux and so gives rise to another inductance, denoted L_{mp}. The calculation of L_{mp} is easy in the high electron density regime, for which the RF current is localized in a narrow skin depth. The magnetic flux is then $\Phi = \mu_0 \tilde{H}_z \pi r_0^2 = L_{mp} \tilde{I}_p$, and the magnetic field is $\tilde{H}_z = \tilde{I}_p/l$, so that

$$L_{mp} = \frac{\mu_0 \pi r_0^2}{l}. \tag{7.42}$$

This expression is not valid in the low electron density limit where the current is not localized in the skin depth. The current is driven by the electric field so that it decays linearly from the edge to the centre, at low density. In this limit, it turns out that L_{mp} is about half the value given by the expression in Eq. (7.42).

Further analysis of the high electron density regime shows that

$$\frac{L_p}{L_{mp}} = \frac{m_e}{n_e e^2 \mu_0 r_0 \delta}, \tag{7.43}$$

i.e., the inductance due to inertia becomes less important at high electron density. Considering $\delta \approx c/\omega_{pe}$, with a discharge radius of 10 cm, then $L_{inertia}/L_{mp} \approx 5.3 \times 10^6/(r_0 n_e^{1/2})$, which gives $L_p/L_{mp} \approx 0.5$ at $n_e = 10^{16}$ m^{-3} and $L_p/L_{mp} \approx 0.17$ at $n_e = 10^{17}$ m^{-3}.

It is also interesting to compare the inductive reactance of the plasma to its resistance. In the high electron density regime, for which the electron inertia inductance can be neglected,

$$\frac{L_{mp}\omega}{R_p} = \frac{r_0}{\delta}\left(\frac{\omega}{\nu_m}\right), \quad \text{if } \omega \gg \nu_m, \tag{7.44}$$

$$\frac{L_{mp}\omega}{R_p} = \frac{r_0}{\delta}\left(\frac{2\omega}{\nu_m}\right)^{1/2}, \quad \text{if } \omega \ll \nu_m, \tag{7.45}$$

where again $\delta \equiv c/\omega_{pe}$ is the collisionless skin depth. It appears that the plasma resistance is small compared to the plasma inductive impedance at low pressure and high frequency, i.e., for $\omega \gg \nu_m$. The opposite limit may only be reached when $\omega \ll \nu_m$, because the skin depth is significantly smaller than the radius r_0.

7.3 The transformer model

A transformer model of an inductive discharge has been proposed by Piejak *et al.* [128]. The coil and the plasma form a transformer – the plasma is regarded as the one-turn secondary coil of an air-cored transformer. The primary has an inductance

7.3 The transformer model

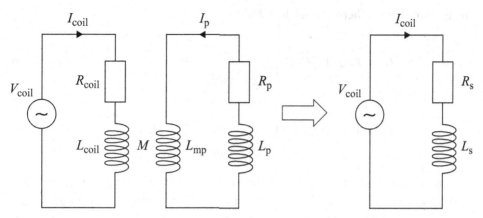

Figure 7.9 The transformer model of an inductive discharge. On the right, the secondary circuit has been transformed into its series equivalent in terms of the primary circuit current.

L_{coil} and a resistance R_{coil}. These two quantities define the Q-factor of the coil, $Q \equiv \omega L_{coil}/R_{coil}$. The coil resistance and the coil inductance, and therefore the Q-factor, may be measured experimentally. It can also be evaluated theoretically. The coil inductance was derived in Section 7.1.6.

Q Calculate the resistance, inductance and Q factor of a coil made of a copper wire 2.75 m long and 6 mm in diameter, and driven at 13.56 MHz. The conductivity of copper is $\sigma_{copper} = 59.6 \times 10^6 \, \Omega^{-1} \, m^{-1}$ and the coil is 0.3 m long, formed from 5 turns of radius 0.08 m.

A The current flows in a cross-section of area $2\pi \times 0.003 \times \delta = 0.0188 \times \delta$, with $\delta = (2/\omega\mu_0\sigma_{copper})^{1/2} = 1.77 \times 10^{-5}$ m. The resistance is therefore $R_{coil} = 2.75/(3.34 \times 10^{-7} \times \sigma_{copper}) = 0.138 \, \Omega$. With $N = 5, l = 0.3$ m and $r_c = 0.08$ m, $L_{coil} = 2.1 \, \mu H$, such that $Q \approx 1300$.

The coil and the one-turn plasma loop are coupled through the mutual inductance M. This takes account of the voltage induced in the secondary by changing current in the primary and *vice-versa*. In this calculation, M is assumed to be real, an assumption that will be discussed later. The coupled circuits shown on the LHS of Figure 7.9 can be transformed into a single circuit composed of a resistance R_s and an inductance L_s, as shown on the RHS of the figure. Applying Kirchoff's laws to the above circuits gives

$$\tilde{V}_{coil} = i\omega L_{coil} I_{coil} + R_{coil} I_{coil} + i\omega M \tilde{I}_p, \tag{7.46}$$

$$\tilde{V}_p = i\omega L_{mp}\tilde{I}_p + i\omega M I_{coil} = -\tilde{I}_p \left[R_p + i R_p \left(\frac{\omega}{\nu_m} \right) \right], \tag{7.47}$$

$$\tilde{V}_{coil} = (i\omega L_s + R_s) I_{coil}. \tag{7.48}$$

The transformation therefore leads to

$$R_s = R_{\text{coil}} + M^2\omega^2 \left(\frac{R_p}{R_p^2 + \left(\omega L_{\text{mp}} + R_p \left(\frac{\omega}{\nu_m}\right)\right)^2} \right), \quad (7.49)$$

$$L_s = L_{\text{coil}} - M^2\omega^2 \left(\frac{L_{\text{mp}} + R_p/\nu_m}{R_p^2 + \left(\omega L_{\text{mp}} + R_p \left(\frac{\omega}{\nu_m}\right)\right)^2} \right). \quad (7.50)$$

In order for the transformer model to describe the inductive discharge accurately, R_s must be set equal to $R_{\text{coil}} + R_{\text{ind}}$, and L_s to L_{ind}, in the entire density range (where R_{ind} and L_{ind} are obtained from the electromagnetic model). In the previous section, by investigating the impedance of the plasma alone, expressions for R_p and L_{mp} have been found so that in principle M remains the only unknown in the problem. From Eq. (7.47) it can be seen that the mutual inductance obeys the following relation:

$$M^2\omega^2 = \left[R_p^2 + \left(\omega L_{\text{mp}} + R_p \left(\frac{\omega}{\nu_m}\right)\right)^2 \right] \frac{|\widetilde{I}_p|^2}{I_{\text{coil}}^2}. \quad (7.51)$$

Substituting this expression into Eqs (7.49) and (7.50) leads to

$$R_s = R_{\text{coil}} + R_p \frac{|\widetilde{I}_p|^2}{I_{\text{coil}}^2}, \quad (7.52)$$

$$L_s = L_{\text{coil}} - \left(L_{\text{mp}} + \frac{R_p}{\nu_m} \right) \frac{|\widetilde{I}_p|^2}{I_{\text{coil}}^2}. \quad (7.53)$$

To provide a global model of an ICP the transformer model must correctly account for the power absorption. This imposes the following relation, $R_p|\widetilde{I}_p|^2 = R_{\text{ind}}I_{\text{coil}}^2$, which leads to

$$R_s = R_{\text{coil}} + R_{\text{ind}}, \quad (7.54)$$

$$L_s = L_{\text{coil}} - L_{\text{mp}} \left(\frac{R_{\text{ind}}}{R_p} \right) - \frac{R_{\text{ind}}}{\nu_m}. \quad (7.55)$$

In this way, the resistance R_s perfectly matches the electromagnetic model, as indeed it must. However, it turns out that L_s is not equal to L_{ind} in the whole electron density range, although they have the same limits at the extremes of electron density. At high electron density, $R_{\text{ind}} = N^2 R_p$, and the inertia term R_{ind}/ν_m is small so that the inductance reduces to

$$L_s \approx L_{\text{coil}} \left(1 - r_0^2/r_c^2 \right), \quad (7.56)$$

7.3 The transformer model

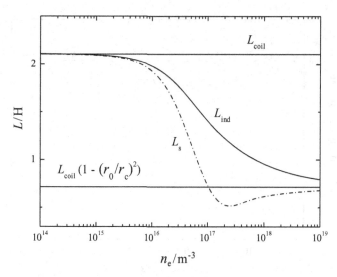

Figure 7.10 Inductance of the transformer model, L_s, along with the inductance derived from the electromagnetic model, L_{ind}, as a function of the electron density for $\nu_m/\omega = 1$. We have chosen $N = 5$, $r_0 = 0.065$ m, $r_c = 0.08$ m and $l = 0.3$ m.

as observed in Figure 7.7. In the low electron density limit, $R_{ind} \to 0$ so that $L_s \approx L_{coil}$, as also observed for L_{ind} in Figure 7.7. The comparison between the transformer model inductance L_s and the electromagnetic model inductance L_{ind} is shown in Figure 7.10 for the entire range of electron density. The discrepancy is significant at intermediate density.

Q (i) What is the implication of the incorrect modelling of the inductance?
(ii) What would be required to fix the discrepancy between L_s and L_{ind}?
A (i) There is no consequence on the global model results presented later because the principle requirement for a correct plasma model is the absorbed power. However, it would be a problem if one needs to evaluate the voltage across the coil, which depends directly upon the reactive part of the impedance.
(ii) To fix this problem, it is necessary to consider a complex mutual inductance, i.e., M should have an imaginary part.

In the following, the definition of M as a real quantity is retained. This mutual inductance is a function of the electron density, which implies that the coupling coefficient of the transformer also varies. The coupling coefficient of the transformer is defined by $M^2/L_{coil}L_{mp}$. It is small when the mutual inductance is weak, and goes to unity for perfect coupling. In the case of a transformer made of two nested, long solenoids, this coefficient is the ratio of the radius of the internal coil,

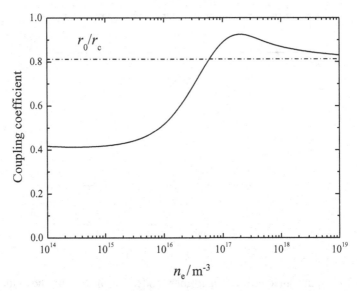

Figure 7.11 Coupling coefficient as a function of the electron density.

r_0, to the radius of the external coil, r_c. Figure 7.11 shows the coupling coefficient of the inductive discharge transformer as a function of the electron density, calculated using the definition of M given in Eq. (7.51). At high electron density, $M^2/L_{coil}L_{mp} \to r_0/r_c$ because the current flows in the skin depth and the plasma indeed behaves like a one-turn internal coil. At low electron density, the coupling is poorer because the current induced in the plasma is distributed across the radius.

Note that at high electron density, $L_s \approx L_{coil}\left(1 - (r_0/r_c)^2\right)$. If the dielectric window were to be infinitely thin, then $(r_0/r_c)^2 \approx 1$ and consequently $L_s \approx 0$. This is the situation of an ideal transformer, for which the inductive reactance of the secondary totally offsets the reactance of the primary such that the primary circuit appears purely resistive. For good coupling, it is therefore necessary to keep the dielectric window thin so the distance between the coil and the plasma is as small as possible and the coupling is maximized.

7.3.1 Review of the transformer model

The inductive discharge may be modelled using the transformer analogy. The plasma is regarded as the one-turn secondary coil of an air-cored transformer, the primary of which is composed of the coil itself. For this model, it is necessary to decompose the full system into two parts: (i) the coil itself and (ii) the plasma loop, in which an RF current flows, distributed in a way that depends on the electron density. The decomposition was accomplished by choosing the mutual inductance M to be purely real. It has been found that:

- The plasma resistance is a continuously decreasing function of the electron density, while the plasma current increases at low electron density to saturate at $I_p = NI_{coil}$.
- Once transformed to the primary circuit, the resistance R_s has to increase at low n_e, go through a maximum, and finally decay at high n_e (like R_{ind} in the electromagnetic model).
- To satisfy the above, the mutual inductance in the transformer model must be a function of the electron density.
- Strictly, a complex mutual inductance (M with real and imaginary parts) is required to model the resistance and the inductance of the transformed circuit in the whole electron density range. Here M was assumed to be purely real at the expense of an approximate form of the inductance. Nevertheless, this is acceptable since the resistance is correct so that the power absorption is correct.
- As a general conclusion, one might say that the transformer analogy is satisfying at high electron density, but should be examined more carefully at low and intermediate electron density.

7.4 Power transfer efficiency in pure inductive discharges

From this point, we go back to R_{ind} and L_{ind} defined in the electromagnetic model. The power delivered by the RF generator is the sum of the power dissipated by the coil, P_{coil}, and the power dissipated by the plasma electrons, P_{abs}:

$$P_{coil} = \frac{1}{2} R_{coil} I_{coil}^2, \tag{7.57}$$

$$P_{abs} = \frac{1}{2} R_{ind} I_{coil}^2. \tag{7.58}$$

A very important quantity to evaluate is the power transfer efficiency, which is defined as

$$\zeta \equiv \frac{P_{abs}}{P_{coil} + P_{abs}} = \left(1 + \frac{R_{coil}}{R_{ind}}\right)^{-1}. \tag{7.59}$$

The maximum power coupling efficiency is reached when R_{ind} is at its maximum. The effective resistance of the ICP, R_{ind}, may be written in terms of the Q-factor of the coil and the structure of the fields as

$$R_{ind} = R_{coil} \left(\frac{2Q}{k_0 r_0} \frac{r_0^2}{r_c^2}\right) \text{Re}\left[\frac{iJ_1(kr_0)}{\sqrt{\varepsilon_p} J_0(kr_0)}\right], \tag{7.60}$$

such that

$$\frac{R_{\text{coil}}}{R_{\text{ind}}} = X\left(\frac{2\, r_c^2}{Q\, r_0^2}\right), \qquad (7.61)$$

where we have introduced the quantity

$$X = k_0 r_0 \left(4\text{Re}\left[\frac{jJ_1(kr_0)}{\sqrt{\varepsilon_p}J_0(kr_0)}\right]\right)^{-1}, \qquad (7.62)$$

which is a function of the electron density. The efficiency may thus be written

$$\zeta = \left[1 + X\left(\frac{2\, r_c^2}{Q\, r_0^2}\right)\right]^{-1}. \qquad (7.63)$$

For a given coil design, the power transfer efficiency depends upon electron density (via X), which in turn depends upon the RF current amplitude. The maximum efficiency is reached when the quantity X reaches its minimum, denoted $X_{\min}$.

In the low-frequency, high-pressure limit ($\nu_m \gg \omega$) typical of fluorescent lamps, it can be shown that $X_{\min} \approx 1$, independent of the ν_m/ω ratio, and consequently the power transfer efficiency becomes

$$\zeta_{m,\text{hp}} = \left[1 + \frac{2\, r_c^2}{Q\, r_0^2}\right]^{-1}. \qquad (7.64)$$

Since the product $Q\, r_0^2/r_c^2$ is usually very large compared to 1, the efficiency can be very high (the efficiency goes to unity when the coil resistance is small and the Q-factor tends to infinity). In addition, it is clear that r_0/r_c has to be as close as possible to unity for high efficiency.

In the lower-pressure inductive discharges used for plasma etching $\nu_m \ll \omega$, so $X_{\min} \approx 2\omega/\nu_m$ and the power transfer efficiency becomes

$$\zeta_{m,\text{lp}} = \left[1 + \frac{4\, r_c^2}{Q\, r_0^2}\left(\frac{\omega}{\nu_m}\right)\right]^{-1}. \qquad (7.65)$$

It is easy to see that $\zeta_{m,\text{lp}} < \zeta_{m,\text{hp}}$. The efficiency of these discharges is typically between 50% and 80% lower than that of those used at higher pressure for fluorescent lamps, which may reach 98%, particularly when enhanced by ferrite cores, as will be described later in this chapter. This can be understood by the fact that when ω/ν_m is large, the ratio of reactive-to-resistive power is high, requiring higher RF currents to maintain a specific level of power absorption in the plasma. This leads to higher dissipation in the coil.

Piejak et al. [128] have developed the full analysis for the maximum power transfer efficiency, obtaining the following formula, which agrees well with the above observations:

$$\zeta_m = \left[1 + \frac{2}{Q}\frac{r_c^2}{r_0^2}\left(\frac{\omega}{\nu_m} + \left(1 + \frac{\omega^2}{\nu_m^2}\right)^{1/2}\right)\right]^{-1}. \qquad (7.66)$$

7.5 Capacitive coupling

As with the previous chapters on RF plasmas, one of the goals is to build a global model that links external currents and voltages to spatially averaged quantities that characterize the plasma. For this it is necessary to consider not only inductive current in the plasma, but also a capacitive component of current. To generate the RF current it is necessary to apply an RF voltage across the coil. This voltage may be large enough for the high-voltage point on the coil to drive a capacitive RF current through the dielectric tube (or window), the sheaths and the plasma and from there to ground.

Q What is the typical voltage amplitude that develops across the coil?
A The voltage amplitude is approximately $V_{coil} \approx \omega L_{ind} I_{coil}$. Considering $L_{ind} \approx L_{coil}$ and using a five-turn coil with $r_c = 0.08$ m and $l = 0.3$ m, gives $L_{coil} = 2.1$ µH (using Eq. (7.30)) and consequently $V_{coil} \approx 1800$ V for $I_{coil} = 10$ A and a driving frequency of 13.56 MHz.

This capacitive coupling is responsible for a fraction of the power deposition. However, it was shown in Chapter 5 that the power deposited in this way decays with $1/n_e$ at the given current. Therefore, capacitive coupling will only be significant at low electron densities. As in very high frequency capacitive discharges, inductive heating will dominate at high electron density. Capacitive discharges are designed to excite the capacitive (or electrostatic E) mode, but may operate in the inductive (or electromagnetic H) mode when driven at very high frequencies, and consequently at high electron densities. By contrast, inductive discharges are designed to operate in the H-mode, at high electron density, but may operate in the E-mode when driven at low power and consequently at low electron density. Both discharges are subject to mode transitions.

The complex geometry of the ICP means that the voltage distribution within the coil is not uniform, so the capacitive coupling is very difficult to model properly, unless performing 3-D numerical calculation of the fields, and is also very design-dependent. To illustrate the physics, the simplified model proposed in [125] is used. This model is shown in Figure 7.12. The inductive branch is modelled with the components discussed above. In parallel with the inductive branch, the capacitive branch is modelled by a capacitance in series with a resistance accounting for ohmic

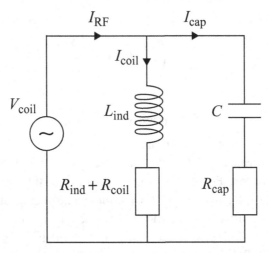

Figure 7.12 Simplified lumped element circuit model of an inductive discharge with capacitive coupling. From [125]. R_{cap} accounts for capacitive heating of electrons (both ohmic and stochastic).

and stochastic heating of electrons. The capacitance is the series combination of the dielectric tube capacitance (a fixed value) and the sheath capacitance, which varies with the plasma parameters. In many instances, the capacitance of the dielectric tube (or window) is smaller than that of the RF sheath and hence it dominates.

The impedance of the capacitive branch is always larger than the impedance of the inductive branch, so $I_{coil} \approx \tilde{I}_{RF} \approx \tilde{V}_{coil}/i\omega L_{ind}$. Then, in almost all operating regimes, $R_{ind} + R_{coil} \ll \omega L_{ind}$ and the resistance in the capacitive branch is small compared to the impedance of the capacitor, so $\omega R_{cap} C \ll 1$. So the power *absorbed by the electrons* in the inductive discharge with capacitive coupling is

$$P_{abs} \approx \frac{1}{2}\left[R_{ind} + (\omega^2 L_{ind} C)^2 R_{cap}\right] I_{coil}^2. \tag{7.67}$$

Q Why is the coil resistance not included in Eq. (7.67)?
A The resistance of the coil should be considered if one needs to evaluate the power dissipated in the system. However, here attention is focused on the power absorbed by the electrons, in preparation for the global model developed in the next section. Thus, the power dissipated by the coil has not been included in Eq. (7.67).

The two resistances in Eq. (7.67) are functions of the electron density. The inductive part, R_{ind}, has been discussed above and is given by Eq. (7.21). The capacitive part, R_{cap}, is not easy to model precisely but is composed of an ohmic

7.5 Capacitive coupling

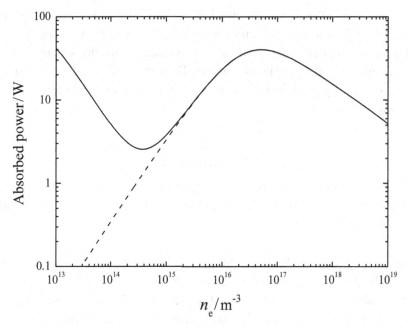

Figure 7.13 Absorbed power as a function of the electron density for a fixed coil current in a pure inductive discharge (dashed line), and in an inductive discharge with capacitive coupling (solid line). $k_B T_e/e = 2.47$ V.

and a stochastic part, and scales with $1/n_e$. From [20, 125]:

$$R_{\text{ohm}} = \frac{m_e \nu_m l_{\text{cap}}}{e^2 n_e A_{\text{cap}}}, \tag{7.68}$$

$$R_{\text{stoc}} = \left(\frac{m_e \bar{v}_e}{e^2 n_e A_{\text{cap}}}\right) \left(\frac{eV_{\text{coil}}}{k_B T_e}\right)^{1/2}, \tag{7.69}$$

$$R_{\text{cap}} = R_{\text{ohm}} + R_{\text{stoc}}. \tag{7.70}$$

Here, l_{cap} and A_{cap} are respectively the length and area in which the capacitive RF current flows. These quantities are not easy to evaluate and are strong functions of the reactor design.

The absorbed power is plotted in Figure 7.13 for a fixed coil current as a function of the electron density for a pure inductive discharge (dashed line), and for an inductive discharge with capacitive coupling (solid line). Here again the situation is that of a five-turn coil of radius $r_c = 0.08$ m around a cylinder of inner radius $r_0 = 0.065$ m and length $l = 0.3$ m. The argon gas pressure is $p = 1.33$ Pa and the 13.56 MHz RF current flowing in the coil is 3 A. The total capacitance (dielectric+sheaths) has been fixed to $C = 10$ pF, which means that sheath size variations with the RF current in the coil and the electron density need not be taken into account. The capacitive coupling parameters are $A_{\text{cap}} = 0.15\,\text{m}^2$, $l_{\text{cap}} = 0.15\,\text{m}$. The inductive power increases linearly with the electron density at low density, passes through a maximum and then decays with the square root of the density, as established earlier in this chapter. The capacitive coupling is dominant at low electron density, but rapidly becomes insignificant since the capacitive power decays with $1/n_e$.

7.6 Global model

To establish the global model of the inductive discharge with capacitive coupling it is necessary to solve simultaneously the particle balance and the power balance for the two variables, n_e and T_e.

> **Q** In previous chapters the RF Child–Langmuir law was used to establish global models. Why is this law not needed here?
> **A** It has already been established that the sheath facing the coil is narrow, and that the physics of this sheath is not dominant in the calculation of the power absorbed by electrons. The sheath size is therefore not a crucial variable for the global model.

The particle balance in cylindrical geometry is

$$n_g K_{\text{iz}} V = 2u_B \left(h_1 \pi r_0^2 + h_{r_0} \pi r_0 l \right), \tag{7.71}$$

where $V = \pi r_0^2 l$ is the plasma volume and the other quantities have their usual meaning. This equation can readily be solved to calculate the electron temperature. Once the electron temperature is found, the loss power is expressed as

$$P_{\text{loss}} = 2n_e u_B \left(h_1 \pi r_0^2 + h_{r_0} \pi r_0 l \right) \varepsilon_T(T_e), \tag{7.72}$$

where all quantities have, again, their usual meaning. The power balance, $P_{\text{abs}} = P_{\text{loss}}$, where the absorbed power P_{abs} is given by Eq. (7.67), allows the equilibrium electron density to be calculated. The full expression for R_{ind} (Eq. (7.21)) is too complicated to yield an explicit analytical expression for n_e. However, as in previous chapters, the solution can be found graphically by plotting the absorbed power and the loss power as a function of the electron density on the same axes. This is done in Figure 7.14 for the same conditions as in Figure 7.13. At a pressure of 1.33 Pa, the electron temperature $k_B T_e/e = 2.47$ V. The equilibrium electron density is located at the intersection of the two power curves, where $n_e \approx 6 \times 10^{16}\,\text{m}^{-3}$.

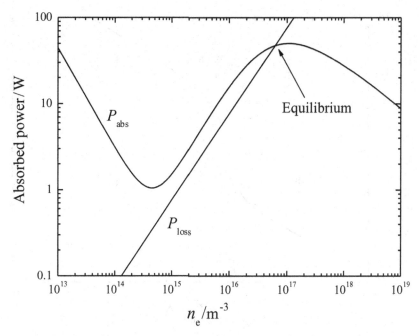

Figure 7.14 Absorbed power and power loss as a function of the electron density for the same conditions as in Figure 7.13.

7.6.1 Electron density as a function of the coil current

Using the procedure described above to solve for the electron density, the equilibrium electron density can be followed while the coil current is scanned. The result is shown in Figure 7.15 for the parameters used in the previous figures. The pressure is kept at 1.33 Pa and consequently the electron temperature remains at $k_B T_e/e = 2.47$ V.

Transition between E and H-modes

One can clearly distinguish three regions. At low coil current, the discharge is in the E-mode, that is the intersection between the absorbed power and the power loss curves occurs before the minimum of the absorbed power. At high coil current, the intersection occurs after the maximum of the absorbed power, i.e., in the inductive mode. In the region delimited in grey (between 1 A and 3 A), the intersection occurs between the minimum and the maximum of the absorbed power; this is the E–H transition region. Note that in this region the increase in the electron density is much faster than in the other two modes because the equilibrium intersection occurs in a region where the slopes of both lines in Figure 7.14 are positive and quite similar.

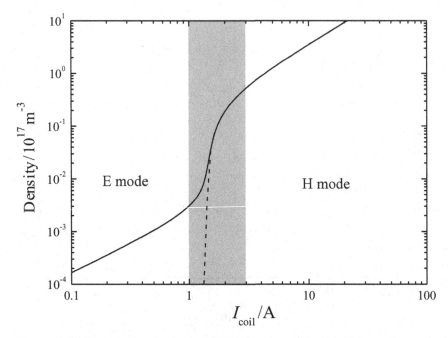

Figure 7.15 Electron density at the equilibrium as a function of the coil current for the inductive discharge with capacitive coupling. The dashed line is the density when capacitive coupling is reduced to zero. The region in grey, between 1 A and 3 A, corresponds to the E–H transition region.

The dashed line in the figure represents the electron density calculated with no capacitive coupling. As already mentioned, there is no equilibrium in this case at low coil current. A current in excess of 1.2 A is required to sustain the inductive mode. The electron density in the inductive mode rapidly exceeds a few 10^{16} m^{-3}. Capacitive coupling plays no role at high electron density, which allows a simplification of the global model to gain further insight into the physics of the inductive mode.

The low-pressure, high-frequency limit at high density

Suppose that the density is high enough for the capacitive coupling to be neglected. It has been shown that in this high-density regime the skin depth is small and the resistance of the inductive branch is $R_{\text{ind}} = N^2 R_p$. Then the power balance becomes

$$\frac{1}{2} R_{\text{ind}} I_{\text{coil}}^2 = \frac{1}{2} N^2 R_p I_{\text{coil}}^2 = 2 n_e u_B \left(h_l \pi r_0^2 + h_{r_0} \pi r_0 l \right) \varepsilon_T(T_e). \quad (7.73)$$

Using the approximate expression of R_p given in Eq. (7.39) leads to the following expression for the electron density as a function of the coil current in the inductive

7.6 Global model

mode:

$$n_e = \left[\frac{\pi r_0 N^2 \nu_m (m/\varepsilon_0)^{1/2}}{4 u_B \left(h_1 \pi r_0^2 + h_{r_0} \pi r_0 l \right) e \varepsilon_T(T_e) l\, c} \right]^{2/3} I_{\text{coil}}^{4/3}. \qquad (7.74)$$

This reveals that at fixed current the electron density increases with the number of turns and with the gas pressure. Indeed, ν_m increases linearly with the gas pressure, and both h_1 and h_{r_0} decrease with the gas pressure. Interestingly, there is no explicit effect of the driving frequency.

The high-pressure, low-frequency limit at high density

In the opposite limit of $\nu \gg \omega$, the resistance R_p is now given by Eq. (7.41). Therefore, when considering the high-density regime with $R_{\text{ind}} = N^2 R_p$, the electron density is

$$n_e = \left[\frac{\pi r_0 N^2 (2 \omega \nu_m)^{1/2} (m/\varepsilon_0)^{1/2}}{4 u_B \left(h_1 \pi r_0^2 + h_{r_0} \pi r_0 l \right) e \varepsilon_T(T_e) l\, c} \right]^{2/3} I_{\text{coil}}^{4/3}. \qquad (7.75)$$

Again at fixed coil current the electron density increases with the number of turns and the gas pressure (with a weaker scaling in pressure). In this high-pressure limit, one can observe a frequency dependence: the electron density increases slightly with the frequency at fixed coil current.

Frequency effect

The scalings derived at high electron density show that the driving frequency plays a very minor role in the physics of the inductive mode. This is very different from what has been demonstrated in capacitive discharges. In order to investigate the effect of frequency in the entire range of electron densities, Figure 7.16 shows the electron density at the equilibrium as a function of the coil current for three different driving frequencies, 4 MHz, 13.56 MHz and finally 60 MHz. The calculation was done with other parameters the same as in Figure 7.15. The scalings predicted above are verified: the frequency has almost no effect at high electron density, where the inductive discharge really works as a transformer. By contrast, the effect of frequency is more marked in the capacitive mode. The capacitive coupling is drastically reduced at low frequency, as might be expected from the two previous chapters.

What experiments show

First of all, many experiments have shown that at moderate power (or density), the coil current does not vary with power or may even decrease when the power (or the density) increases [129]. This is in apparent contradiction with the scaling derived

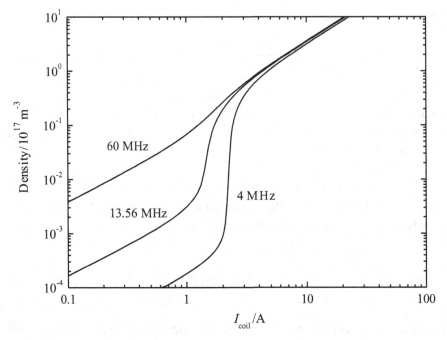

Figure 7.16 Electron density at the equilibrium as a function of the coil current for three different driving frequencies.

above. In fact, Eqs (7.74) and (7.75) describe the large current limit of Figure 7.15. In this regime, the RF current in the coil increases with the power delivered by the power supply. It turns out that many experiments actually operate in the regime corresponding to the grey area in Figure 7.15. When capacitive coupling is ignored (the dashed line in the figure), the electron density increases drastically (because the power increases) but the RF current remains almost unchanged. In fact, in this regime $R_{\text{ind}} \propto n_e$, as does the power loss, so that the power balance requires that the RF current is independent of the electron density, as observed in experiments.

7.6.2 Power transfer efficiency

When capacitive power coupling is included, the power transfer efficiency becomes

$$\zeta = \frac{P_{\text{abs}}}{P_{\text{coil}} + P_{\text{abs}}} \approx \frac{R_{\text{ind}} + (\omega^2 L_s C)^2 R_{\text{cap}}}{R_{\text{coil}} + R_{\text{ind}} + (\omega^2 L_s C)^2 R_{\text{cap}}}. \tag{7.76}$$

The power transfer efficiency is plotted in Figure 7.17 for a 1.33 Pa discharge, with the same conditions as before. The coil inductance is $L_{\text{coil}} = 2.1\,\mu\text{H}$, the coil resistance is $R_{\text{coil}} = 0.137\,\Omega$, the angular frequency is $\omega = 2\pi \times 13.56\,\text{MHz}$, and

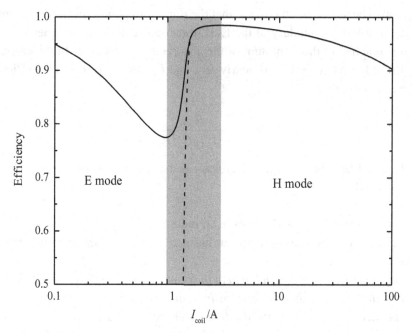

Figure 7.17 Power transfer efficiency as a function of the coil current for the inductive discharge with capacitive coupling. The dashed line is the efficiency when capacitive coupling is reduced to zero. The region in grey, between 1 A and 3 A, corresponds to the E–H transition region.

$Q \approx 1300$ and $r_0 = 0.81\, r_c$. Again marked in grey is the region of E–H transition, the region of capacitive coupling being to the left (at low RF current) and the region of inductive coupling to the right. The maximum power transfer efficiency occurs at the beginning of the inductive mode and thereafter decreases at higher RF current (or RF power). The maximum efficiency occurs when the equilibrium is reached at the maximum of the inductive discharge resistance R_{ind}. The maximum efficiency predicted by Eq. (7.65) is $\zeta_m = 0.988$, in good agreement with the value observed in Figure 7.17.

Comment: The efficiency calculated above is unrealistically high, when compared with experiments, mostly because the Q-factor has been overestimated. In reality, R_{coil} is larger than the estimations presented here because of so-called proximity effects between the turns of coil. More realistic values of Q lie between 100 and 300. Taking $Q = 200$ leads to a maximum efficiency of $\zeta_m = 0.927$, much closer to experimentally measured values. The general form of the dependence of the efficiency on the electron density remains similar and is observed in experiments.

It can also be seen from the figure that the minimum power transfer efficiency is located at the lower boundary of the E–H transition region, that is when the equilibrium is reached at the minimum of the absorbed power curve. In the capacitive mode, the inductive resistance is nearly zero and $L_{\text{ind}} \approx L_{\text{coil}}$, so that the efficiency may be approximated by

$$\zeta \approx \left[1 + \frac{R_{\text{coil}}}{(\omega^2 L_{\text{coil}} C)^2 R_{\text{cap}}}\right]^{-1}. \tag{7.77}$$

Since $R_{\text{cap}} \propto 1/n_{\text{e}}$, the efficiency decreases as the electron density increases in the capacitive mode.

Q Is any power dissipated by ions in an inductive discharge?
A There must indeed be some power dissipated by ions in the sheath in front of the coil.
Comment: This was not included in the power transfer efficiency calculations presented here, mainly because it is relatively small compared with that in capacitive discharges. Note also that the energy of ions arriving at a substrate can be controlled by including a third electrode that is biased by, for instance, an independent RF power supply (cf. Chapter 4).

Q Is there any need for a match-box in inductive discharges?
A A match-box is indeed required since the plasma impedance is not 50 Ω. The match-box can be incorporated in the circuit, in the same way as for capacitive discharges.
Comment: Losses in the match-box can be large, in particular when the coupling efficiency is small.

7.7 Summary of important results

- Inductive discharges are generated by RF current in a coil separated from the plasma by a dielectric window. They can be modelled starting from Maxwell's equation to calculate the electromagnetic fields, leading to an equivalent circuit model based on the Poynting theorem. They are more commonly modelled as a transformer, in which the plasma loop current is the secondary.
- Inductive discharges may have a very high power transfer efficiency, specifically when $\nu_{\text{m}} \gg \omega$. In the limit of $\nu_{\text{m}} \ll \omega$, the efficiency is reduced because the reactive power is larger and consequently higher coil currents (leading to more losses) are required to maintain the same plasma density. For high

coupling efficiency, the distance between the plasma and the coil, i.e., the thickness of the dielectric window, must be small.
- Although inductive discharges are designed to excite the electromagnetic H-mode, they may operate in the E-mode at low RF current (or power). Consequently, they are subject to E to H-mode transitions. These transitions are more pronounced than in very high frequency capacitive discharges.
- The effect of the driving frequency is not important at high electron density, when the inductive discharge works as a transformer. However, the frequency has a strong effect on capacitive coupling.
- In principle, ICPs allow ion energy and ion flux to be varied quasi-independently, because the plasma is generated by a coil and the substrate holder is biased independently.

7.8 Further considerations

The discussion has so far missed out some very important aspects of inductive discharges which are considered in this last section. Technological aspects will be dealt with first and a discussion of more subtle physical mechanisms will conclude this chapter.

7.8.1 Strategies to minimize capacitive coupling

The reduction of capacitive coupling is useful for many reasons. It may avoid instabilities at the E–H transition (see Chapter 9) and it reduces the sputtering of the dielectric window by ions accelerated across the sheaths. From a scientific point of view, it reduces the RF fluctuations of the plasma potential, which complicates electrical diagnostics of the discharge.

There are several ways to reduce the capacitive coupling. One may drive the coil at lower frequency. An alternative is to introduce a capacitor between the coil and the ground, as shown in Figure 7.18. In the figure, the coil inductance has been artificially divided into two parts for the sake of the demonstration. One can see that the resistance of the coil and that due to plasma load have been ignored to simplify the argument, and because they do not dominate the total impedance. The voltage that develops across the capacitor introduced between one end of the coil and the ground is 180° out-of-phase with the voltage that develops across the coil itself. Therefore, if the capacitor is chosen so that

$$L_{\text{coil}} C \omega^2 = 2, \tag{7.78}$$

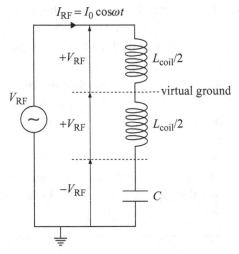

Figure 7.18 Schematic of a design with a capacitor between the coil and the ground. The coil inductance has been artificially divided into two parts for the sake of the demonstration.

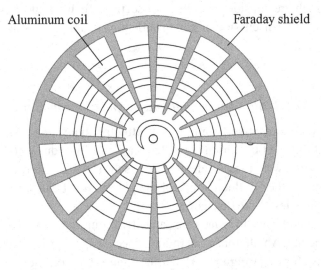

Figure 7.19 View of a Faraday shield for a planar coil, seen from the side of the dielectric window. From [130].

then there is a voltage node (a virtual ground) in the centre of the coil, as shown in the figure. Consequently, for the same current in the coil, the voltage at both ends of the coil is half that in the absence of a capacitor.

Another classical way of reducing capacitive coupling is to introduce a 'Faraday shield' between the coil and the plasma, as seen in Figure 7.19. The rôle of this

7.8 Further considerations

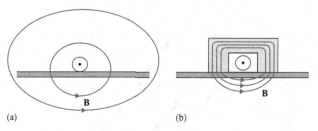

Figure 7.20 Schematic diagram describing the effect of the use of a ferromagnetic core on the magnetic field lines. (a) Large RF field loop created by the coil alone, (b) RF field channelled by the ferromagnetic core. From [131].

shield is to localize the electrostatic field between the coil and the shield, without affecting the induced electromagnetic field. The shield is grounded to provide a path for the capacitive current that does not pass through the plasma. The spacing and openings in the shield shown in the figure are designed to prevent the circulation of azimuthal RF current. The induced electromagnetic field is consequently weakly affected. Faraday shields are very efficient in reducing the capacitive coupling. Actually, in plasma processing, a strong reduction of capacitive coupling may be problematic because discharge ignition then becomes almost impossible.

7.8.2 Enhanced inductive coupling with ferromagnetic cores

The coupling efficiency of inductive discharges may be increased by using ferromagnetic cores. While electrical transformers have ferromagnetic cores and operate at low frequency (industrial or audio frequencies), conventional inductive discharges used in plasma processing usually do not have them and operate at higher frequency (typically 13.56 MHz). In contrast, many compact fluorescent RF lamps are based on ICPs with internally located coils, enhanced by ferrite cores. These operate at 2.65 MHz, a frequency especially allocated for lighting.

Ferromagnetic cores have a high magnetic permeability, which acts as a magnetic flux concentrator, as shown in Figure 7.20, reproduced from [131]. The ferromagnetic core concentrates magnetic flux where it enters the plasma load. In terms of the transformer analysis developed in this chapter, the mutual inductance between the coil and the plasma is thereby increased. The transformer model has been revisited by Lloyd *et al.* [132] to include ferromagnetic core enhancement. The main effect is a reduction of the coil current, and consequently of the losses, as shown in Figure 7.21. The relative inductor loss with a ferrite core is an order of magnitude lower and, at large discharge power, the power transfer efficiency reaches 99%.

The benefit of ferromagnetic cores is mostly visible at low and moderate frequencies (typically below 4 MHz), because the magnetic permeability of these materials

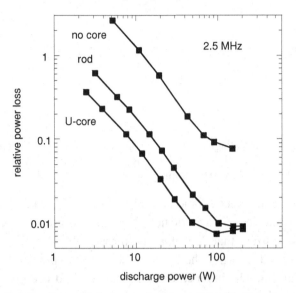

Figure 7.21 Relative inductor loss as a function of the discharge power with and without ferrite core. From [133].

drops at higher frequencies. The trend in inductive discharges, unlike in capacitive discharges, may therefore be frequency reduction rather than frequency increase. This has the great advantage of avoiding the standing wave effects described in the previous chapter.

7.8.3 Anomalous skin depth and collisionless heating

As in capacitive discharges, there is a regime at low pressure in ICPs where ohmic power absorption is not the dominant process. In capacitive discharges, electrons pass within the RF sheath in a time that is shorter than the period of the sheath motion. Similarly here, the induced electric field is localized in a skin layer, so if an electron traverses this layer in a time that is shorter than the RF period, then it will gain net energy from the field. The related condition is [134, 135]:

$$\omega \delta \leq \left(\frac{k_B T_e}{m_e} \right)^{1/2}. \tag{7.79}$$

Under such conditions, a change in the electric field at one particular location, and time, affects the current everywhere in the plasma, and at all later times, because of the rapid thermal motion of electrons. Consequently, the current density in the plasma *is not* related to the electric field by the usual law, i.e., $\tilde{J}_\theta \neq i\omega\varepsilon_0\varepsilon_p \tilde{E}_\theta$. The spatial distribution of the electric field and of the RF current is not the usual exponential decay within the traditional skin depth, and the phase between these

7.8 Further considerations

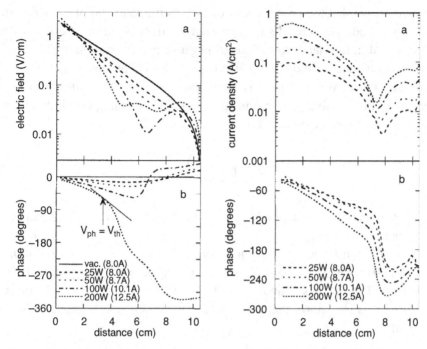

Figure 7.22 Spatial distribution of the electric field (left) and the RF current density (right) in a 0.133 Pa ICP in argon. Also shown is the phase distribution (the phase reference here is that of electric field in vacuum).

two quantities may virtually take any value. This 'anomalous skin effect' implies a variety of interesting phenomena such as collisionless heating or negative power absorption, which have been studied intensively in the late 1990s and more recently. A good review of these effects has been published by Godyak [136].

In Figure 7.22 the spatial distributions of the electric field and the RF current density are shown, as measured by Godyak and Piejak [137, 138] in a 0.133 Pa inductive discharge in argon. The measurement method involved mapping out the magnetic field changes detected by a small loop of wire. This experiment used a planar ICP (see Figure 7.1b). As expected from earlier discussion of ICPs, the electric field decays exponentially away from the dielectric window in the absence of plasma; i.e., there is a geometric decay length of the field. When a plasma is struck, the decay length becomes shorter, as expected from the classical plasma skin effect. However, at powers above 100 W, the electric field decays rapidly to reach a first minimum (a little after 6 cm for 100 W), and then increases again on going deeper into the plasma. This is a typical signature of anomalous field penetration, also observed by Cunge et al. [139] and recently calculated by Hagelaar using a fluid approach [140]. The RF current (shown on the right) also exhibits an anomalous structure, with different positions for the minima and different phase evolution. It therefore appears that the propagations of the electric field and of the RF current are not correlated, which results in somewhat arbitrary phases between these two quantities, and in turn leads to regions of negative power absorption [141] which means electromagnetic fields take power out of the local plasma.

Although the field penetration is anomalous, and therefore the decay is not exponential, it is possible to define an anomalous skin depth, which is an effective penetration length, given by [2, 135]

$$\delta_{\text{eff}} \approx \left(\frac{\bar{v}_e c^2}{\omega \omega_{\text{pe}}^2} \right)^{1/3}. \quad (7.80)$$

Q Is the anomalous skin depth larger or smaller than the collisionless skin depth defined as $\delta \equiv c/\omega_{\text{pe}}$? Show that it depends on the driving frequency.

A The condition for $\delta_{\text{eff}} \leq \delta$ is equivalent to $\bar{v}_e/\omega \leq c/\omega_{\text{pe}}$. For excitation at 13.56 MHz $\omega = 8.5 \times 10^7$ s^{-1}, a plasma having $n_e = 10^{17}$ m^{-3} and $k_B T_e/e = 3$ V (so $\bar{v}_e \approx 10^6$ m s^{-1} and $\omega_{\text{pe}} \approx 1.8 \times 10^{10}$ s^{-1}) has $\delta_{\text{eff}} \leq \delta$. Doing the full calculation shows that $\delta = 1.67$ cm while $\delta_{\text{eff}} = 1.48$ cm, a small but significant difference.

The anomalous field penetration regime is often a regime in which collisionless heating is dominant (this may not be true at very low frequency, where the field penetration may be anomalous and yet collisionless heating may be negligible). This has been demonstrated experimentally by Godyak *et al.* [142], who measured the power deposited to the plasma and compared it with the calculated ohmic power. It appears that at low pressure, the collisionless power may easily be an order of magnitude larger than the ohmic power. A convenient way of accounting for collisionless heating is to define an effective collision frequency due to stochastic interaction between the plasma electrons and the skin depth electric field. This stochastic frequency has been evaluated by Lieberman and Lichtenberg [2], and is

$$\nu_{\text{stoc}} \approx \frac{\bar{v}_e}{4\delta_{\text{eff}}}. \quad (7.81)$$

It is then easy to define an effective collision frequency, $\nu_{\text{eff}} = \nu_{\text{stoc}} + \nu_m$. The high electron density plasma resistance under these circumstances becomes

$$R_p = \frac{\pi r_0}{\sigma_{\text{eff}} l \delta_{\text{eff}}}, \quad (7.82)$$

where $\sigma_{\text{eff}} \equiv n_e e^2 / m_e \nu_{\text{eff}}$ is the new expression for the conductivity. Finally, for completeness, it is noted that the shape of the electron energy distribution function can be distorted profoundly by collisionless heating in inductive discharges [143].

7.8.4 Non-linear effects

The final topic in this chapter concerns the non-linear effects that occur in inductive discharges. The non-linearities mostly come from the Lorentz force on the electron fluid, which is produced by the RF magnetic field. Momentum conservation for electrons has to include this force and becomes

$$nm \left[\frac{\partial \mathbf{u}}{\partial t} + (\mathbf{u} \cdot \nabla) \mathbf{u} \right] = nq \left(\mathbf{E} + \mathbf{u} \times \mathbf{B} \right) - \nabla p - m\mathbf{u} \left[n\nu_m + S - L \right]. \quad (7.83)$$

It is clear that the Lorentz force, which is proportional to the product of the electron drift velocity and the magnetic field, $F_L \propto \mathbf{u} \times \mathbf{B}$, introduces a non-linear response which manifests itself by the generation of second-harmonic currents, and a DC component known as the ponderomotive force. It has also been suggested that the RF-induced field reduces the plasma conductivity leading to a non-linear skin depth [144], though it was shown later that this effect does not in fact exist [145].

> Q How many different skin depths have been introduced in this chapter?
> A Many! The classical skin depth has two expressions depending on pressure: Eq. (2.57) at low pressure and Eq. (2.58) at high pressure. Then, at low pressure, non-local (collisionless) effects lead to anomalous skin penetration given by Eq. (7.80). Finally, one should also note that even in the absence of plasma, the electric field always decays away from the antenna due to geometric effects.

It has been shown that the second-harmonic currents do not contribute significantly to the electron heating. The DC ponderomotive force has a considerable effect on the plasma density profile. This force pushes cold electrons away from the skin layer. Since it acts differently on cold and hot electrons, the EEDF is affected; the EEDF tends to be depleted of cold electrons within the skin layer.

These non-linear effects become important at low pressure and at low driving frequency, when the Lorentz force becomes comparable to the electric force. Godyak [136] showed that for a flat coil, the main component of the magnetic field is directed radially while the RF drift velocity is directed azimuthally: $B_r = -E_\theta/\delta\omega$ and $u_\theta \approx eE_\theta/m(\omega^2 + \nu_{\text{eff}}^2)^{1/2}$. Consequently, the Lorentz force, which is proportional to the product $B_r u_\theta$, scales in the following way: $F_L \propto E_\theta^2/\delta\omega(\omega^2 + \nu_{\text{eff}}^2)^{1/2}$. Since the electric field is only a weak function of the driving frequency and the gas pressure, the Lorentz force increases when these quantities decrease. Typically, non-linear effects are not significant for 13.56 MHz discharges at a few Pa. However, they are considerable at the lower frequency of 0.45 MHz [136].

8
Helicon plasmas

Adding a static magnetic field to an RF-excited plasma has two major consequences. Firstly, the plasma transport is reduced in the direction perpendicular to the magnetic field lines; this will be discussed in the next chapter. It will be shown that the magnetic field reduces the transverse plasma flux and may therefore be used to increase the plasma density at given power. More generally, the addition of a static magnetic field can be used to adjust the uniformity of the plasma flux, and to modify the electron temperature or the electron energy distribution function. This is achieved by changing the magnetic field topology. Some of these properties are used in magnetically enhanced reactive ion etching (MERIE) reactors, which are capacitively coupled reactors with a magnetic field parallel to the electrodes. In some instances, this magnetic field is designed to rotate at low speed in order to average out modest asymmetries of the plasma parameters.

Secondly, a static magnetic field enables the propagation of electromagnetic waves at low frequencies, that is at $\omega \ll \omega_{pe}$; a class of such waves, known as 'helicons', are of particular importance in plasma processing and in space plasma propulsion. Helicons are part of a bigger group of waves called 'whistlers'. The first report of whistlers, that is whistling tones descending in frequency from kilohertz to hundreds of hertz in a few seconds, was in the early twentieth century. A possible origin of these signals was given later when the first theories of propagating waves in a magnetized plasma were proposed by Hartree [146] and Appleton [147]. The atmospheric lightning flashes at one location of the Earth generates a localized impulse of electromagnetic disturbance. A broad spectrum of electromagnetic waves subsequently propagates along the Earth's magnetic field lines at a speed that depends on the frequency (lower frequencies propagate at slower speed). The signals received at the other end of the magnetic field line arrive over a period of a few seconds and, when converted into acoustic waves, mimic a descending whistle as lower-frequency components arrive later.

The word helicon was first proposed by a French scientist, Aigrain, to describe an electromagnetic wave propagating in the free electron plasma within a solid metal [148]. The propagation of equivalent waves in gaseous plasmas was subsequently studied in the 1960s. In 1970, Boswell proposed using them as a source of energy to sustain a plasma [149]. The content of this chapter is largely inspired by a review published in two papers by Boswell and Chen [150] and Chen and Boswell [151], charting the subsequent development of the field. The name 'helicon' comes from the fact that the wave rotates during its propagation in the z-direction, carrying the electrons in a helical motion. The electric and magnetic fields of the wave have the following form:

$$E, B \sim \exp j\,(\omega t - k_z z - m\varphi) \qquad (8.1)$$

where m is the azimuthal mode number, k_z is the longitudinal wavenumber, and φ is the azimuthal angle. Throughout the analysis in this chapter, the static magnetic field B_0 is along z. The radial structure of the helicon wave fields will be discussed in Section 8.2.2.

In a plasma produced by a helicon wave, energy is transferred from it to the plasma electrons to produce heating by collisional or collisionless mechanisms. The propagating character of the wave implies that heating penetrates deeper in the plasma than inductive heating (localized in the skin depth) or capacitive heating (mostly localized in the RF sheaths). This achieves high ionization efficiency in large plasma volumes and/or long plasma columns. Since the antenna is excited by an RF voltage, helicon plasmas may also operate in capacitive (E) mode at low power. In addition, the RF current flowing in the antenna induces fields near the antenna that tend to excite an inductive (H) mode – cf. Chapter 7. The H-mode usually dominates at intermediate power. The plasma eventually operates in the W-mode (where W signifies the propagating helicon wave mode) when the power is large enough to provide the required plasma density to support helicon wave propagation. Therefore, helicon plasmas are subject to E–H–W transitions. Further mode transitions are also observed within the W-mode, because of resonant coupling to the antenna. All these phenomena lead to abrupt variations of the electron density with the input power, which is inconvenient for plasma processing, unless the source is appropriately designed to control these mode jumps.

Q Capacitive discharges are driven by parallel plates while inductive discharges are driven by coils. Explain why the word 'antenna' is used here.

A An antenna has a specific design in order to launch propagating electromagnetic waves. The shape can be chosen to select specific wavelengths and modes.

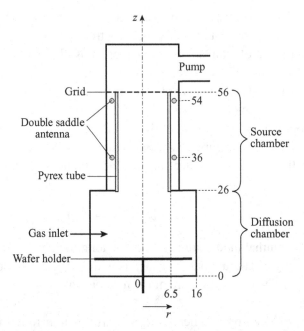

Figure 8.1 A schematic of a helicon plasma processing reactor where the source tube, surrounded by the helicon antenna, sits on top of an expansion chamber at the bottom of which is placed a wafer holder. The numbers are radial and axial distance markers in cm.

The combination of efficient wave heating and increased plasma confinement make helicon reactors attractive for highly ionized plasmas [152,153], with applications in plasma thrusters [15–17]. Helicon reactors have also been used in various plasma processing applications, such as silicon dioxide deposition [154] or fast etching of silicon [155] and silicon carbide [9, 11]. A typical configuration of a helicon processing reactor is shown in Figure 8.1. The source tube, surrounded by a helicon antenna, sits on top of an expansion chamber at the bottom of which a wafer holder is placed. When the amplitude of the static magnetic field (along the z-axis) is maximum in the source region, the plasma is mostly generated in that region and expands into the bottom chamber with a strong decay of the plasma density in the diverging magnetic field. The wafer-holder area exposed to the expanding plasma may be significantly larger than the cross-section of the source tube. The uniformity across the wafer may also be adjusted by varying the magnetic field shape and/or by adding a multipole confinement based on arrays of strong permanent magnets. Alternatively, the magnetic field amplitude may be kept constant along z, or even larger in the bottom chamber to concentrate the field lines, in which case the plasma density is very high at the wafer-holder position. This, however, is at the expense of uniformity. In the case of a diverging magnetic field, it has been found that the plasma expansion may be associated with complex transport phenomena including double-layer formation and instabilities. This will be described in the next chapter.

In this chapter, the properties of helicon waves will be studied in a growing level of complexity. The propagation parallel to the static field B_0 is first studied in an infinite plasma with uniform density. The general dispersion relation is given, various waves are described and helicon waves are defined. These waves are compared to the simple case of electromagnetic waves in non-magnetized plasmas, discussed in Chapter 2. Off-axis propagation and waves contained in a cylinder are then studied, leading to boundary conditions and eigenvalues for the wavenumbers; the assumption of uniform plasma density is retained and its validity is discussed in a dedicated section. The antenna coupling is treated in a simple and idealized way that then allows the conditions for the existence of helicon modes to be derived. Having defined the wave properties, the absorption of wave energy is discussed, leading to wave heating of the electron population. The E–H–W transitions are described at the end of the chapter.

Warning:

- The word 'mode' is used to describe several phenomena in this chapter. Careful attention should be paid to the context in which this word is used. The wave rotates as it propagates, introducing an azimuthal structure to the wave fields described by a mode number m. There are transitions between the various energy coupling mechanisms: the capacitive (E) mode, the inductive (H) mode, and the helicon (W) modes. Finally, resonant coupling between the antenna and the helicon wave occurs at discrete longitudinal wavelengths (or equivalently densities). Therefore, within the W-mode, there are several longitudinal wave modes described by the mode number χ.
- Since this chapter is about waves, Boltzmann's constant is written k_B so that k can be used as a wavenumber.
- Since m is used as an azimuthal mode number, the electron mass will be written m_e to avoid any confusion.

Unlike in previous chapters, a global model of helicon plasma processing reactors will not be developed. The reason for this is twofold: (i) the power balance is complicated and the heating of plasma electrons by helicon waves is still not fully understood, (ii) the particle balance is complicated because of the geometry of the reactor which involves complex transport phenomena. A global model has been proposed by Lieberman and Boswell [156], in which collisionless power absorption was ignored and the geometry was simplified.

8.1 Parallel propagation in an infinite plasma

The subject of waves in magnetized plasmas (see, for example, Stix [157]) will not be treated here in detail. The principal aim is to obtain the electromagnetic modes in a magnetized plasma, following the approach in Chapter 2 for the simpler non-magnetized case. The linearization of the fluid equations for electrons and ions leads to an expression for plasma permittivity. When a static magnetic field is applied to the plasma, the response of the medium to the fields becomes anisotropic and one has to define a *plasma permittivity tensor*. The dispersion of electromagnetic waves then depends on the direction of the wave vector with respect to the magnetic field. The anisotropy comes from the Lorentz force that acts in a direction perpendicular to the particle motion. Charged particles therefore tend to rotate around magnetic field lines at their respective cyclotron frequencies, defined by $\omega_{ce} \equiv eB_0/m_e$ (electron cyclotron frequency) and $\omega_{ci} \equiv eB_0/M$ (ion cyclotron frequency). Note that ions rotate at a much slower frequency since $\omega_{ci}/\omega_{ce} = m_e/M \ll 1$.

Q Calculate the cyclotron frequency for electrons and for argon ions in a field of 5 mT.
A $\omega_{ce} = 8.9 \times 10^8 \, \text{s}^{-1}$ and $\omega_{ci} = 1.2 \times 10^4 \, \text{s}^{-1}$.

To account for the anisotropy, it is convenient to divide the waves into two classes: waves propagating along the magnetic field, and waves propagating perpendicular to the magnetic field. Helicon waves propagate mostly along the magnetic field lines. Suppose the magnetic field is along the z-axis. For the collisionless (low-pressure, non-resistive) case with the wave vector parallel to the direction of the static field B_0 (so that the wavenumber is $k \equiv k_z$), considering the cold plasma approximation ($T_e = T_i = 0$) and neglecting terms of order m_e/M compared to unity, it turns out that there are two types of waves having the following dispersion relations:

$$n_{\text{ref},R}^2 = \frac{k^2 c^2}{\omega^2} = 1 + \frac{\omega_{pe}^2}{\omega \omega_{ce} \left(1 + \frac{\omega_{ci}}{\omega} - \frac{\omega}{\omega_{ce}}\right)}, \tag{8.2}$$

$$n_{\text{ref},L}^2 = \frac{k^2 c^2}{\omega^2} = 1 - \frac{\omega_{pe}^2}{\omega \omega_{ce} \left(1 - \frac{\omega_{ci}}{\omega} + \frac{\omega}{\omega_{ce}}\right)}. \tag{8.3}$$

The first wave (dispersion relation given by Eq. (8.2)) is called the right-hand polarized (RHP) wave, since the wave electric field rotates clockwise when the static magnetic field B_0 is seen from behind. The second wave (dispersion relation given by Eq. (8.3)) is called the left-hand polarized (LHP) wave, since the wave electric

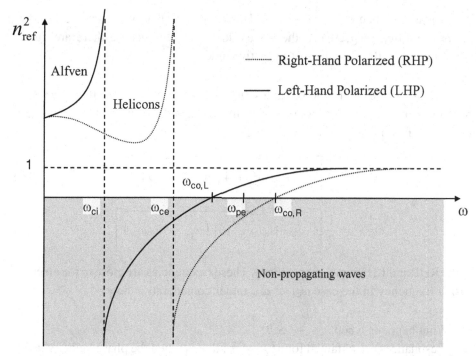

Figure 8.2 Square of the refractive index as a function of frequency for the LHP and RHP waves propagating parallel to the direction of B_0 in an infinite magnetized plasma; $\omega_{ce} \ll \omega_{pe}$.

field rotates anti-clockwise when the static magnetic field B_0 is seen from behind. The dispersion diagram is sketched in Figure 8.2 for the case where $\omega_{ce} \ll \omega_{pe}$.

Q In Chapter 2, the plasma permittivity, ε_p, and the associated plasma refractive index, $n_{ref} = \sqrt{\varepsilon_p}$, were obtained in the case of a non-magnetized plasma. It was found that the plasma refractive index was isotropic (independent of the direction in space) and that it was given by $n_{ref}^2 = 1 - \omega_{pe}^2/\omega^2$ (when dissipations due to electron–neutral collisions are neglected). Subsequently, ω_{pe} was shown to be a cut-off frequency since waves cannot propagate for $\omega < \omega_{pe}$. First, check that the non-magnetized case is recovered by setting $B_0 = 0$ in Eqs (8.2) and (8.3). Second, identify cut-off frequencies in Figure 8.2. Will waves ever propagate below those frequencies?

A When $B_0 = 0$ in Eqs (8.2) and (8.3), $\omega_{ce} = \omega_{ci} = 0$ and consequently $n_{ref,R}^2 = n_{ref,L}^2 = 1 - \omega_{pe}^2/\omega^2$; the non-magnetized case is recovered. Two cut-off frequencies, $\omega_{co,R}$ and $\omega_{co,L}$, are identified for the RHP wave and the LHP wave, respectively. The RHP wave does not propagate (it becomes evanescent) when $\omega_{ce} \leq \omega \leq \omega_{co,R}$, and similarly the LHP wave does not propagate when $\omega_{ci} \le \omega \le \omega_{co,L}$. However, note that at frequencies below ω_{ce} and ω_{ci}, respectively, the waves do propagate because there are other solutions with real values of refractive index.

Imaginary values of refractive index ($n_{ref}^2 < 0$) indeed indicate that the waves are evanescent, i.e., they are not propagating waves, and the cut-off frequencies are at $n_{ref}^2 = 0$. From Eqs (8.2) and (8.3), the cut-off frequencies are

$$\omega_{co,R} = \frac{1}{2}\left[\omega_{ce} + \sqrt{\omega_{ce}^2 + 4\left(\omega_{pe}^2 + \omega_{ce}\omega_{ci}\right)}\right], \quad (8.4)$$

$$\omega_{co,L} = \frac{1}{2}\left[-\omega_{ce} + \sqrt{\omega_{ce}^2 + 4\left(\omega_{pe}^2 + \omega_{ce}\omega_{ci}\right)}\right], \quad (8.5)$$

for the RHP and LHP wave, respectively. These frequencies are close to the electron plasma frequency in the case $\omega_{ce} \ll \omega_{pe}$ under consideration here.

Q What happens when $n_{ref}^2 \to \infty$?
A Resonances are obtained for $n_{ref}^2 \to \infty$, that is when the phase velocity goes to zero. At the resonance, charged particles rotate at the same frequency as the waves so that they experience a quasi-constant field. This leads to resonant energy absorption.

The electric field for the LHP wave rotates in the same direction as ions around the magnetic field, thus the resonance is at ω_{ci}. By contrast, the electric field for the RHP wave rotates in the same direction as electrons around the magnetic field, and therefore its resonance is at ω_{ce}. As shown on the diagram, the waves become evanescent above their respective resonance frequencies. As ω approaches ω_{pe}, the waves become propagating again and when ω goes to infinity, the phase velocity approaches the speed of light in vacuum (since $n_{ref}^2 \to 1$). In this limit, charged particles do not respond to the wave fields and the medium behaves as a dielectric (eventually as a vacuum when $n_{ref} \approx 1$). Note again that since $n_{ref} \le 1$ the phase speed is greater than the speed of light.

A helicon is in fact the low-frequency RHP wave; we therefore restrict further discussion to Eq. (8.2) in the domain $\omega < \omega_{ce}$. Then taking $\omega_{ce} \ll \omega_{pe}$ and $\omega \ll \omega_{pe}$ gives the following dispersion relation:

$$n_{ref,R}^2 = \frac{\omega_{pe}^2}{\omega\omega_{ce}\left(1 + \frac{\omega_{ci}}{\omega} - \frac{\omega}{\omega_{ce}}\right)}. \quad (8.6)$$

8.1 Parallel propagation in an infinite plasma

In this frequency domain, there are three types of waves that are important in various subfields of plasma physics: space science, magnetic fusion energy science and plasma processing science. The frequency domain can be refined further by examining different dominant terms in Eq. (8.6). From now on, n_{ref} will be used in place of $n_{\text{ref,R}}$ for simplicity.

8.1.1 Alfvèn waves

Considering frequencies below ω_{ci} so that $\omega/\omega_{ce} \ll 1$, the dispersion relation becomes

$$n_{\text{ref}}^2 = \frac{\omega_{pe}^2}{\omega_{ce}(\omega_{ci} + \omega)}. \tag{8.7}$$

At very low frequencies ($\omega \ll \omega_{ci}$) this gives the so-called Alfvèn waves that are non-dispersive since the phase velocity is independent of frequency. The phase velocity is called the Alfvèn velocity and is given by

$$v_\varphi = \frac{c}{n_{\text{ref}}} = v_A = \frac{c}{\omega_{pe}}\sqrt{\omega_{ci}\omega_{ce}} = c\frac{\omega_{ci}}{\omega_{pi}}. \tag{8.8}$$

Alfvèn waves are used in Tokamak reactors for ion heating and they are also observed in the Earth's magnetosphere.

8.1.2 Electron cyclotron waves

For frequencies near ω_{ce}, the dispersion relation with $\omega_{ci}/\omega \ll 1$ becomes

$$n_{\text{ref}}^2 = \frac{\omega_{pe}^2}{\omega\omega_{ce}\left(1 - \dfrac{\omega}{\omega_{ce}}\right)}. \tag{8.9}$$

The index n_{ref}^2 passes through a minimum at $\omega = 0.5\,\omega_{ce}$ (where the phase speed is maximum), as shown in Figure 8.2. The wave nature is different depending on whether $\omega > 0.5\,\omega_{ce}$ or $\omega < 0.5\,\omega_{ce}$ (see [150, 151, 157] for details). The electron cyclotron waves (at $\omega > 0.5\,\omega_{ce}$) are used for electron heating in Tokamaks. At the resonance ($\omega = \omega_{ce}$), the electrons rotate around the magnetic field line in synchronism with the wave electric field and consequently experience a quasi-constant field that accelerates them over many cyclotron orbits. This results in a very efficient resonant heating. The ECR (electron cyclotron resonance) is also used in some plasma processing reactors. The excitation frequency used is usually 2.45 GHz, which requires a magnetic field of 0.0875 T for resonance. From a plasma processing point of view, there are several drawbacks with ECRs. Firstly, the required static magnetic field is rather high. Secondly, the electron energy distribution function becomes anisotropic with a tail of very energetic electrons (those efficiently interacting with the wave). These energetic electrons are problematic in etching because they induce charge-effect damages in the microelectronic 'ultra large scale integrated' circuits. Finally, the operating pressure must remain fairly low because collisions inhibit the resonance.

> **Q** The electron–neutral elastic collision for an argon plasma having an electron temperature of 5 eV is about $\nu_m \simeq 1.5 \times 10^{-13} \times n_g\,\mathrm{s}^{-1}$, where n_g is the density of neutrals. What is the condition on gas pressure for efficient ECR operation?
>
> **A** The pressure must be such that the electron–neutral collision frequency is much smaller than the electron cyclotron frequency, $\nu_m \ll \omega_{ce}/2\pi = 2.45\,\mathrm{GHz}$. Then, the condition is $n_g \ll 1.6 \times 10^{22}\,\mathrm{m}^{-3}$; the related condition on pressure is $p \ll 60\,\mathrm{Pa}$ at 300 K. Typically, plasma processing ECR reactors operate at $p \leq 1\,\mathrm{Pa}$.

8.1.3 Helicon waves

Helicon waves are at the low-frequency limit of electron cyclotron waves (i.e., $\omega < 0.5\,\omega_{ce}$). The frequency is sufficiently high so that ions do not respond to the field, and sufficiently low so that electron inertia is small, i.e., $\omega_{ci} \ll \omega \ll \omega_{ce}$. The dispersion relation of helicon waves is given by

$$n_{\mathrm{ref}}^2 = \frac{\omega_{pe}^2}{\omega \omega_{ce}}. \tag{8.10}$$

Helicon reactors are designed to allow helicon wave propagation with a frequency of 13.56 MHz and the typical conditions of operation in argon are $n_e = 10^{18}\,\mathrm{m}^{-3}$ and $B_0 = 5\,\mathrm{mT}$. The important frequencies are consequently

$$\omega_{ci} = 1.2 \times 10^4\,\mathrm{s}^{-1},$$
$$\omega = 8.5 \times 10^7\,\mathrm{s}^{-1},$$
$$\omega_{ce} = 8.9 \times 10^8\,\mathrm{s}^{-1},$$
$$\omega_{pe} = 5.7 \times 10^{10}\,\mathrm{s}^{-1},$$

such that the conditions $\omega_{ci} \ll \omega \ll \omega_{ce} \ll \omega_{pe}$ are satisfied.

8.2 Helicon wave propagation in a cylinder

In a helicon reactor, the plasma is spatially limited and the wave propagation is more complex than has been described above. The boundary conditions impose

8.2 Helicon wave propagation in a cylinder

both standing waves and off-axis propagation, that is the wave does not propagate parallel to the direction of B_0. In this section the wave propagation is analysed at an angle $\theta \neq 0$ relative to the magnetic field. The boundary conditions imposed on the fields, and the way the waves are launched by various types of antenna, are also discussed.

8.2.1 Off-axis propagation in an infinite plasma

The dispersion relation for helicon waves propagating at an angle θ relative to the magnetic field in an infinite magnetized plasma [150] is

$$n_{\text{ref}}^2 = \frac{\omega_{\text{pe}}^2}{\omega(\omega_{\text{ce}}\cos\theta - \omega)}. \tag{8.11}$$

Note that with $\theta = 0$ and $\omega \ll \omega_{\text{ce}}$ the dispersion relation established earlier (see Eq. (8.10)) is recovered. It appears that when $\theta \neq 0$ there exists a limiting angle for propagation, defined as follows:

$$\theta_{\text{res}} = \arccos\left(\frac{\omega}{\omega_{\text{ce}}}\right), \tag{8.12}$$

at which a resonance occurs (since $n_{\text{ref}}^2 \to \infty$). The wave vector is therefore restricted to a cone of angles $\theta < \theta_{\text{res}}$. Using the above example with a driving frequency of 13.56 MHz and an electron cyclotron frequency of $\omega_{\text{ce}} = 8.9 \times 10^8 \text{ s}^{-1}$ defines a phase velocity resonance cone at $\theta_{\text{res}} = 1.4748$ rad (or equivalently $\theta_{\text{res}} = 84.5°$). The refractive index is plotted as a function of θ in Figure 8.3. As θ approaches θ_{res} the refractive index goes to infinity (the resonance). Above θ_{res} the index becomes imaginary and the waves are evanescent. However, as discussed below, another condition constrains the wave energy to an even smaller angle.

The energy flow in a lossless medium propagates along the group velocity vector, which does not necessarily coincide with the direction of the wave vector [157]. The direction of the group velocity vector is called the 'ray direction'. As with the wave vector direction, there is a limiting angle for the ray direction, defining a cone in which the wave energy will be restricted (Figure 8.4). To find the relationship between the various angles, first let ψ be the angle between the ray direction and the static magnetic field. Stix ([157], chapter 4) shows that the angle between the wave vector direction and the ray direction, denoted α, is defined by

$$\tan\alpha = \frac{1}{n_{\text{ref}}}\frac{\partial n_{\text{ref}}}{\partial \theta}. \tag{8.13}$$

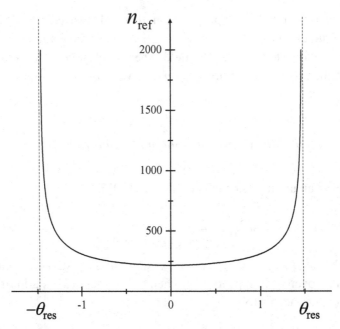

Figure 8.3 Refractive index as a function of θ in radians; the propagation is restricted within the phase velocity resonance cone of angle θ_{res}. The following conditions were used: $f = 13.56\,\text{MHz}$, $B_0 = 0.005$ tesla and $n_e = 10^{18}\,\text{m}^{-3}$.

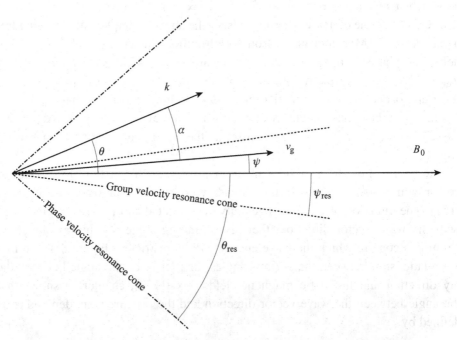

Figure 8.4 Schematic representation of the wave and group velocity vectors, limited in their respective resonance cones, of angle θ_{res} for the phase velocity and ψ_{res} for the group velocity.

8.2 Helicon wave propagation in a cylinder

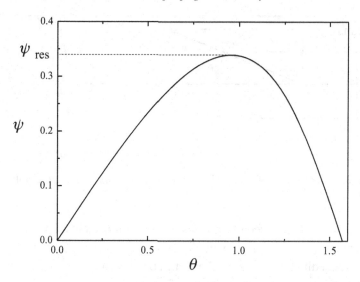

Figure 8.5 The angle between the ray direction (group velocity vector direction) and the magnetic field direction, ψ, as a function of θ. A maximum angle $\psi_{res} \approx 0.33$ is reached when $\theta \approx 0.95$. The angular limitation of the ray direction accounts for the tendency of helicon waves to follow the magnetic field lines.

Q Show, with reference to Figure 8.4, that with the approximation $\omega_{ce} \cos \theta \gg \omega$ we get

$$\psi = \theta - \arctan\left[\frac{\tan \theta}{2}\right].$$

A In this limit,

$$n_{ref}^2 = \frac{\omega_{pe}^2}{\omega \omega_{ce} \cos \theta},$$

so that

$$\tan \alpha = \frac{1}{n_{ref}} \frac{\partial n_{ref}}{\partial \theta} = \frac{1}{2} \tan \theta.$$

Using $\psi = \theta - \alpha$ gives the appropriate result.

The ray direction angle ψ is plotted as a function of θ in Figure 8.5. This angle passes through a maximum $\psi_{res} \approx 0.33$ rad (which corresponds to approximately 20°), at $\theta \approx 0.95$ rad. This defines a resonance cone for the group velocity. Therefore, the direction of the wave energy flow is limited to small angles, $\psi \lesssim 20°$. It has indeed been observed that helicon waves (and more generally whistlers) tend to propagate along the magnetic field lines. Figure 8.4 summarizes the above; the

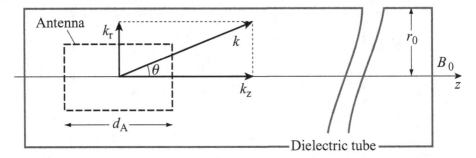

Figure 8.6 Schematic of the helicon antenna around the source tube, with wavenumbers.

wave and group velocity vectors are restricted within their respective resonance cones, of angle θ_{res} for the phase velocity and ψ_{res} for the group velocity.

Before proceeding, note that in the limit considered above, $\omega_{ce}\cos\theta \gg \omega$, the dispersion relation may also be written as follows:

$$kk_z = \frac{e\mu_0 n_e \omega}{B_0}, \tag{8.14}$$

if the electron plasma frequency and the electron cyclotron frequency are expressed as functions of the electron density n_e, and if $k = k_z/\cos\theta$. This form will be used later in the chapter.

8.2.2 Fields and boundary conditions in a cylinder

In a finite system, k and k_z must satisfy boundary conditions on the electromagnetic fields; at a given plasma density, the propagation angle depends on the size of the system. Maxwell's equations have been solved to obtain the radial structure of the helicon wave fields in a cylinder of radius r_0, for a uniform plasma density and a constant static magnetic field along the z-axis. The results are as follows (see for instance [153]):

$$\tilde{B}_r = A\left[(k + k_z) J_{m-1}(k_r r) + (k - k_z) J_{m+1}(k_r r)\right], \tag{8.15}$$

$$\tilde{B}_\varphi = iA\left[(k + k_z) J_{m-1}(k_r r) - (k - k_z) J_{m+1}(k_r r)\right], \tag{8.16}$$

$$\tilde{B}_z = -2iA J_m(k_r r), \tag{8.17}$$

where A is an arbitrary amplitude, m is the azimuthal mode number, k is the magnitude of the wave vector, k_z and k_r are the magnitude of the axial (longitudinal) and radial wave vectors, respectively (see Figure 8.6) and

$$k_r^2 + k_z^2 = k^2; \tag{8.18}$$

8.2 Helicon wave propagation in a cylinder

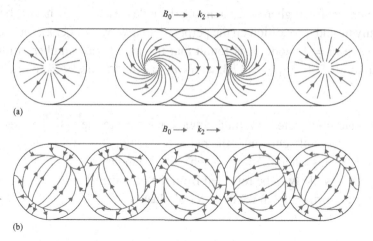

Figure 8.7 Transverse electric field lines as they evolve during the propagation for (a) $m = 0$ and (b) $m = +1$. Reproduced from Lieberman and Lichtenberg [2], after [153].

J_m is the mth-order Bessel function. As shown in Figure 8.7, reproduced from [153], the transverse structure of the electric field is very complicated and depends strongly on the azimuthal mode m. For the $m = 0$ azimuthal mode the field structure evolves during propagation as sketched in Figure 8.7(a). The first pattern, on the left, shows a phase when the electric field lines are purely radial; the field is purely electrostatic. By contrast, at the later phase shown in the third pattern the field lines are circular, indicating that the electric field is purely electromagnetic. In between these two cases, the electric field lines follow spirals. Note that the electrostatic field changes sign between the first pattern (on the left) and the second pattern (on the right), separated by half a wavelength. The pattern of the $m = +1$ mode (Figure 8.7b) is even more complicated, but note that this one does not change form as it propagates, though it does rotate. There is a strong electrostatic radial component in the centre that changes sign at half a wavelength.

The boundary conditions at $r = r_0$ depend on the electrical nature of the chamber walls. An insulator would require the tangential component of the magnetic field to be zero, while at a conducting boundary the tangential electric field is zero, $\tilde{E}_\varphi = 0$. It turns out that in this problem, both conditions are equivalent to $\tilde{B}_r = 0$, which subsequently imposes the following relation [153]:

$$mkJ_m(k_r r_0) + k_z J'_m(k_r r_0) = 0. \qquad (8.19)$$

This sets eigenvalues for k_z and k_r. In this equation J'_m is the first derivative of J_m with respect to its argument. It turns out that for the $m = 0$ mode, the condition defines a unique value for the radial wavenumber, $k_r r_0 = 3.83$, whatever the value of k_z. However, for higher-order azimuthal modes, Eq. (8.19) has to be solved numerically to find a condition on the perpendicular wavenumber k_r, for each value of k_z. For the $m = 1$ mode, it is found that the radial wavenumber is limited to values satisfying $2.4 < k_r r_0 < 3.83$, with $k_r r_0 \approx 3.83$ when $k_z \ll k_r$ and $k_r r_0 \approx 2.4$ when $k_z \gg k_r$.

> **Q** The wave propagates at a finite angle compared to the cylinder axis so that it must eventually encounter the radial boundary, at least for a long cylinder. What happens when the wave reaches the boundary?
> **A** The wave is likely to be reflected from the cylindrical boundary and to then propagate further down the cylinder of plasma.

8.2.3 Non-uniform plasma density

> **Q** Is the assumption of uniform plasma density realistic?
> **A** Probably not. It has been shown in Chapter 3 that in a confined, actively sustained plasma, the density is usually maximum in the centre and decays towards the edges.

The effect of non-uniform plasma density on the fields and on the dispersion of helicon modes was first considered by Blevin and Christiansen [158] and more recently revisited by Chen *et al.* [159] and Breizman and Arefiev [160]. As might be expected, it was found that the wave tends to be guided by the density gradients, even before the boundaries are reached, i.e., the fields are more concentrated in the centre, where the density is peaked. Breizman and Arefiev [160] have also proposed that this wave localization is responsible for enhanced collisional power absorption.

8.2.4 Antenna coupling

Several types of antennas have been used in laboratory experiments and plasma processing reactors. A schematic of the antennas most commonly used for launching helicon waves is presented in Figure 8.8. The simplest is a loop that excites the azimuthally independent $m = 0$ mode. The azimuthal electromagnetic field produced by the one-turn antenna (very similar to that of inductive discharges) then couples to the wave electromagnetic field (the circular pattern in Figure 8.7a). In practice, there is also a quasi-electrostatic field generated at the ends of the one-turn antenna (the ends are not shown in the idealized drawing of Figure 8.8) that may also couple to the radial field of the wave (the radial pattern in Figure 8.7a). The

8.2 Helicon wave propagation in a cylinder

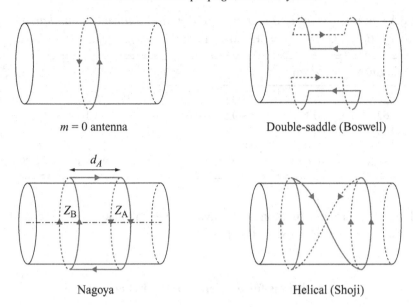

Figure 8.8 Schematic of the most common types of helicon antennas. The top left is a single-loop $m = 0$ antenna whereas the other three antennas are designed to excite the $m = 1$ mode.

three other antennas shown in the figure are designed to excite the $m = \pm 1$ modes. It has been found that the $m = -1$ mode is poorly coupled and does not propagate far out of the forcing region (near the antenna). By contrast, the $m = +1$ mode has excellent coupling and leads to very efficient plasma production. The double-saddle antenna was introduced by Boswell in 1970 [149], while the plane-polarized Nagoya antenna is a simplified scheme introduced in 1978 [161]. A twisted version of this antenna was used by Shoji at Nagoya University in the 1980s [162].

Although antennas are not easy to model in detail, one characteristic can be simply anticipated: the length of the antenna is critical in selecting discrete longitudinal wavelength or wavenumbers k_z. For example, to impose an axial wavelength ($\lambda = 2\pi/k_z$), the $m = 0$ antenna can be excited by a pair of single-loop coils with equal, but oppositely directed currents in each loop, axially separated by half the chosen wavelength. The axial lengths, d_A, of the various $m = 1$ structures in Figure 8.8 are chosen in a similar way to set k_z [2]. Higher-order modes that match antenna length to any odd number of half wavelengths are also possible, so there is a longitudinal mode number, χ, defined by

$$k_z = (2\chi + 1)\frac{\pi}{d_A}. \tag{8.20}$$

As mentioned at the beginning of this chapter, χ is different from m, the azimuthal mode number.

Table 8.1 *Characteristics of the longitudinal χ modes for the azimuthal $m = 0$ mode, with $B_0 = 0.005$ T, $f = 13.56$ MHz and $r_0 = 6.5$ cm*

χ	k_z/m^{-1}	λ_z/m	k_r/m^{-1}	k/m^{-1}	$\theta/°$	n_e/m^{-3}
0	20.9	0.3	58.92	62.53	70.43	3.82×10^{17}
1	62.8	0.1	58.92	86.14	43.16	1.58×10^{18}
2	104.7	0.06	58.92	120.16	29.37	3.67×10^{18}
3	146.6	0.043	58.92	158	21.9	6.76×10^{18}
4	188.8	0.033	58.92	197.49	17.36	1.09×10^{19}

All the necessary information is now available to calculate the characteristics of each longitudinal mode χ excited by the antenna.

8.3 Conditions for existence of the helicon modes

In the following, the characteristics of each mode χ will be calculated for the following reference conditions: a static magnetic field of $B_0 = 0.005$ tesla, a driving frequency of 13.56 MHz, a plasma radius of $r_0 = 0.065$ m and an antenna length of $d_A = 0.15$ m.

The external parameters that are at the disposal of the designer are B_0, ω, r_0 and d_A. The 'unknowns' are k, k_z, k_r, n_e, which can be determined using the dispersion relation Eq. (8.14), the relationship between magnitudes of the wave vectors Eq. (8.18), the cylindrical boundary condition at $r = r_0$ Eq. (8.19) and the wavenumber of the axial mode k_z Eq. (8.20).

8.3.1 The $m = 0$ case

First of all, k_z is imposed by the antenna length, according to Eq. (8.20). For the given parameters, $k_z = 20.9 \text{ m}^{-1}$ for $\chi = 0$, $k_z = 62.8 \text{ m}^{-1}$ for $\chi = 1$, etc. For the $m = 0$ azimuthal mode, the radial wavenumber is independent of k_z and then Eq. (8.19) is satisfied by $k_r r_0 = 3.83$, leading to $k_r = 58.92 \text{ m}^{-1}$. The value of the total k then follows from Eq. (8.18). Finally, the dispersion relation, Eq. (8.14), is used to calculate the electron density. The characteristics of the different χ modes are summarized in Table 8.1. The first helicon mode appears for an electron density of $n_e = 3.82 \times 10^{17} \text{ m}^{-3}$, and higher densities are required for higher-order modes ($\chi \geq 1$). Note the large jump in density between the $\chi = 0$ and $\chi = 1$ modes. For electron densities below $n_e = 3.82 \times 10^{17} \text{ m}^{-3}$, the discharge will most likely work in the inductive H-mode because the conditions for helicon wave propagation are not met. It should, however, be noted that this density depends upon the static

Table 8.2 *Characteristics of the longitudinal χ modes for the azimuthal $m = 1$ mode, with $B_0 = 0.005$ T, $f = 13.56$ MHz and $r_0 = 6.5$ cm*

χ	k_z/m^{-1}	λ_z/m	k_r/m^{-1}	k/m^{-1}	$\theta/°$	n_e/m^{-3}
0	20.9	0.3	53.3	57.2	68.5	3.5×10^{17}
1	62.8	0.1	47.4	78.7	37	1.44×10^{18}
2	104.7	0.06	46.1	114.4	23.7	3.5×10^{18}
3	146.6	0.043	45.7	153.3	17.3	6.6×10^{18}
4	188.8	0.033	45.5	194	13.6	1×10^{19}

magnetic field B_0. If one wants to excite the first helicon mode at lower density, then B_0 should be smaller. On the other hand, to operate at higher density, one should increase B_0. The value of θ given in the table is the angle between the wave vector and the static magnetic field. Note that the wave vector is better aligned with the magnetic field for higher plasma densities (higher χ). As indicated in Section 8.2.1, the wave energy propagates within a different angle because the direction of the group velocity vector is different from that of the wave vector. It was shown that the angle of propagation of the wave energy is always smaller than 20°.

8.3.2 The $m = 1$ case

In this case, the only difference lies in the cylindrical boundary condition Eq. (8.19), which now constrains the relationship between k_z and k_r. Table 8.2 summarizes the characteristics of each mode χ for $m = 1$.

Having defined the condition for helicon wave propagation, it is necessary to consider how the wave energy may be absorbed by the electron population. Efficient electron heating is required for efficient ionization and consequently for efficient plasma production. When the characteristic length for wave absorption is short, the heating efficiency can be described as being particularly high.

8.4 Wave power absorption: heating

It was reported in earlier chapters that in capacitive and inductive discharges at low neutral gas pressure the heating of electrons is predominantly by collisionless mechanisms. The same is true in plasmas sustained by helicon waves at low gas pressure and moderate plasma density (typically the conditions for the $\chi = 0$ mode), since the collision frequencies are too low for efficient ohmic heating of electrons [150, 151]. Before discussing possible heating mechanisms, the characteristic absorption length along the z-axis, α_z, will be calculated for an effective collision frequency ν_{eff} that incorporates all possible dissipations (collisional and collisionless).

Q For a 0.5 m long plasma column (cf. Figure 8.1) to be sustained by the helicon waves, what can be said about the absorption length?
A To produce efficient electron heating, the wave energy must be absorbed in a characteristic length that is comparable to, or shorter than, the experimental arrangement: $\alpha_z \leq 0.5$ m.

Q What would happen if the absorption length were much larger than the system size and the end boundaries did not absorb wave energy?
A A standing wave could become established if the wave were reflected at the ends.
Comment: This has indeed been observed by Boswell [149]; see also [163] for a more recent observation of this phenomenon.

8.4.1 Characteristic absorption length of the wave

To calculate α_z, one needs to include an effective collision frequency, ν_{eff}, in the helicon wave dispersion relation. This can be done by modifying Eq. (8.11) in the same way as in unmagnetized plasmas, giving

$$n_{\text{ref}}^2 = -\frac{\omega_{\text{pe}}^2}{\omega(\omega - \omega_{\text{ce}} \cos\theta - i\nu_{\text{eff}})}. \tag{8.21}$$

The unmagnetized case, given by Eq. (2.52), is easily recovered for $\omega \ll \omega_{\text{pe}}$ by setting $B_0 = 0$, i.e., $\omega_{\text{ce}} = 0$. When considering the regime of helicon waves, for instance with $\omega, \nu_{\text{eff}} \ll \omega_{\text{ce}} \cos\theta$, Eq. (8.21) reduces to

$$n_{\text{ref}}^2 = \frac{\omega_{\text{pe}}^2}{\omega \omega_{\text{ce}} \cos\theta}\left(1 - \frac{i\nu_{\text{eff}}}{\omega_{\text{ce}} \cos\theta}\right), \tag{8.22}$$

which may also be written

$$kk_z = \frac{e\mu_0 n_e \omega}{B_0}\left(1 - \frac{i\nu_{\text{eff}} k}{\omega_{\text{ce}} k_z}\right). \tag{8.23}$$

Because of the dissipation term, the wavenumbers will now be complex quantities so let $k_z = k_{\text{real}} - ik_{\text{imag}}$. The characteristic absorption length along the z-axis can then be identified with the imaginary part: $\alpha_z \equiv k_{\text{imag}}^{-1}$. It can be expected that the wave will be absorbed over a distance of several wavelengths, so the absorption

8.4 Wave power absorption: heating

length can be presumed to be large compared to the wavelength, $\alpha_z \gg \lambda_z$, which is equivalent to $k_{imag} \ll k_{real}$.

To calculate α_z, one should substitute the complex wavenumbers into the dispersion relation and solve for real and imaginary parts. This is difficult in the general case but it is relatively easy in asymptotic regimes. First consider the case where $k_r \ll k_z$, such that $k \approx k_z$. Then

$$k_{real}^2 - k_{imag}^2 = \frac{e\mu_0 n_e \omega}{B_0} \tag{8.24}$$

and

$$2k_{real}k_{imag} = \frac{e\mu_0 n_e \omega}{B_0}\left(\frac{\nu_{eff}}{\omega_{ce}}\right), \tag{8.25}$$

which, using $k_{imag} \ll k_{real}$ and consequently $k_{real} \approx k_z$, leads to

$$\alpha_z \equiv k_{imag}^{-1} = \frac{2\omega_{ce}}{k_z \nu_{eff}}. \tag{8.26}$$

Using the same approach in the opposite limit of $k_r \gg k_z$ leads to

$$\alpha_z = \frac{\omega_{ce}}{k_r \nu_{eff}}. \tag{8.27}$$

Not surprisingly, it appears that the characteristic absorption length is mainly governed by the effective collision frequency; the larger the collision frequency, the shorter the absorption length. It is the purpose of the next two sections to evaluate the relative contribution of collisional and collisionless processes to the effective collision frequency ν_{eff}.

> **Exercise 8.1: Effective collision frequency** Use the reference conditions given at the start of Section 8.3 and Tables 8.1 and 8.2 to estimate the order of magnitude of ν_{eff} that is required for efficient helicon wave heating in a plasma column that is 0.5 m long.

The example shows that $\nu_{eff} \gtrsim 2 \times 10^7 \text{ s}^{-1}$ is the necessary condition for efficient wave absorption in a half-metre long system. The electron–neutral elastic collision for an argon plasma having an electron temperature of 5 eV is about $\nu_m \simeq 1.5 \times 10^{-13} \times n_g \text{ s}^{-1}$, where n_g is the neutral argon gas density in m^{-3}. At a neutral gas pressure of 0.133 Pa at room temperature, $\nu_m = 4.8 \times 10^6 \text{ s}^{-1}$, which is significantly smaller than the requirement for efficient heating by helicon waves so unless other mechanisms are present, wave modes will not be effective. It turns out that the helicon modes are effective here, so the next task is to try to understand why this is so.

Table 8.3 *Characteristic absorption length for the χ modes of Table 8.2*

χ	n_e/m^{-3}	ν_c/s^{-1}	ξ	ν_w/s^{-1}	$\nu_{\text{eff}}/\text{s}^{-1}$	α_z/m
0	3.5×10^{17}	5.7×10^6	6.1	$\ll 1$	5.7×10^6	2.9
1	1.44×10^{18}	8.55×10^6	2.03	4×10^7	4.9×10^7	0.38
2	3.5×10^{18}	1.39×10^7	1.22	1.24×10^8	1.37×10^8	0.12
3	6.6×10^{18}	2.2×10^7	0.87	9.4×10^7	1.15×10^8	0.1
4	1×10^{19}	3.25×10^7	0.68	5.9×10^7	9.2×10^7	0.1

8.4.2 Collisional wave absorption

Q Are there other types of collisions to consider in high-density plasmas in addition to the usual electron–neutral collisions considered so far?

A In high-density plasmas, the collisions between pairs of charged particles (electrons colliding with ions) will be more frequent than collisions between charged particles and the neutral gas.

Comment: Collisions between pairs of charged particles are called 'Coulomb collisions'.

A calculation of the frequency of electron–ion collisions may be found in many plasma physics textbooks (see for instance [2]) and will not be detailed here. Chen [153] gives a simple result that applies for singly charged ions:

$$\nu_{ei} \approx 2.9 \times 10^{-11} n_e \left(\frac{k_B T_e}{e}\right)^{-\frac{3}{2}} \text{s}^{-1}, \quad (8.28)$$

where the quantities are to be specified in SI units. Choosing again $k_B T_e/e = 5$ V leads to $\nu_{ei} \simeq 2.6 \times 10^{-12} \times n_e$ s^{-1}, while the electron–neutral collision frequency in the same conditions is about $\nu_m \simeq 1.5 \times 10^{-13} \times n_g$ s^{-1}. The ratio of electron–ion to electron–neutral collisions is therefore

$$\frac{\nu_{ei}}{\nu_m} \simeq 17.3 \times \frac{n_e}{n_g}, \quad (8.29)$$

which shows that electron–ion collisions dominate as soon as the ionization fraction n_e/n_g is greater than 6%. This condition can be met in helicon-sustained plasmas, in particular because low-pressure, high-density plasmas are subject to depletion of the neutral gas, that is a lowering of the particle density, n_g, owing to gas heating, as discussed in Chapter 9. The total collision frequency, $\nu_c = \nu_m + \nu_{ei}$, which includes both electron–neutral and electron–ion collisions, can be calculated for each of the χ modes, for instance those listed in Table 8.2. This is done in Table 8.3, with

8.4 Wave power absorption: heating

$\nu_m = 4.8 \times 10^6 \, \text{s}^{-1}$. For the first longitudinal mode, $\chi = 0$, the electron–neutral collisions still dominate, because the electron density is still too low for electron–ion collisions to be significant. For higher-order modes, the electron–ion collisions take over; for instance, for $\chi = 4$, they are almost six times more frequent than electron–neutral collisions.

Q Check to see if the assumption that the absorption length is large compared to the wavelength, $\alpha_z \gg \lambda_z$, is valid.
A It is indeed the case, as can be checked by inspecting Tables 8.2 and 8.3.

The introduction of electron–ion coulomb collisions does not make a difference at moderate electron densities (typically the $\chi = 0$ mode). Collisionless energy exchange mechanisms have therefore been invoked to explain the experimentally observed wave absorption. In the next section, a wave–particle interaction mechanism is considered as a means of transferring energy into the electron population without collisions between particles.

8.4.3 Collisionless wave absorption

When a helicon mode passes through the plasma, charged particles are oscillated by the electric field of the wave as the disturbance moves by. The wave propagates at the phase speed, $v_\phi = \omega/k_z$, which depends on the static magnetic field and the electron density. For the electrons, the periodic displacement adds to the background thermal motion at speeds characterized by $v_e = \sqrt{k_B T_e/m_e}$. Under typical conditions, $v_\phi \sim v_e$. It is tempting therefore to imagine that any electrons moving at exactly the same speed as a helicon wave, and in the same direction, are barely affected by it. Furthermore, electrons moving slightly faster than the wave will drive into the back edge of the crests while those moving more slowly will be swept forward by the leading edge of the crests. Although that image is a poor visualization of the interaction, it does suggest that energy could be transferred between the wave and the electrons and that the energy transfer is likely to be a function of particle velocity. So, to take this wave–particle interaction into account one must integrate the interaction over the electron velocity distribution. The most significant contributions to the result will come from those particles that have a speed close to the phase speed of the wave.

Q In a velocity distribution that falls off monotonically at higher velocity (as with a Maxwellian, for instance), what will be the net effect of the wave–particle interaction?

> **A** The bell shape of the velocity distribution means that the group of particles that have a velocity close to the phase velocity of the wave will contain more electrons that have a lower speed than have a higher speed. The net effect is therefore a damping of the wave and a heating of the particle distribution.
> *Comment: This process is sometimes referred to as 'Landau damping', but that term was originally used to name a related but non-dissipative phenomenon, so it has been avoided here.*

The wave–particle damping mechanism is 'collisionless' but it is nevertheless convenient to find a means of including the effect in the algebra as an additional 'equivalent' collision frequency ν_w. Chen [153] has carried out the necessary integrations over a Maxwellian velocity distribution to derive the equivalent frequency for this mechanism:

$$\nu_w = 2\sqrt{\pi}\xi^3\omega\exp\left(-\xi^2\right), \tag{8.30}$$

with $\xi = \sqrt{2}\,v_\phi/v_e$. This expression is an approximation, valid when $\xi > 1$. In Table 8.3, ν_w is calculated for the different χ modes, and the sum of the collision frequency and the equivalent frequency $\nu_{\text{eff}} = \nu_c + \nu_w$ is then used to calculate the characteristic absorption length α_z for each mode.

This procedure shows that the $\chi = 0$ mode is still not efficiently absorbed, even when damping by wave–particle interaction is introduced, since the typical absorption length of 2.9 m is much larger than any experimental arrangements. In fact, the wave–particle mechanism is completely negligible in this condition ($\nu_w \approx 0$) because the phase speed is much larger than the electron thermal speed. Consequently, the number of electrons interacting with the wave is extremely small. For higher-order modes, $\chi \geq 1$, the phase speed of the wave becomes comparable to the electron thermal speed.

In the previous section it was concluded that collisional damping is inefficient for $\chi = 0$ because both the plasma density and the neutral gas density are too small. It becomes gradually efficient for $\chi \geq 1$, as the electron density increases (due to electron–ion collisions). Neither collisional nor collisionless energy damping seem to explain the observed absorption of the $\chi = 0$ mode. Exactly how the energy is transferred from the $\chi = 0$ mode remains a puzzle. Degeling *et al.* [164] have shown that electron trapping in the helicon wave may be responsible for the wave energy absorption. Other mechanisms, such as the excitation of other wave modes, have also been proposed [151]. Breizman and Arefiev [160] have proposed that the strong electron density gradient in the radial direction leads to wave guiding and enhanced heating in the longitudinal direction.

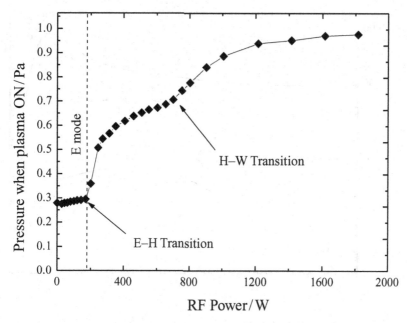

Figure 8.9 Gas pressure measured at the wall of a SF_6 plasma produced in a helicon reactor. The E–H and H–W transitions are identified as the power increases.

8.5 E–H–W transitions

It has been shown that helicon wave propagation requires a minimum plasma density. However, a helicon plasma reactor can operate at low electron density (or low injected power), in a regime where the helicon wave is not launched. There is nevertheless a substantial voltage across the antenna, so a capacitive current is driven in the plasma and a fraction of the discharge power is therefore deposited capacitively. Moreover, the RF current flowing in the antenna (which behaves as a non-resonant inductive coil) results in plasma generation in the vicinity of the antenna by the induced RF electric field. The discharge can therefore exist in three different modes: the capacitive mode (E-mode), at low power, the inductive mode (H-mode), at intermediate power and finally the helicon mode (W-mode), at high power. As the power is increased, transitions from capacitive to inductive to helicon modes (E–H–W) are observed.

Q Figure 8.9 displays the pressure measured by a capacitance manometer (an absolute pressure gauge) located at the wall of the diffusion chamber of the helicon plasma processing reactor shown in Figure 8.1. The feedstock gas was SF_6. Explain why an abrupt increase in pressure is an indication of mode transition.

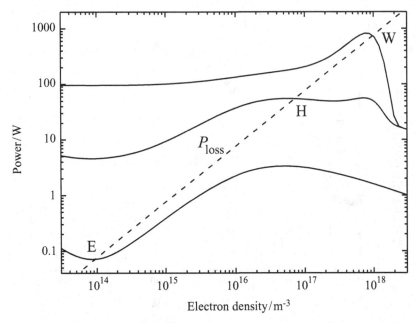

Figure 8.10 Schematic power representation of the E–H–W transitions in a helicon discharge.

A   From the previous chapters, one expects E–H–W mode transitions to result in an abrupt increase in the electron density. In a molecular gas like SF_6, an increase in the electron density will lead to higher dissociation of the molecules and consequently to an increase in total pressure (if the pumping speed remains constant, which was the case in this experiment). Another effect arises: depletion of neutral gas in the centre, which in some instances may lead to neutral gas accumulation at the wall and consequently to an increase in the measured pressure.
Comment: Neutral depletion in high-density plasmas is treated in the next chapter.

The mode transitions in a helicon reactor may be understood by looking at the power–electron density space, as done in the previous chapter and shown in Figure 8.10. The solid lines represent the absorbed power curves, whereas the dashed line is the power loss curve which, as discussed extensively, is a straight line (for an electropositive gas with single-step ionization only). The lower curve of absorbed power, for low RF current in the antenna, intersects the loss curve in the capacitive branch at very low electron density, around $n_e \approx 10^{14}$ m^{-3}: the discharge operates in E-mode. As the current in the antenna increases, the inductive peak

8.5 E–H–W transitions

takes over and the discharge runs in the inductive H-mode, with an intersection at $n_e \approx 8 \times 10^{16}$ m^{-3} in this example. Finally, at higher current in the antenna, a new peak in the absorbed power, centred around $n_e \approx 10^{18}$ m^{-3}, arises. This peak corresponds to the density at which the first longitudinal (χ) helicon mode propagates. This corresponds to operation in the $\chi = 0$ helicon mode, which in this example occurs at higher electron density than that in Tables 8.1 and 8.2 because the parameters are slightly different. The discharge therefore experiences E–H–W transitions as the RF current in the antenna (or the output power of the generator) increases.

Q Is it possible to control the position of the helicon peak and consequently to control the abruptness of the H–W transition?
A It is indeed possible to change the position of the helicon peak by changing the magnitude of the static magnetic field. From the dispersion relation it appears that the longitudinal wavelength is mostly governed by the ratio B_0/n_e. Thus, to keep the same wavelength selected by the antenna, an increase in B_0 will result in the same increase in n_e. If one chooses a modest magnetic field, the helicon peak will merge with the inductive peak and the H–W transition will be smooth. However, the helicon mode will run at moderate density. If one chooses a strong magnetic field, the H–W transition will be more abrupt and the helicon mode will run at higher density. The first case may be preferable for plasma processing, whereas the second case may be more appropriate for plasma thrusters.

Q Why is there only one peak in the helicon mode in Figure 8.10?
A There should be several peaks corresponding to higher χ modes, as shown in Tables 8.1 and 8.2. The higher-order modes will appear successively as the power is increased.
Comment: Lieberman and Boswell have proposed a simplified global model of a helicon discharge able to describe E–H–W transitions, including higher-order χ modes [156].

8.6 Summary of important results

- The addition of a static magnetic field to an RF plasma has two major consequences: (i) the plasma transport is modified (see next chapter), (ii) electromagnetic waves can propagate at low frequencies; the helicon waves studied in this chapter are low-frequency waves such that $\omega_{ci} \ll \omega \ll \omega_{ce} \ll \omega_{pe}$.
- Helicon waves tend to propagate along the static magnetic field line. When confined in a cylinder, eigenvalues of the radial and longitudinal wavenumbers are obtained.
- Antennas may be designed to select specific longitudinal wavenumbers (or wavelengths). For a given magnetic field, this in turn defines a typical electron density for efficient helicon wave propagation.
- The helicon wave energy is absorbed efficiently by the plasma electrons by collisional and collisionless mechanisms. This leads to efficient ionization in a long plasma column.
- Helicon reactors are subject to E–H–W transitions. They operate in the E (capacitive) mode at low power, in the H (inductive) mode at intermediate power, and eventually in the W (helicon) mode at high power. The W-mode encompasses several azimuthal (m) and longitudinal (χ) modes.

9
Real plasmas

The plasma systems treated in this book have been simplified to enable analysis and insight. So far the discussion of plasmas and sheaths has considered an ideal low-temperature plasma that contains only singly ionized species formed from an atomic gas. Etching plasmas, deposition plasmas and plasmas in thrusters all involve more complex phenomena than have been included in the sheath and transport models.

> **Q** With reference to Sections 1.2.1 and 2.1.3, identify three species that can be expected in fluorocarbon plasmas (for semiconductor processing) that were not included in Chapter 3.
> **A** (i) The (fluorocarbon) gas is not atomic and so a fluorocarbon plasma may contain *radical species* as well as the parent molecules; (ii) fluorine and fluorinated radicals are electronegative so in addition to molecular species one can anticipate that *negatively charged ions* might also occur in fluorocarbon plasmas; (iii) *various positive ions of fragmented molecules* will also be present.

Here are some of the issues. When plasmas are formed in molecular gases, electron–molecule collisions, chemical reactions in the gas phase and interactions of reactive species with surfaces (the reactor walls or the substrate) together determine the plasma composition. In some instances, the reactions in the gas phase lead to the formation of macro-molecules which may then agglomerate to make fine particles and hence 'dusty plasmas'. As more and more energy is coupled into a plasma, the fraction of the gas that is ionized rises. Eventually this leads to another class of complication when the plasma pressure becomes comparable with that of the gas, in which case the plasma dynamics and the neutral gas dynamics are coupled. When a static magnetic field is also present, such as in helicon systems, the transport of charged particles is modified by the field. High-density plasmas used for surface treatments may be generated in a source region from where they expand into a larger interaction chamber, but the expansion is sometimes found to be associated with non-linear structures in the plasma profile. Finally, electrical discharge plasmas may be unstable or chaotic.

This chapter will examine the impact of some of these non-ideal issues on plasma transport and plasma/sheath boundaries, as well as on plasma stability. First, the effect of neutral gas depletion in high-density plasmas will be analysed. Then, attention moves to the effect of a static magnetic field on the fluxes of particles crossing a plasma, before re-examining theories of sheaths and transport in the presence of negative ions, i.e., in electronegative plasmas. After considering plasmas that are allowed to expand from a source region into expanding plasmas, the final topic concerns instabilities observed at the E–H transition of inductive discharges. Not all of these scenarios are thoroughly understood.

9.1 High-density plasmas

In high-density plasma sources, such as inductively coupled and helicon reactors, at high power per unit volume the plasma density, n, may be sufficiently high for the plasma pressure $nk(T_e + T_i)$ to become comparable to that of the neutral gas $n_g k T_g$. In this situation, the density of the neutral gas cannot be taken to be a constant value everywhere, set by the original gas pressure. Instead, the fluid models of Chapter 3 require additional conservation equations for a third fluid that represents the neutral gas. In experiments, high plasma pressure has been found to cause depletion of neutral gas in the reactor centre. In the following, attention will be focused on the effect of this on plasma transport, as analysed by Fruchtman et al. [165] and by Raimbault et al. [44]. In this section, for the sake of simplicity, neutral dynamics will be added only to the Schottky solution, first assuming that the gas temperature remains fixed and then with gas heating included. The analysis of the intermediate and low-pressure cases can be found elsewhere [44].

9.1.1 Neglecting gas heating: isothermal gas

The conservation equations, including neutral dynamics for a high-pressure, isothermal gas, become

$$(nu)' = nn_g K_{iz}, \tag{9.1}$$

$$(n_g u_g)' = -nn_g K_{iz}, \tag{9.2}$$

$$0 = neE - nuMn_g K_g, \tag{9.3}$$

$$0 = -kT_g n_g' + nuMn_g K_g, \tag{9.4}$$

$$0 = -neE - kT_e n', \tag{9.5}$$

where u_g is the speed of the fluid that represents the neutral gas, T_g is the gas temperature and $K_g \equiv \sigma_i \bar{v}_i$, with σ_i the cross-section for ion–neutral momentum transfer collisions. The ionization rate depends on the electron temperature; see Eq. (2.27).

Q Equation (9.1) is a steady-state continuity equation for the ions. It balances the divergence of flux out of a vanishingly small volume against the ionization happening within it. Ionization transforms gas atoms into ions at a rate that depends on gas density, electron density and a rate constant for the process. Account for the second equation in a similar way.

A Equation (9.2) is a continuity equation for the gas. The divergence of the flux is locally balanced by the loss of gas atoms caused by the ionization process. *Comment: Notice that the loss of density from the neutral gas fluid is equal to the gain in density of the ion fluid.*

Equation (9.3) is a force balance equation for the ions. The electric force is balanced by a friction term arising from collisions with the gas that transfer momentum out of the ion fluid into the neutral gas. Compared with these two terms, the rate of change of momentum arising from the pressure gradient in the ion fluid is negligible. Equation (9.4) is the force balance for the neutrals. Since there can be no direct electric force on the gas, the momentum gained from collisions with the ions must be balanced by a pressure gradient. Finally, Eq. (9.5) is the force balance for the electrons. The electric force is balanced by a pressure gradient; integrating this equation leads to the Boltzmann equilibrium.

Compared to the three transport equations used to solve the standard Schottky model, the inclusion of gas temperature and gas density requires two further transport equations and a specification of the constant gas temperature. The above system of equations can be integrated analytically, with different boundary conditions for neutrals. The case treated here will fix the neutral density at the reactor walls, which means that the total number of neutrals is not conserved. This is appropriate when the gas pressure is controlled at the reactor wall. Another limiting case would be to consider a fixed number of neutrals (see Fruchtman et al. [165]). Before looking at the full solution, note that simply adding the force balance equations for the three fluids and then integrating from the wall where $n_g = n_{gw}$ and $n = 0$, leads to

$$n(x)kT_e + n_g(x)kT_g = n_{gw}kT_g. \tag{9.6}$$

Since the electron (plasma) pressure is maximum at the reactor centre and decays towards zero at the wall, this shows that the neutral pressure has to fall towards the centre. The neutral gas density depletion at the plasma centre is therefore

$$\frac{n_{g0}}{n_{gw}} = 1 - \frac{n_0 k T_e}{n_{gw} k T_g}. \tag{9.7}$$

Since $T_e \gg T_g$, only a few percent of ionization leads to severe neutral depletion.

Q Show that taking an ionization fraction in the centre of only $n_0/n_{g0} = 0.01$, given a temperature ratio $T_e/T_g = 100$, leads to a neutral gas depletion of $n_{g0}/n_{gw} = 0.5$.

A The pressure ratio in Eq. (9.7) can be written

$$\frac{n_0 k T_e}{n_{gw} k T_g} = \left(\frac{n_0}{n_{g0}} \frac{T_e}{T_g}\right) \frac{n_{g0}}{n_{gw}}.$$

Inserting the given values and combining with Eq. (9.7), gives $n_{g0}/n_{gw} = 0.5$.

Setting $\gamma_g \equiv T_e/T_g$ and $N_0 \equiv n_0/n_{gw}$, Raimbault et al. [44] obtained the following condition for the electron temperature (which is now disguised in γ_g and in the functional dependence of K_{iz} and u_B):

$$n_{gw} l = \frac{4 u_B}{(K_g K_{iz})^{1/2}} \frac{1}{(1 - (\gamma_g N_0))^{1/2}} \arctan \left[\frac{(1 - (\gamma_g N_0))^{1/2}}{1 - \gamma_g N_0}\right]. \tag{9.8}$$

When the degree of ionization is low, $\gamma_g N_0 \to 0$ and this reduces to

$$n_{gw} l = \frac{\pi u_B}{(K_g K_{iz})^{1/2}}; \tag{9.9}$$

there is no gas depletion, so $n_g(x) = n_{gw} = n_g$. Noting that $D_a \equiv u_B^2/n_g K_g$, one can see that Eq. (9.9) is identical to Eq. (3.72), as indeed it must be. The full expression for the plasma density profile in this situation is found to be

$$n(x) = n_0 \frac{\tan^2\left(K\sqrt{1 - (\gamma N_0)^2}\right) - \tan^2\left(2K\sqrt{1 - (\gamma N_0)^2} x/l\right)}{\tan^2\left(K\sqrt{1 - (\gamma N_0)^2}\right) + \tan^2\left(2K\sqrt{1 - (\gamma N_0)^2} x/l\right)}, \tag{9.10}$$

with $K \equiv (n_{gw} l)(K_i K_g)^{1/2}/(4 u_B)$. The neutral gas density profile can then be deduced from pressure conservation and the plasma density profile.

Q Noting that when $\gamma N_0 \to 0$ then Eq. (9.9) applies, show that the plasma density profile reduces to $n(x) = n_0 \cos(\pi x/l)$, as obtained in the classical Schottky model.

> A When $\gamma N_0 \to 0$, Eq. (9.9) gives $K \to \pi/4$. Putting both these limits simultaneously into Eq. (9.10) gives
>
> $$n(x) = n_0(\cos^2 \pi x/2l - \sin^2 \pi x/2l)$$
>
> and using standard trigonometric identities recovers the expected result.

Equation (9.8) sets a condition on the electron temperature, but unlike in the simpler cases of earlier chapters, this condition now depends on the plasma density through N_0. The consequence of the coupling between the plasma fluid and the neutral gas fluid is therefore that the electron temperature depends on the plasma density, i.e., on the amount of power deposited into the plasma. It turns out that increasing the plasma density increases the neutral depletion and increases the electron temperature. In other words, the particle and power balances are no longer independent.

9.1.2 Including gas heating: non-isothermal gas

The isothermal neutral approximation is not satisfactory because of the coupling of energy between the plasma and the neutral gas that leads to significant gas heating [26, 166–168]. Gas heating results from collisions between charged particles and neutrals. In the preceding analysis, this energy transfer has not been included. To do so requires one further equation. This has been included in sophisticated numerical simulations (see for example [169, 170]), and was studied more specifically by Liard et al. [171], who added the neutral energy conservation to the model of Raimbault et al. In place of a set gas temperature, a heat flux balance is needed to link its value to the local energy density and collisions. Taking account of the fact that in binary collisions between disparate masses, $m \ll M$, no more than $2m/M$ of the incident kinetic energy can be transferred from the lighter to the heavier species, the energy flux balance can be approximated by

$$\left(\kappa T_g'\right)' = nn_g k \left[3\frac{m}{M}K_e(T_g - T_e) + \frac{3}{4}K_g(T_g - T_i)\right], \quad (9.11)$$

where κ is the thermal conductivity of the gas (which in turn depends on gas temperature), K_e is the rate coefficient for electron–neutral elastic collisions and T_i is the ion temperature. Within this model, the gas heating term on the RHS is mostly dominated by electron–neutral collisions, so it is reasonable to set ion and gas temperatures equal and thereby avoid the further complexity of following the development of the ion thermal energy. In molecular gases some other mechanisms, such as vibrational excitation followed by vibration–translation transfer, also need to be included and may even dominate the gas heating [26].

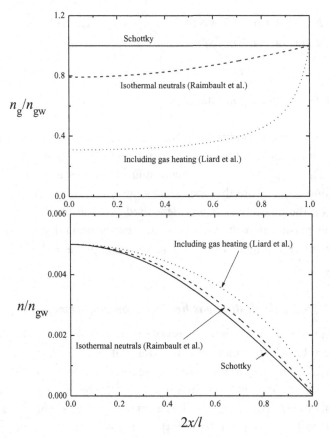

Figure 9.1 Neutral and plasma density profiles for the classical Schottky model, the neutral depletion model of Raimbault *et al.* and the model of Liard *et al.* that includes gas heating.

Results of the different models are shown in Figure 9.1, where the neutral gas density profile $n_g(x)$ and the plasma density profile $n(x)$, normalized to the gas density at the wall, are plotted as a function of x. Each figure has three curves: the solid line represents the classical Schottky model (i.e., with uniform gas density and gas temperature), the dashed line represents the result of the model of Raimbault *et al.* (which includes neutral dynamics but within the isothermal neutral approximation) and finally the dotted line represents the model by Liard *et al.*, i.e., including neutral gas heating. All these curves have been calculated for the reduced parameters $Pl = 3.9$ Pa m and $N_0 = n_0/n_{gw} = 0.005$ (which corresponds to a few percent of ionization at the discharge centre). The neutral gas depletion at the centre is about 25% if gas heating is ignored, but reaches 70% when gas heating is included. The neutral gas temperature is high, about 1000 K at the centre, in good qualitative agreement with previously published experimental results.

The depletion of neutral gas at the discharge centre and the flatter plasma density profiles are due to enhanced plasma transport, and consequently to higher edge-to-centre plasma density ratio, the so-called h_1 factors [172]. This enhanced transport (or plasma deconfinement) leads to an increase in the electron temperature.

9.2 Magnetized plasmas

Q Charged particles rotate around magnetic field lines at the cyclotron frequency. What is the radius of gyration (or 'Larmor radius') for electrons with the average speed $\bar{v}_e = (8kT_e/\pi m)^{1/2}$? Compare the Larmor radius for electrons with that for ions.

A Since charges tend to rotate around field lines at the cyclotron frequency $\omega_{ce} \equiv eB_0/m$, the Larmor radius is $r_{Le} = \bar{v}_e/\omega_{ce} \equiv m\bar{v}_e/(eB_0)$.
The ratio of ion-to-electron Larmor radius is: $r_{Li}/r_{Le} \equiv \sqrt{MT_i/mT_e} \gg 1$.

Charged particles tend to spiral around magnetic field lines. This motion must perturb the plasma transport, though as for wave propagation, the effect will be different for motion parallel and perpendicular to the magnetic field lines.

Q Describe the process of charged particle transport in a magnetized plasma in the absence of collisions.

A In the absence of collisions the motion perpendicular to the magnetic field cannot extend farther than a Larmor radius, because the particles are trapped in the cycloidal orbits. Therefore, there should be no plasma flux in the perpendicular direction, resulting in perfect plasma confinement. Along the magnetic field lines, the charged particles move freely, resulting in the same type of transport as that described by Tonks and Langmuir.

The transport along the magnetic field lines being essentially unaffected, the discussion focuses on transport perpendicular to the field, which requires collisions. The schematic motion of a charged particle perpendicular to the magnetic field lines is shown in Figure 9.2. A collision allows the particle to shift the cycloidal motion from line to line, globally resulting in perpendicular transport. On average, the shift is of one Larmor radius. In the following, this is implemented in the fluid model of plasma transport studied in Chapter 3. The particle conservation equation remains unchanged, but the momentum conservation equation needs to include the Lorentz force, proportional to $\mathbf{u} \times \mathbf{B}$:

$$nm\left[\frac{\partial \mathbf{u}}{\partial t} + (\mathbf{u} \cdot \nabla)\mathbf{u}\right] = nq\,(\mathbf{E} + \mathbf{u} \times \mathbf{B}) - \nabla p - m\mathbf{u}\,[n\nu_m + S - L]. \quad (9.12)$$

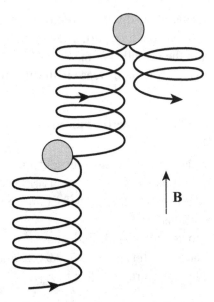

Figure 9.2 Schematic of electron transport perpendicular to magnetic field lines. The large grey sphere symbolizes neutral atoms while the small black sphere symbolizes electrons. Without collisions, electrons cannot travel farther than a Larmor radius.

The magnetic field reduces the flux perpendicular to the field lines, resulting in better confinement. It consequently decreases the edge-to-centre density ratio, h_l, and the electron temperature required to sustain the plasma.

9.2.1 Ambipolar diffusion with transverse magnetic field

This section is about how the static magnetic field modifies the ambipolar diffusion of Schottky. The plasma is assumed to be infinite in the z-direction (all quantities are uniform in z) and the static magnetic field is along the y-axis. In the steady state, neglecting the small contribution of the source and loss terms, the conservation of momentum, Eq. (9.12), written for electrons and projected on the x and z-axes becomes

$$0 = -n_e e \left(E - u_{ez} B_y\right) - kT_e \frac{dn_e}{dx} - n_e u_{ex} m \nu_e, \tag{9.13}$$

$$0 = -n_e e u_{ex} B_y - n_e u_{ez} m \nu_e, \tag{9.14}$$

where ν_e is the electron–neutral collision frequency. Note that in the z-direction the gradients and the electric field are zero. Eliminating the components along z,

$$-n_e e E - kT_e \frac{dn_e}{dx} - n_e u_{ex} m \nu_e \left(1 + \frac{\omega_{ce}^2}{\nu_e^2}\right). \tag{9.15}$$

9.2 Magnetized plasmas

It therefore appears that the equation takes the same form as the non-magnetized case, with an effective collision frequency defined as

$$v_e^* = v_e \left(1 + \frac{\omega_{ce}^2}{v_e^2}\right). \tag{9.16}$$

Using quasi-neutrality, $n_e = n_i = n$, and doing the same calculation for the ions leads to the following system of equations:

$$nu = -n\mu_e^* E - D_e^* n', \tag{9.17}$$

$$nu = n\mu_i^* E - D_i^* n', \tag{9.18}$$

where the diffusion and mobility coefficients have their usual definition with $v_{e,i}^*$ instead of $v_{e,i}$. The rest of the analysis is straightforward and follows from that developed in Chapter 3. The effective ambipolar diffusion coefficient,

$$D_a^* = \frac{\mu_i^* D_e^* + \mu_e^* D_i^*}{\mu_i^* + \mu_e^*}, \tag{9.19}$$

will be used to evaluate the flux exiting the discharge and in turn the electron temperature.

In the absence of magnetic field, the ion mobility and diffusion coefficients are much lower than those for electrons, leading to an approximate form of the ambipolar diffusion coefficient given by Eq. (3.67). However, in the presence of magnetic field, this approximation is no longer satisfactory. It is even the case that the electron mobility becomes smaller than the ion mobility because the 'magnetization' (a term used to describe the strong inhibition of the perpendicular flux due to the cycloidal motion) of electrons is much more efficient.

Q Why are the electrons more easily 'magnetized' than the ions?

A As mentioned previously, the Larmor radius is much smaller for electrons than for ions, which suggests the above statement.

Comment: A more precise analysis consists of comparing the ratios ω_{ce}/v_e and ω_{ci}/v_i, which will be done later.

Q Show that $\omega_{ce}/v_e \equiv \lambda_e/r_{Le}$ and $\omega_{ci}/v_i \equiv \lambda_i/r_{Li}$.

A Combining the previous definitions for the Larmor radius, $r_{Le,Li} \equiv \bar{v}_{e,i}/\omega_{ce,ci}$ and for the collision frequency $v_{e,i} \equiv \bar{v}_{e,i}/\lambda_{e,i}$ gives the required result. According to Figure 9.2, the particles should not be magnetized when the mean free path for elastic collisions is much smaller than the Larmor radius, i.e., $\lambda_{e,i}/r_{Le,Li} \ll 1$.

A very common regime of operation of helicon plasma processing reactors is such that electrons are magnetized ($\omega_{ce}^2/\nu_e^2 \gg 1$) while ions are not ($\omega_{ci}^2/\nu_i^2 \ll 1$). In that case, the following approximations apply:

$$\nu_e^* = \nu_e \left(1 + \frac{\omega_{ce}^2}{\nu_e^2}\right) \approx \frac{\omega_{ce}^2}{\nu_e}, \tag{9.20}$$

$$\nu_i^* = \nu_i \left(1 + \frac{\omega_{ci}^2}{\nu_i^2}\right) \approx \nu_i, \tag{9.21}$$

which leads to

$$\mu_e^* = \frac{e\nu_e}{m\omega_{ce}^2}, \quad D_e^* = \frac{kT_e \nu_e}{m\omega_{ce}^2}, \quad \mu_i^* = \frac{e}{M\nu_i}, \quad D_i^* = \frac{kT_i}{M\nu_i}.$$

Note that $\mu_i^* D_e^* \gg \mu_e^* D_i^*$ because $T_e \gg T_i$, such that, after some algebra, the ambipolar diffusion coefficient takes the following approximate form:

$$D_a^* \approx \frac{kT_e}{M\nu_i} (1 + \delta_B)^{-1} \tag{9.22}$$

with

$$\delta_B \equiv \frac{m\omega_{ce}^2}{M\nu_i \nu_e} = \frac{\omega_{ci}\omega_{ce}}{\nu_i \nu_e}. \tag{9.23}$$

When there is no magnetic field, i.e., $\delta_B = 0$, this expression reduces to Eq. (3.67) established in Chapter 3. As the magnetic field increases, δ_B increases and therefore D_a^* decreases. The flux perpendicular to the magnetic field is therefore reduced. The electron temperature and the edge-to-centre density ratio are obtained straightforwardly by replacing D_a by D_a^* in Eqs (3.72) and (3.75), respectively.

Q In a plasma at $p = 13.3$ Pa with $B_0 = 0.2$ T, the important frequencies for transport are

$$\omega_{ci} = 2.4 \times 10^5 \text{ s}^{-1},$$
$$\nu_i = 1.8 \times 10^6 \text{ s}^{-1},$$
$$\omega_{ce} = 1.8 \times 10^{10} \text{ s}^{-1},$$
$$\nu_e = 3.7 \times 10^8 \text{ s}^{-1}.$$

Verify the hypothesis that electrons are magnetized and evaluate the effect of the magnetic field on the transport.

A The electrons are magnetized because $\omega_{ce}^2/\nu_e^2 \gg 1$ and the ions are not because $\omega_{ci}^2/\nu_i^2 \ll 1$. On evaluating the magnetization parameter given in Eq. (9.23), it is found that $\delta_B \approx 6.4 \geq 1$. The ambipolar diffusion coefficient

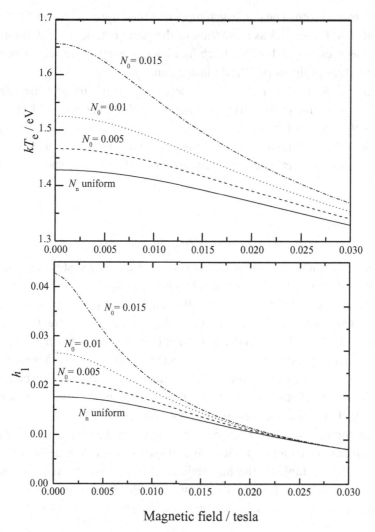

Figure 9.3 Electron temperature (left) and edge-to-centre density ratio (right) as a function of the magnetic field amplitude and for increasing electron densities, normalized to gas density at the wall. This figure is from Liard et al. [173], in which the pressure was fixed at 13.3 Pa and the discharge length at $l = 0.15$ m.

is therefore likely to be significantly reduced compared to the non-magnetized case. The flux, proportional to D_a^*, is reduced accordingly and consequently both T_e and h_1 are significantly smaller.

Liard et al. [173] have recently improved the above model to include neutral dynamics and therefore account for neutral depletion, which was discussed in the previous section. The details will not be presented here but instead the key results of their analysis are given. The electron temperature and the edge-to-centre density ratio are plotted in Figure 9.3 as a function of the static magnetic field amplitude, for a neutral gas pressure of 13.3 Pa. Each figure has several curves corresponding to increasing values of the normalized plasma density.

First observe that, as stated before, the electron temperature and the edge-to-centre density ratio (and so the flux) decrease when the magnetic field increases. Second, note the effect of neutral depletion: both quantities increase at a given magnetic field as the plasma density increases, i.e., when neutral depletion occurs. However, the neutral depletion effect becomes less and less important as the magnetic field increases.

9.2.2 Limitations of the above theory

The focus on perpendicular transport implies that it is as if the system were infinite parallel to the field, in the z-direction. However, because electrons travel very fast along the magnetic field lines, the finite extent of any real system must be considered. A two-dimensional calculation is then required to correctly treat this problem (see for instance Lieberman and Lichtenberg [2]). In addition, the problem has different solutions depending on the conductivity of the walls. Whereas with insulating walls the electron and ion fluxes to any small portion of surface must, on average, be equal, DC current can flow in conducting boundaries and therefore a local balance of electron and ion fluxes is not necessary.

Another important aspect that has been presumed in the above considerations is that the diffusion of charged particles is a stable process. Magnetized plasmas become inherently unstable as the magnetic field increases, and the transport of charged particles is then a turbulent process [23]. Plasma turbulence is a major research topic in the study of fusion plasmas and for plasmas in space.

9.3 Electronegative plasmas

Plasma processing applications often require feedstock gases that contain atoms with a large electron affinity (or 'electronegativity'). It is particularly the case in plasma etching, where the common feature of the typical gas mixtures such as $HBr/Cl_2/O_2$ or $Ar/C_4F_8/O_2$ is inclusion of halogen-containing components. Fluorine and chlorine in particular have large electronegativities, so one can expect the plasma formed in these gases to contain negative ions. The term 'electronegative plasma' implies one containing a significant fraction of negative ions – enough to profoundly modify the equilibrium and the dynamics of the plasma. Among the new phenomena that arise are spontaneous instabilities with temporal cycles

9.3 Electronegative plasmas

of negative ion creation and destruction, and spatial structures including double layers of charge-separating regions of different ionic composition.

There are then three types of charged particles to consider, and conservation equations for negative ions must be introduced and coupled with those already established for positive ions and electrons. Because of their negative charge, negative ions experience a confining electric field, as electrons do. However, unlike electrons, their temperature is small and their mass is large so that they barely penetrate the positive ion sheaths at the boundary and so they tend to be confined in the plasma.

Q Under what circumstances would negative ions *not* be confined in a plasma?
A Plasma boundaries are usually places where the positive space charge sheaths arise. Immediately adjacent to these sheaths there will usually be quasi-neutral plasmas in which a weak electric field accelerates positive charge out of the plasma and into the sheath.
(i) This situation may be different near surfaces that are taken to a potential that is more positive than that of the local plasma, such as obtains when a Langmuir probe is taken into its electron saturation region. Electrons and negative ions will both be attracted to such a surface.
(ii) Should negative ions be formed on, or close to, surfaces that are biased below the floating potential, then these ions will be accelerated into the plasma, gaining sufficient energy to escape across other sheaths formed on surfaces that are at a less negative potential.

It is commonly the case that negative ions are produced in the plasma volume, by electron attachment to molecules, and lost in the plasma volume, by detachment or by recombination with positive ions. Under these circumstances, the flux of negative ions to the reactor walls may be presumed to be zero. Then, the simplest way, though not necessarily the most realistic, to treat the negative ions is to suppose that they are in a Boltzmann equilibrium with the local potential. This may seem reasonable since, like electrons, negative ions will tend to be confined in the potential structure. This naturally arises in plasmas to expel positive charge and retain negative charge (some of which is markedly more mobile), thereby balancing production and loss processes. However, as shown in Chapter 2, the Boltzmann equilibrium can be derived from the fluid momentum equation only when inertial terms like $Mu\,n_g K_g$ and $Mu\,u'$ are insignificant compared with the isothermal pressure gradient and the electric force. Thus, high gas pressure and gradients in drift speed both threaten the validity of the Boltzmann assumption. Notwithstanding these observations, the convenience of the Boltzmann factor is sufficient to justify its use in examining the characteristic behaviour of electronegative plasmas, so in what follows it will be presumed to hold for negative ions unless otherwise stated.

9.3.1 Debye length in electronegative plasmas

The Debye length is modified when there are two separate populations of negative charge characterized by different temperatures. In an electronegative plasma, assuming Boltzmann negative ions of central density n_{n0} and temperature T_n, small changes in potential, $e\phi \ll kT_{e,n}$, give rise to a small space charge perturbation described by

$$\rho(x) = e(n_p - n_e - n_n) \approx e \left(n_{p0} - n_{e0} \left(1 - \frac{e\phi(x)}{kT_e} \right) - n_{n0} \left(1 - \frac{e\phi(x)}{kT_n} \right) \right). \tag{9.24}$$

The two following parameters are appropriate: the 'central' electronegativity, $\alpha_0 \equiv n_{n0}/n_{e0}$, and the electron-to-negative ion temperature ratio, $\gamma \equiv T_e/T_n$. Quasi-neutrality applies in the unperturbed plasma $n_{p0} = n_{e0} + n_{n0}$. The linearized Poisson equation is then

$$\phi''(x) = -\frac{\rho}{\varepsilon_0} = \frac{e^2 n_{e0} \phi(x)}{\varepsilon_0 kT_e} (1 + \gamma \alpha_0) \tag{9.25}$$

and the Debye length is the scale length of the spatially decaying, exponential solution (cf. Eq. (3.6)):

$$\lambda_D^* = \sqrt{\frac{\varepsilon_0 kT_e}{n_{e0} e^2}} \sqrt{\frac{1}{1 + \gamma \alpha_0}} = \lambda_{De} \sqrt{\frac{1}{1 + \gamma \alpha_0}}. \tag{9.26}$$

The Debye length in an electronegative plasma is therefore smaller than that in an electropositive plasma, though the difference is not significant at very low electronegativity ($\alpha_0 \ll 1$).

9.3.2 Bohm criterion in electronegative plasmas

The Bohm criterion for electropositive plasmas sets a minimum speed for ions entering a positive ion sheath. The criterion needs to be revisited for electronegative plasmas. Boyd and Thompson in 1959, and more recently Braithwaite and Allen [174], have done this analysis. They showed that the Bohm speed for an electronegative plasma becomes

$$u_B^* = u_B \left(\frac{1 + \alpha_s}{1 + \gamma \alpha_s} \right)^{1/2}, \tag{9.27}$$

9.3 Electronegative plasmas

where α_s is evaluated at the sheath edge. It appears that the Bohm speed is reduced as the electronegativity increases. The potential drop (from the centre to the sheath edge), which is necessary to accelerate the ions to this reduced Bohm speed, is also reduced, and ignoring collisions in the approach to the plasma/sheath boundary this is

$$\frac{e\phi_s}{kT_e} = \frac{1}{2}\left(\frac{1+\alpha_s}{1+\gamma\alpha_s}\right). \qquad (9.28)$$

Q Using the same approach as in Section 3.2.1, establish Eq. (9.27) when the negative ion density follows a Boltzmann exponential.

A For a positive ion sheath, the positive ion density falls slower than that of the negative species as the potential becomes increasingly negative, so

$$\frac{d\rho}{d\phi} < 0,$$

where

$$\rho = e\left[n_{ps}\left(1 - \frac{2e\phi}{Mu_s^2}\right)^{-1/2} - n_{es}\exp\left(\frac{e\phi}{kT_e}\right) - n_{ns}\exp\left(\frac{e\phi}{kT_n}\right)\right]$$

and the subscript 's' signifies the value at the plasma/sheath boundary, where quasi-neutrality holds: $n_{ps} = n_{es}(1+\alpha_s)$. Carrying out the differentiations:

$$\frac{e^2 n_{es}(1+\alpha_s)}{Mu^2}\left(1 - \frac{2e\phi}{Mu^2}\right)^{-3/2} < \frac{e^2 n_{es}}{kT_e}\left(\exp\left(\frac{e\phi}{kT_e}\right) + \gamma\alpha_s\exp\left(\frac{e\phi}{kT_e}\right)\right).$$

As with the electropositive case (Section 3.2.1), Taylor expansion shows that the inequality is assured for $\phi < 0$ if the terms on the LHS exactly equal those on the RHS when $\phi = 0$, whereupon $u = u_B^*$, as defined by Eq. (9.27).

The analysis is in fact more subtle than it appears above. The negative ion fraction at the sheath edge, α_s, differs from the electronegativity in the discharge centre, α_0, because the negative ion temperature is different from the electron temperature. In terms of the central densities, the ratio of the two Boltzmann populations at the boundary ($\phi = \phi_s$) shows that

$$\alpha_s = \alpha_0 \exp\left(\frac{e\phi_s}{kT_e}(1-\gamma)\right). \qquad (9.29)$$

Equations (9.28) and (9.29) can be solved together to show how the negative ion fraction, α_s, and the potential at the sheath/plasma boundary, ϕ_s, vary with the central electronegativity, α_0. Figure 9.4 shows the results for a fixed temperature

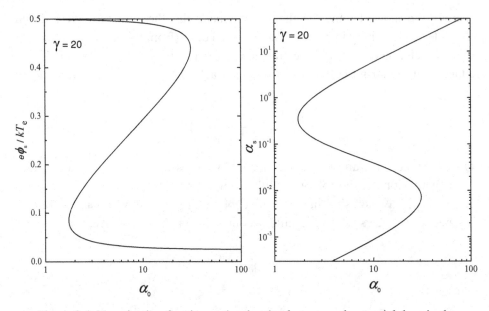

Figure 9.4 Negative ion fraction at the sheath edge, α_s, and potential drop in the presheath, ϕ_s, as a function of the negative ion fraction at the centre, α_0.

ratio of $\gamma = 20$. Both curves are multi-valued in the intermediate-α_0 regime. Three regimes can be identified:

- At low electronegativity, $\alpha_0 \leq 2$, the negative ions do not reach the sheath edge (α_s is almost zero) and the potential drop in the presheath is nearly unperturbed compared to the electropositive solution. The Bohm speed remains almost unchanged. This regime is called the stratified regime, where the centre of the discharge contains negative ions while the presheaths remain electropositive, i.e., free of negative ions.
- In the opposite limit of large electronegativity, $\alpha_0 \geq 30$, the negative ions occupy almost all the plasma volume and reach the sheath edge in significant numbers, and the potential drop in the presheath is small, on the order of $e\phi_s \approx kT_n/2$. As a result, the positive ion speed at the sheath edge is reduced and is $u_B^* \approx (kT_n/M)^{1/2}$ (obtained by taking the limit $\alpha_s \to \infty$ in Eq. (9.27)). This regime is called uniform, because the electron density is fairly uniform in all the plasma, there being almost no potential drop from the centre to the sheath edge.
- In the intermediate regime, both curves are multi-valued, and the two branches described above co-exist with a third solution to the equations. The appropriate physical solution will be discussed briefly in the following.

It turns out that for $\gamma \gtrsim 10$ the curves of α_s and ϕ_s are always triple-valued in the region of $\alpha_0 \sim 1$, so it is necessary to determine which conditions prevail at

9.3 Electronegative plasmas

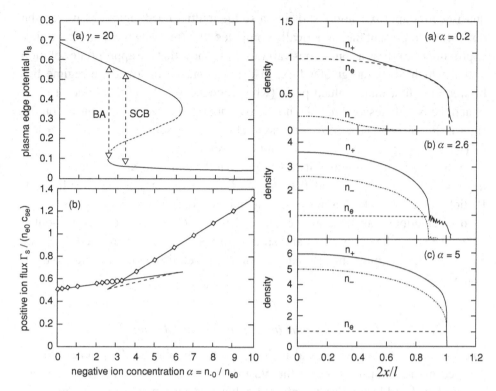

Figure 9.5 On the left, the potential drop in the presheath and normalized flux at the wall as a function of α_0. On the right, the density profiles in the three regimes. From results of numerical simulations in [175].

the plasma/sheath boundary. Sheridan and coworkers have analysed the problem of intermediate electronegativities using numerical solutions of the fluid equations [175], kinetic theories [242] and particle-in-cell simulations [176]. They did not impose quasi-neutrality and they solved Poisson's equation self-consistently. Their findings can be summarized by examining the curves shown in Figure 9.5, that were obtained with the fluid calculations [175]. First, on the left, is shown the potential drop in the presheath and the normalized positive ion flux at the wall as a function of α_0. Note that the curve is somewhat different from that in Figure 9.4, although the multi-valued feature appears. The difference lies in the fact that Sheridan *et al.* included the ionization term in the momentum equation, which was not the case in the calculation by Braithwaite and Allen; note that for $\alpha_0 = 0$, the potential drop is $e\phi_s/kT_e \approx \ln 2$ rather than $e\phi_s/kT_e \approx 1/2$, as already discussed in Section 3.3.1. This difference is only quantitative and does not change the physical meanings discussed here.

The positive ion flux to the wall is plotted below the potential curve in Figure 9.5. The lines represent the flux associated with the potential variations within the quasi-neutral approximation, showing also the multi-valued region. However, the dots, which represent the numerically computed flux (relaxing the quasi-neutrality approximation) show a continuous variation, and show that the appropriate solution is that which gives the greater flux. In conclusion, the intermediate regime lies between the first multi-valued point (when increasing α_0) and the value of α_0 at which the two fluxes associated with the two branches are equal. As shown on the right in Figure 9.5 and already discussed above, the discharge is stratified at low α_0 (with electropositive presheaths) and uniform at high α_0.

In the intermediate regime, it was found that the solutions are oscillatory when approaching the sheaths, and that double layers form before the sheath itself. Particle-in-cell simulations have shown that the oscillations are an artefact of fluid calculations, but have also confirmed that double layers do arise. Double layers have also been observed and studied in a somewhat different context (see Section 9.4). The following references contain further discussions relevant to these issues: [177–181].

9.3.3 Transport in electronegative plasmas

The inclusion of negative ions also significantly complicates the problem of the charged-particle transport within the plasma. As already discussed, negative ions are produced and lost within the plasma volume because they cannot escape from the potential well for negative charge that is sustained in the plasma bulk by the continued escape of the more energetic and more mobile electrons. It turns out that the mechanisms for production and loss may be very different depending on the gas mixture and the gas pressure. In some cases, negative ions are lost by mutual recombination with positive ions, while in other situations they are lost by detachment on collision with excited neutral atoms or molecules. On top of this, it should be noted that the simplicity of a plasma with only one type of positive ion and only one type of negative ion almost never exists: negative ions are formed in certain *molecular* gases (hydrogen, oxygen, halogens, fluorocarbons, etc.) and inevitably the plasma will contain a rich mixture of charged species. These issues are very important for careful modelling of electronegative plasmas and for appreciating the limitations of the models. Nevertheless, as in earlier sections, electrons and a single negative ion species are both assumed here to be in Boltzmann equilibrium with the potential:

$$n_e = n_{e0} \exp\left(\frac{e\phi}{kT_e}\right), \tag{9.30}$$

$$n_n = n_{n0} \exp\left(\frac{e\phi}{kT_n}\right), \tag{9.31}$$

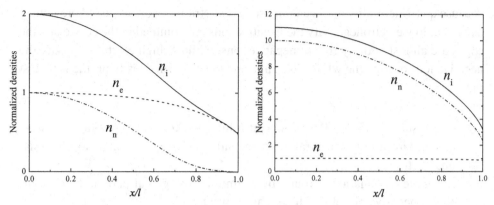

Figure 9.6 Charged particle density profiles for $\alpha_0 = 1$ (left) and $\alpha_0 = 10$ (right). The pressure was chosen such that $l/\lambda_i = 10$. The electron temperature is 3.1 eV for $\alpha_0 = 1$ and 3.6 eV for $\alpha_0 = 10$.

where $\phi = 0$, $n_e = n_{e0}$ and $n_n = n_{n0}$ at the discharge centre. Taking only a single (dominant) species of positive ion requires two equations for conservation of number and momentum:

$$\frac{d(n_i u_i)}{dx} = v_{iz} n_e, \qquad (9.32)$$

$$e n_i E = n_i m_i v_{in} u_i. \qquad (9.33)$$

Combining the above equations leads to the three following equations for a quasi-neutral plasma, for the three variables n_i, u_i and ϕ:

$$\frac{d(n_i u_i)}{dx} = v_{iz} n_{e0} \exp\left(\frac{e\phi}{kT_e}\right), \qquad (9.34)$$

$$-e\frac{d\phi}{dx} = m_i v_{in} u_i, \qquad (9.35)$$

$$n_i = n_{e0} \exp\left(\frac{e\phi}{kT_e}\right) + n_{n0} \exp\left(\frac{e\phi}{kT_n}\right). \qquad (9.36)$$

These may be numerically integrated to obtain the density and potential profiles. As discussed in Chapter 3, the procedure is to choose an electron temperature and a trial value of the electronegativity at the centre (α_0), and then integrate from the centre to the boundary where a condition must be specified, iterating on the central electronegativity until the specified condition is reached. The appropriate condition is to set the ion fluid speed to the sound speed in electronegative plasmas, defined by Eq. (9.27). The profiles obtained by integration of Eqs (9.34)–(9.36) are shown in Figure 9.6, and they are similar to the results of particle simulations shown in Figure 9.5 (note that the sheath is not resolved in the present model because quasi-neutrality has been imposed). The stratification discussed previously occurs: at low electronegativity the negative ions are confined in the centre and the edges are almost entirely free of negative ions, while at high electronegativity the negative ions occupy the whole discharge and the electron density profile is almost flat.

Q The results in Figure 9.6 show that a higher electron temperature is associated with a higher central electronegativity. Give an explanation for this observation.

A A simple explanation is found by examining the global particle balance in electronegative plasmas, which may be written

$$n_e n_g (K_{iz} - K_{att}) = h_1 n_i u_B^* \frac{A}{V},$$

where u_B^* is the modified Bohm speed and h_1 is the edge-to-centre positive ion density ratio. Using quasi-neutrality, $n_e + n_n = n_i$, then leads to

$$K_{iz} - K_{att} = \frac{h_1 u_B^* (1 + \alpha) A}{n_g V}.$$

Since $K_{iz} - K_{att}$ is a strongly increasing function of T_e, it follows that the electron temperature increases moderately when α_0 increases. Intuitively, it seems reasonable that a smaller fraction of electrons must have a higher average energy to maintain the plasma.

Comment: *Given the very complicated spatial structure of the densities shown in Figure 9.6, one should wonder if the flux leaving the plasma can still be written in the form $h_1 n_i u_B^*$, and if so, what is the appropriate expression for h_1. It turns out that it is a very complicated problem. Monahan and Turner [182] have offered a thorough analysis of global models of electronegative discharges and tested them against particle-in-cell simulations. In particular, they have discussed the h_1 formulas proposed by Kim et al. [183].*

Further insight into the subtleties of electronegative plasmas can be found in a large section of Lieberman and Lichtenberg [2] (second edition, chapter 10) and in the related articles by the same authors [184, 185]. Another extensive study of electronegative plasmas has been reported by Franklin [186]. One very important point raised is the fundamental difference between the recombination-dominated case (see [187]), in which a minimum electron temperature is required to satisfy the particle balance, and the detachment-dominated case, for which such a criterion does not arise (see [188]).

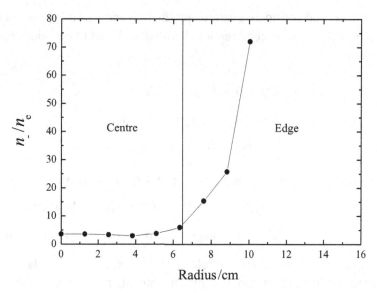

Figure 9.7 Negative ion fraction as a function of the radius in a magnetized helicon plasma in SF_6. The solid line at 6.5 cm separates the central region (where the helicon wave energy is absorbed) from the diffusion region near the edge. From [190].

9.3.4 Transport in magnetized electronegative plasmas

The effect of a static magnetic field is to increase the plasma confinement, chiefly because electron motion across the magnetic field is strongly impeded. In most cases, the mobility of positive ions moving perpendicular to the magnetic field is unaffected and may be larger than that of the electrons. In electronegative plasmas there will also be negative ions and since these are slow, heavy particles, their Larmor radius is of the same order as that of positive ions. Consequently, one may expect that the electrons will be 'filtered' out of a plasma flow across a magnetic field, while positive and negative ions diffuse together to maintain quasi-neutrality. This has indeed been observed experimentally by several authors [189, 190]; a so-called ion–ion plasma is formed. In Figure 9.7, the negative ion fraction is shown as a function of the radius in a magnetized helicon plasma in SF_6 (reproduced from [190]). The static magnetic field is along the z-axis so that the radial direction is perpendicular to it. The electronegativity increases drastically from the centre towards the edge, essentially because the electron density falls rapidly in the outer regions where there is little ionization. The formation of an ion–ion plasma at the edge is at the basis of the dual-ion (PEGASES) thruster (Section 1.3). Ion–ion plasmas have also been considered for charge-free etching in microelectronics.

Franklin and Snell [191] have proposed a fluid model to quantify the electron filtering described above. They reduced the problem to one dimension (an infinite slab or cylinder) and considered various production and loss mechanisms for negative ions. When ion–ion recombination dominates, the set of equations to solve is the following:

$$\nabla(n_e v_e) = (K_{iz} - K_{att})n_g n_e, \qquad (9.37)$$

$$\nabla(n_i v_i) = K_{iz} n_g n_e - K_{rec} n_i n_n, \qquad (9.38)$$

$$\nabla(n_n v_n) = K_{att} n_g n_e - K_{rec} n_i n_n; \qquad (9.39)$$

$$m(K_{iz} n_g - K_{att} n_g + v_e) n_e \mathbf{v}_e + e n_e \mathbf{E} + kT_e \nabla n_e + e n_e \mathbf{v}_e \times \mathbf{B} = 0 \qquad (9.40)$$

$$kT_i \nabla n_i - e n_i \mathbf{E} + M_i \nabla(n_i \mathbf{v}_i \mathbf{v}_i) + M_i v_i n_i \mathbf{v}_i = 0 \qquad (9.41)$$

$$kT_n \nabla \mathbf{n}_n + e n_n \mathbf{E} + M_n \nabla(n_n \mathbf{v}_n \mathbf{v}_n) + M_n v_n n_n \mathbf{v}_n = 0 \qquad (9.42)$$

The first three equations are the particle conservation equations, where electrons are produced by ionization and lost by attachment, positive ions are produced by ionization and lost by recombination with negative ions, and finally negative ions are produced by attachment and destroyed by the recombination with positive ions. The other three equations balance the forces on the three charged fluids. The key issue lies in the boundary conditions. Franklin and Snell [191] have investigated the case when the negative ion flux at the edge is zero (for example, in the case when a positive ion sheath forms). In one dimension, the charge fluxes are balanced at each position, $\Gamma_e + \Gamma_n = \Gamma_i$, so that the Franklin and Snell condition leads to $\Gamma_e(R) = \Gamma_i(R)$. However, with strong electron filtering one also requires $\Gamma_e(R) \to 0$ (no electrons reach the edge) so that all the particle fluxes must end up zero at the edge, meaning that the charged particles have to be produced and destroyed in the volume. Leray et al. [192] have revisited the problem in order to allow for ion extraction at the edge. They showed that the problem cannot be solved in one dimension because the condition of no electron flux at the edge imposes the following condition, obtained from integration of Eq. (9.37):

$$\Gamma_e(R) = (K_{iz} - K_{att})n_g \int_0^R n_e(x) dx = 0, \qquad (9.43)$$

which can only be achieved if $K_{iz} = K_{att}$; i.e., again electrons have to be produced and lost in the volume. To get around this, they proposed modelling the region of ionization as a finite cylinder for which there were no radial (cross-field) losses of electrons but losses along the axis (in which direction the motion is unimpeded by the field) were modelled as an effective volume (integrated) loss, and then replaced Eq. (9.37) by

$$\nabla(n_e v_e) = (K_{iz} - K_{att})n_g n_e - \nu_L n_i, \qquad (9.44)$$

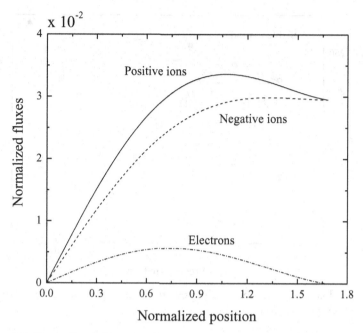

Figure 9.8 Normalized fluxes as a function of normalized radius showing the decay of the electron flux to form a flux of ion–ion plasma at the edge of a magnetized cylinder; in the chosen normalization scheme the radial wall was located at 1.8. From [192].

where ν_L accounts for electron losses in the axial direction. In this way, they were able to obtain solutions in which the electron flux (and the electron density) vanishes to zero at the edge, with finite and equal positive and negative ion fluxes, as shown in Figure 9.8.

9.3.5 Instabilities at the E–H transition in electronegative gases

In electronegative plasmas, widely used in plasma etching (for example, O_2, SF_6, CF_4, Cl_2, etc.), the E–H transition in inductive discharges has been found to be unstable [19, 20, 124–126, 193–195].

Experimental observations

Figure 9.9 shows the instability window in the power/pressure space, as measured in a CF_4 inductive discharge. At low power, the discharge operates in a stable capacitive (E) mode, while at high power the discharge operates in a stable inductive (H) mode. In the region marked in grey in the figure, the discharge parameters (electron and ion densities, electron temperature, etc.) undergo large relaxation oscillations. The light emitted by the plasma therefore fluctuates (see insert) with

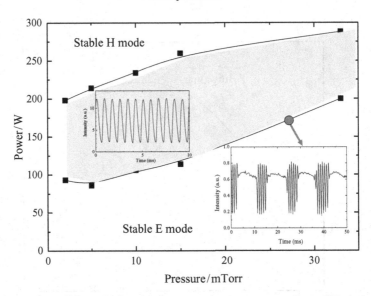

Figure 9.9 Instability window in the power/pressure space for a CF_4 inductive discharge. Originally from [124].

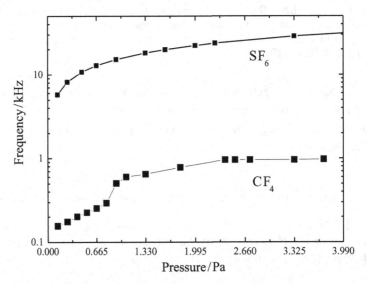

Figure 9.10 Frequency of the oscillations measured in CF_4 and SF_6.

a clearly defined frequency; near the edges of the unstable region more complex behaviour is observed with intermittent bursts of instability.

The frequency of these relaxation oscillations varies over a wide range, from a few hundred Hz to several tens of kHz. The frequency depends mostly on the gas composition and the gas pressure, but it is also affected by other parameters such as the RF power and the match-box settings. To illustrate this, Figure 9.10 shows

9.3 Electronegative plasmas

the frequency of the instability as a function of the gas pressure, for SF_6 and CF_4 as feedstock gases. The frequency increases with the gas pressure for both gases, and the frequency is much higher in SF_6. More generally, it has been shown that the frequency increases with the electronegativity of the discharge, which increases with the gas pressure, and is much higher in SF_6 than in CF_4.

Finally, it has also been shown that when the frequency is not too high (typically in CF_4), the gas chemistry (including dissociation, chemical reactions at surfaces and gas heating) is modulated during the instability [26, 124, 196].

Global model of the E–H instability

A global model has been proposed to explain the instability mechanism [19, 20]. This model is based on an inductive discharge with capacitive coupling, as in Chapter 7. However, in electronegative gas mixtures the plasma is composed of electrons, positive ions and negative ions. Moreover, to model the instability one has to keep time-dependent electron and negative ion terms in the global model equations. Using the quasi-neutral approximation, $n_e + n_n = n_i$, the electron and negative ion particle balance equations are

$$\frac{dn_e}{dt} = n_e n_g (K_{iz} - K_{att}) + n_n n_g^* K_{det} - \Gamma_e \frac{A}{V}, \qquad (9.45)$$

$$\frac{dn_n}{dt} = n_e n_g K_{att} - n_n n_g^* K_{det} - n_n n_i K_{rec} - \Gamma_n \frac{A}{V}. \qquad (9.46)$$

Electrons are generated by ionization and detachment of negative ions, with the reaction coefficients K_{iz} and K_{det}, and are lost by attachment to molecules having a density n_g, with a rate coefficient K_{att}. Electrons are also lost at the wall with a flux Γ_e. Negative ions are produced by attachment, and lost via three mechanisms: (i) recombination with positive ions (with the rate coefficient K_{rec}); (ii) detachment by collisions with metastable species of fixed density n_g^*; (iii) lost at the wall. The third mechanism is unlikely since there is usually a sheath in front of walls, which prevents negative ions from escaping the plasma. There is no need to consider positive ion balance explicitly, because this is assured by the quasi-neutrality assumption.

There must be flux balance at the wall to ensure that quasi-neutrality can be maintained, $\Gamma_e + \Gamma_n = \Gamma_i$. Since $\Gamma_n \approx 0$, it follows that $\Gamma_e \approx \Gamma_i$. Chabert et al. [124] used the following heuristic form for the flux, valid for all values of electronegativity $\alpha = n_n/n_e$:

$$\Gamma_i = \left[\frac{h_{l0} - h_{l\infty}}{(1+\alpha)^{3/4}} + h_{l\infty} \right] n_i u_B = h_l n_i u_B, \qquad (9.47)$$

where

$$h_{l\infty} = \frac{3}{2} \left[1 + \frac{l}{\sqrt{2\pi \lambda_i}} \right]^{-1}. \qquad (9.48)$$

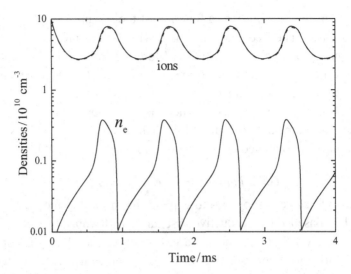

Figure 9.11 Densities during the relaxation oscillations as calculated by the global model for the case of SF_6.

The power balance is

$$\frac{d}{dt}\left[\frac{3}{2}n_e kT_e\right] = P_{abs} - P_{loss}, \qquad (9.49)$$

where P_{abs} is given by a relation very similar to Eq. (7.67), although the authors used somewhat simplified forms of R_{ind}. When the negative ion flux to the wall is neglected, the loss power is given by

$$P_{loss} = n_e n_g e\left(K_{iz}\varepsilon_{iz} + K_{att}\varepsilon_{att} + K_{exc}\varepsilon_{exc}\right) + \Gamma_e\left(e\phi_f + 2kT_e\right)\frac{A}{V}, \qquad (9.50)$$

where $\phi_f \approx 5kT_e/e$, and the flux of negative ions to the wall is neglected. The constants ε_{iz}, ε_{att} and ε_{exc} are typical threshold energies for the related processes.

The three balance equations Eqs (9.45), (9.46) and (9.49) may be solved numerically to calculate the time variations of n_e, n_n and T_e. This can be done for various RF currents in the coil. When doing this, the authors found, as in the experiments, that the discharge was stable (no time fluctuations of the plasma parameters) at low power, and at high power, but experienced relaxation oscillations at intermediate powers. A typical example of calculated densities in the unstable regime is shown in Figure 9.11, for the case of SF_6. The electronegativity is always very large, so that it is almost impossible to distinguish between positive and negative ion densities. This has also been observed experimentally. The frequency of the instabilities is around 1.2 kHz. In the experiment, the frequency was larger at around 10 kHz.

In addition, the window of instability predicted by the model was smaller than the experimental window.

The dynamics of the instability within this model has been reported theoretically in [125]. Some of the discrepancies between theory and experiments have been explained. The general conclusion is that the instability is sensitive to the amount of capacitive coupling. Strategies to suppress the instability by controlling the capacitive coupling have therefore been proposed.

9.4 Expanding plasmas

9.4.1 Electropositive plasmas

Helicon (or cylindrical inductively coupled) plasma processing reactors are generally composed of a cylindrical source region sitting on top of a larger expansion chamber. When the ionization is mainly localized in the source, this geometry leads to a gradient in the plasma density, decreasing from the source to the bottom of the expansion chamber. Because electrons are generally in (or very close to) Boltzmann equilibrium, this gradient of electron density is accompanied by a gradient in (DC) plasma potential. Consequently, there is a weak electric field that accelerates positive ions out of the source region while confining, to some extent, the electrons. In general, the ion acceleration is modest and the plasma remains quasi-neutral during the expansion. However, Charles and co-workers [197–199] found experimentally that by adding a strongly diverging magnetic field it is possible to obtain conditions in which ions become supersonic, i.e., exceed the speed of sound (Bohm speed). Chen [200] has used an analytical derivation of a plasma expansion in a diverging magnetic field based on classical sheath theory to show that ions will reach the speed of sound at a position where the plasma radius has expanded by 28%. When ions become supersonic near a surface, the quasi-neutral plasma goes over into a sheath region where there is net positive space charge adjacent to the surface. For an expanding plasma the supersonic condition is generally reached in open space, and in this case a region of positive space charge develops immediately adjacent to a layer of negative space charge, forming a so-called 'double layer'. The double layer is a sort of electrostatic shock across which there is an abrupt potential step between two quasi-neutral plasmas. The positive space charge layer is located on the high potential side, usually called the 'upstream' side, while the negative charge is located on the low potential, 'downstream' side.

In Figure 9.12, the plasma potential is plotted as a function of the axial distance. The symbols are retarding field analyser measurements (see Chapter 10) taken by Charles and co-workers [199], whereas the solid line is the prediction of a model proposed by Lieberman *et al.* [201]. This model couples particle balance upstream

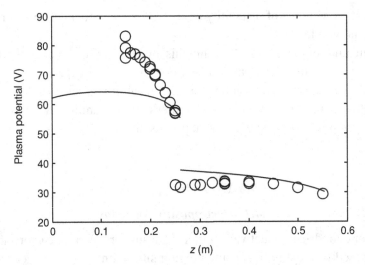

Figure 9.12 Potential as a function of axial distance in a helicon plasma (argon 0.03 Pa) under conditions where a double layer forms at $z = 0.25$ m; the circles are measurements and the solid line comes from a model. From [199].

and downstream to the requirements of the double layer. The double layer is marked by an abrupt change in potential of about 25 V at around $z = 0.25$ m.

Positive ions enter the double layer from upstream, at the Bohm speed, because the Bohm criterion has to be fulfilled in order to form the space charge. They are subsequently accelerated to a much larger speed when falling across the 25 V double-layer potential drop.

Q At what speed do argon ions leave the double layer downstream?
A Applying energy conservation, that is neglecting collisions within the double layer, the speed is $v_i = \sqrt{2e\Delta V/M + u_B^2}$.
With $\Delta V = 25$ V, $M = 40$ amu and $kT_e/e = 5$ eV, the result is $v_i = 11\,500$ m s^{-1}.

The acceleration of ions across the double layer forms a positive ion beam downstream that has been observed in experiments [199] – see Figure 9.13. The ion velocity distribution function, measured by a rotating retarding field analyser (RFA – see Section 10.3), exhibits two peaks when the RFA faces the source (the solid line), and only one peak when the RFA is rotated by 90° (dashed–dotted line). The other curves represent intermediate angles. The high-energy peak is the signature of an ion beam, that is ions that were born upstream and accelerated through the double layer. The peak at low energy (around 25–30 V) is due to ions generated downstream and accelerated by the sheath that forms in front of the RFA. This peak

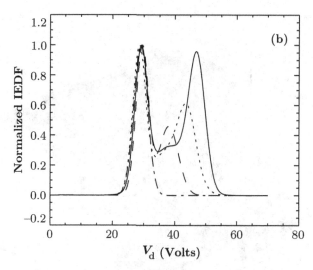

Figure 9.13 Ion velocity distribution downstream measured with a rotating retarding field analyser and plotted on an energy scale (see Section 10.3.2). A high-energy peak, signature of an ion beam, is detected when the RFA faces the source (the solid line). The beam does not appear when the RFA is rotated by 90° (dashed–dotted line). Reproduced from [199].

sits at the local plasma potential. As mentioned in Chapter 1, Charles's group at the Australian National University has proposed taking advantage of the positive ion beam acceleration to generate thrust in satellites. The concept is called HDLT (for helicon double-layer thruster). A theoretical analysis of the momentum transferred to ions within the double layer has been performed by Fruchtman [202].

9.4.2 Electronegative plasmas

Double layers have also been observed in similar reactors operating with electronegative gases [203, 204]. It has been shown that in electronegative media the diverging magnetic field is not necessary to observe double layers, partly because the Bohm speed is much lower in electronegative plasmas. Therefore, the electric field amplitude required to accelerate ions to supersonic speed is lower. In the absence of a magnetic field, double layers were not observed for purely electropositive gases (typically argon), but when between 5% and 15% SF_6 was added to the argon, stable double layers were formed at the junction between the source tube and the expansion chamber. These double layers had weaker amplitudes of only 6–8 V. As the SF_6 percentage was increased further, the double layers become unstable [204]. This is illustrated in Figure 9.14, which shows the plasma potential as a function of space and time measured in the reactor. The abrupt drop in plasma potential

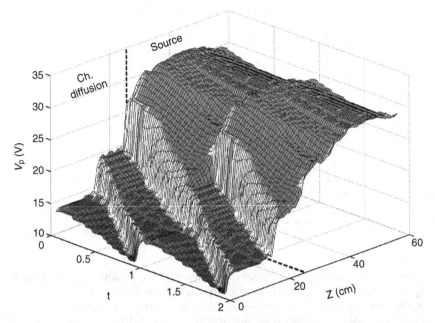

Figure 9.14 Periodic formation and propagation of double layers in the expanding chamber of an inductive/helicon plasma source working with Ar/SF$_6$ mixtures.

resembles cliffs in this representation; this structure forms at the interface between the source and the expansion chamber and subsequently travels downstream as time evolves. The double layers propagate relatively slowly, at about 150 m s^{-1}. It was found that a second double layer forms before the first one has reached the bottom of the chamber, so that two double layers co-exist at any particular time during the cycle.

The unstable behaviour of expanding, electronegative, inductive discharges was first reported by Tuszewski [205]. Later, Tuszewski and Gary [206] demonstrated that the expanding plasma becomes linearly unstable if the difference between the positive ion and the negative ion drift velocities exceeds some threshold (positive and negative ions are streaming in opposite directions). Although that is certainly a consequence of double layers, the exact connection between the kHz-level relaxation oscillations and slowly drifting double layers remains unclear at the time of writing.

9.4.3 Reflections

This chapter has considered a few scenarios that are more realistic than the simplified systems that were the subject of earlier chapters. It should be clear that the inclusion, for instance, of magnetic fields or electronegativity from the outset

9.4 Expanding plasmas

would have obscured the quantitative scalings and classifications that have been revealed for E, H and W-modes of plasma excitation. There are many other realities that apply in the wide variety of technological applications of plasmas.

Q Give a preliminary comment on the following realities.
  (i) Plasma sources often operate in molecular gases.
  (ii) There is usually more than one species of positive ion.
  (iii) The pressure is often between the collisionless and collisional regimes.
  (iv) The interaction of plasmas with surfaces means that the plasma composition is not uniform.
  (v) $\omega_{pi} \ll \omega < \omega_{pe}$.

A The following responses are intended to be relevant comments but are not comprehensive answers.
  (i) Electron energy will be diverted from ionization into molecular excitations – the RF power may have to be higher than simple models suggest to sustain the plasma at a given density.
  (ii) One species of positive ion may dominate but if not, then one must start to consider where each species is produced and whether there are mechanisms that couple the different species together – an effective (average) ion mass could be considered, but there will be times when a more careful approach is required.
  (iii) There has been some attention given to this.
  (iv) Depletion of reactants may occur in time and space and etch products may 'contaminate' the plasma. These phenomena may severely perturb the parameter space established by the simple, chemistry-free models.
  (v) Ions will respond to lower frequencies, so the simple mono-energetic response to the mean field needs to be reconsidered, especially if $\omega \lesssim \omega_{pi}$.

Whether the behaviour of a plasma is or is not close to the predictions of a model must be judged on the basis of measurements or the key plasma parameters such as electron density, mean electron energy, potential of the plasma, etc. There are various ways to proceed with such measurements using optical, electrical or indeed electro-optical methods, and the topic of plasma diagnostics is worthy of a book in its own right. The final chapter looks at just a few, concentrating on techniques that fall within the electrical category and that are modelled using the same equations and data sets that have already been used to model capacitive, inductive and helicon plasma sources.

10

Electrical measurements

As a simple means of probing the charge composition of a plasma, one of the more immediate temptations for an experimental plasma physicist is to insert some kind of small, refractory, electrically conducting material, such as a bare wire. Applying a potential to this conductor might then be supposed to enable it to act as a rudimentary collector of charged particles.

> **Q** Explain why a refractory material has been specified for the collection of electrons or positive ions from a plasma, bearing in mind the nature of bounded plasmas as set out in earlier chapters and the difficulty of taking heat out of a small wire probe immersed in an ionized gas at low pressure.
>
> **A** It has already been established in earlier chapters that the mean thermal energy of the electron population in a low-pressure plasma is typically a few eV. It has also been found that ions are naturally expelled with at least the Bohm speed, and may pick up additional energy in the sheath, so the mean ion energy at a surface may be several eV; furthermore, the neutralization of an ion on a surface will liberate the ionization energy, which is also several eV. Compared with the molecules of any residual gas, the charged particles are hundreds of times more energetic than the gas, so one should anticipate the possibility of a small, thermally isolated surface becoming heated.

In the 1920s, Langmuir was one of the first to develop an electrostatic probe method based on the insertion of a small, charge-collecting surface. There are various forms including planar, cylindrical and spherical geometries for so-called single, double and triple probes. There are also probes that use electrical resonances and others that launch and detect waves. All types can be used in steady and transient plasmas, though there is always an upper frequency limit. Special schemes have been devised for RF plasmas, using techniques that compensate for the RF potential fluctuations with passive or active circuitry. Magnetized plasmas pose further challenges. Each configuration is accompanied by assumptions that constrain both the applicability and the analytical methods that translate the measured currents and voltages variously into charge densities, space potentials, particle fluxes, energy distributions and measures of collisionality.

This chapter will take a broad look at the options and opportunities for electrical probes, suitable for the environment of the RF plasmas. First, the traditional electrostatic (Langmuir) probe will be analysed in sufficient detail to recognize the main benefits and limitations of the method. Next, a relatively simple development of the electrostatic probe is introduced, adding an electrostatic filter to form a retarding field energy analyser. In the next section, a range of high-frequency probes will be described. Finally, the topic will be broadened to include global methods such as wave transmission and impedance analysis. It should be noted that data on plasma behaviour can also be extracted non-invasively by optical methods that are both localized and time-resolved, but at the expense of much more sophisticated apparatus and analysis – optical measurements are not discussed in this book.

10.1 Electrostatic probes

A simple electrostatic probe can be made by placing a short bare wire projection from a coaxial cable directly into the plasma volume and biasing it with respect to some other conducting surface in contact with the plasma. The cable must be enclosed and sealed so that it is compatible with vacuum and plasma. In this section, current–voltage relationships will be established for the two most common configurations namely symmetrical (double) probes and highly asymmetrical (single) probes. Planar and cylindrical geometries are considered.

10.1.1 Planar geometry

To form a device that can be used locally to sample the charged particle fluxes, one needs to consider a small area of surface that can be independently biased at a particular potential (Figure 10.1). The equations developed in Chapter 3 when considering what were imagined to be *external* plasma boundaries can be applied here for the *internal* boundaries that are formed around objects inserted in the plasma. For a planar surface, area A, at a potential ϕ which is below that of the plasma (ϕ_p), the electron and ion fluxes can be deduced from Eqs (3.29) and (3.31). The net current to the surface is then

$$I = eA\left[-\frac{n_0\bar{v}_e}{4}\exp\left(\frac{e(\phi-\phi_p)}{kT_e}\right) + n_s u_B\right] \qquad \phi < \phi_p. \qquad (10.1)$$

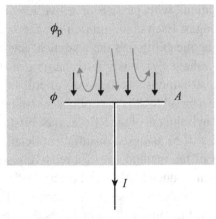

Figure 10.1 Particle fluxes to a surface (area A, potential ϕ) immersed in plasma (potential $\phi_p > \phi$) – electron flux is indicated in grey, ion flux in black. Imbalance between the fluxes forms a net current I. It is supposed that the lower surface is insulated from the plasma, so current is collected only on the upper surface.

The electron flux has been written in terms of the density of the 'undisturbed' plasma, n_0, some way from the surface, whereas the ion flux is written in terms of the plasma density at the sheath plasma boundary, n_s. It was shown in Chapter 3 that when an external plasma boundary actively determines the balance between production and loss, n_s and n_0 are related by h_l factors – when small probes are inserted in confined plasmas, that is the plasma boundary is internal, n_s at the boundary and n_0 in the bulk plasma are not necessarily related by h_l. Notice that if the surface were biased very negatively with respect to the plasma, it would collect virtually no electrons and could then be said to collect an 'ion saturation current', which in planar geometry is $n_s e u_B$ per unit area. Also note that if it were biased at or above the DC plasma potential, the current density would be $-n_0 e \bar{v}_e/4$, as the ions would no longer be attracted. Clearly, Eq. (10.1) is not strictly correct at the plasma potential because it still contains the ion current (based on there being an ion sheath), but the error is in most cases negligibly small.

It is useful to catalogue the assumptions that underpin Eq. (10.1).

(i) All incident charge is absorbed by the surface.
(ii) The arrival and recombination of charge does not initiate any secondary processes through which charge is emitted from the surface.
(iii) Any other secondary processes, such as photo-emission, do not release a significant quantity of charge.
(iv) The electron population is in equilibrium with the electric field (that is the use of the Boltzmann factor is appropriate).
(v) The ion flux into the sheath around a small planar probe surface is governed by factors similar to those that apply at large external boundaries.

10.1 Electrostatic probes

(vi) One species of singly charged positive ions exists in the plasma.
(vii) The distributions of particle speeds far from the surface are isotropic.
(viii) The surface is sufficiently small that it does not significantly affect the overall particle and energy balances and therefore does not perturb the factors that determine electron temperature and plasma density.
(ix) Unless otherwise stated, the motion of charged particles near probes is collisionless.

The first three assumptions can be ensured by the use of clean, refractory, metals with a high work function, provided the incident energy of ions is no more than a few 10s of eV. The others need to be borne in mind before drawing conclusions from specific probe data.

Q If a probe were to draw a steady current of electrons from a plasma, what would happen to the potential of the plasma, supposing the normal production and loss processes to remain in balance?
A The plasma potential would tend to become more positive if there were a net removal of negative charge.

In practice, for a low-pressure plasma that is generated by volume ionization and lost by recombination at the walls, a steady loss of electrons to a small surface can be sustained indefinitely without changing the plasma potential provided there is an equivalent loss of positive ions to a second 'probe' surface, not too far away. In effect, there must be an electrical current in an external circuit that ultimately unites positive and negative charge arriving from the plasma by these two separate routes. Within the plasma the same current must also flow between the collection points. For the plasma to be able to remain an equipotential the density of probe current in the plasma must be much less than the random thermal current density.

So now suppose that there are two probe surfaces, joined by a battery that fixes the potential between them so that $\phi_2 - \phi_1 = V$ and the current in the external circuit is I, as illustrated in Figure 10.2. To derive a single expression for the current–voltage relationship, $I(V)$, that could then be measured, first use Eq. (10.1) written in terms of A_2 and ϕ_2 to define a term containing the 'unknown' plasma potential, ϕ_p, and density, n_0, as

$$-\frac{n_0 \bar{v}_e}{4} \exp\left(\frac{-e\phi_p}{kT_e}\right) = \left(\frac{I_2}{eA_2} - n_s u_B\right) \exp\left(\frac{-e\phi_2}{kT_e}\right).$$

Then, using Eq. (10.1) in terms of A_1 and ϕ_1 can be written

$$I_1 = eA_1 \left[\left(\frac{I_2}{eA_2} - n_s u_B\right) \exp\left(\frac{-e\phi_2}{kT_e}\right) \exp\left(\frac{e\phi_1}{kT_e}\right) + n_s u_B\right].$$

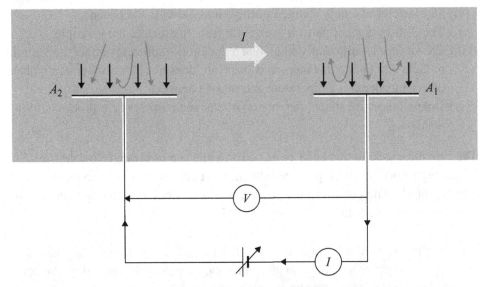

Figure 10.2 Two surfaces linked by a battery – the potential is distributed in order that the current is continuous. It is supposed that the lower surfaces are insulated from the plasma, so current is collected only on the upper surfaces. $I(V)$ is given by Eq. (10.2).

Matching the currents so that $I = +I_1 = -I_2$, this then rearranges to

$$I\left(1 + \frac{A_1}{A_2}\exp\left(\frac{-eV}{kT_e}\right)\right) = n_s u_B e A_1 \left(-\exp\left(\frac{-eV}{kT_e}\right) + 1\right).$$

The general current–voltage relationship for planar surfaces is then

$$I = n_s u_B e A_1 \left[\exp\left(\frac{eV}{kT_e}\right) - 1\right]\left[\exp\left(\frac{eV}{kT_e}\right) + \frac{A_1}{A_2}\right]^{-1}. \qquad (10.2)$$

10.1.2 Symmetrical double probe

In this subsection it will be supposed that there are two identical collecting surfaces, $A_1 = A_2 = A$, exposed to the same plasma, for instance, placed side by side, a few tens of λ_{De} apart. This arrangement forms a so-called symmetrical double probe. Then a further simplification of the expression for the current follows on taking $\exp(eV/2kT_e)$ out of the two sets of square brackets in Eq. (10.2), giving

$$I = I_i \tanh\left(\frac{eV}{2kT_e}\right), \qquad (10.3)$$

where $I_i = n_s u_B e A$ is the ion saturation current to one of the probe surfaces. Since the voltage applied to a double probe does not need to be referenced to ground, it is sometimes termed 'a floating double probe'. As the potential difference between the probes is swept from a large negative to a large positive voltage, the current changes symmetrically between the ion saturation current being collected by one surface through zero to the ion saturation current being collected by the other surface. In each case the net electron current at the other surface is just sufficient to maintain current continuity. No matter how hard one tries, neither probe surface can be taken close to plasma potential because once a surface is drawing the ion saturation current, all additional applied voltage is accommodated across the adjacent sheath, which widens accordingly.

Q Use Eqs (3.7) and (3.90) to estimate the width of a sheath on a small planar probe across which there is a potential difference of 25 V when immersed in a plasma having 10^{16} electrons m^{-3} with a temperature equivalent to 2 eV, and hence suggest a minimum diameter for a disk-shaped planar probe surface in this plasma.

A From Eq. (3.7), the Debye length for 10^{16} electrons m^{-3} and 2 eV is 10^{-4} m, so Eq. (3.90) gives

$$s = \sqrt{2eV_0/kT_e}\, \lambda_{De} = 5 \times 10^{-4} \text{ m}.$$

A disk probe would need to have a diameter several times this, say ~5 mm, to ensure that the sheath forms a thin, conformal layer, with negligible effect of the edges on current collection.

Comment: The ion matrix model tends to underestimate sheath dimensions, but it is adequate for the present estimate; a denser plasma would lead to a smaller limit on the minimum diameter.

It is instructive also to evaluate the change in potential of the more positive probe that is required for the more negative side to draw the ion saturation current. When the applied voltage is zero, both probes float with respect to the plasma. That is,

$$0 = eA \left[-\frac{n_0 \bar{v}_e}{4} \exp\left(\frac{-e\phi_p}{kT_e}\right) + n_s u_B \right], \quad (10.4)$$

which specifies the potential difference between a floating probe and the plasma, ϕ_p. When $eV \gg kT_e$ the saturation current is drawn. At the more negative probe this is due entirely to the *arrival* of positive ions that account for a current I_i into the external circuit. At the more positive probe since the same flux of ions also arrives here and this must be offset, so twice that amount of electron flux must be

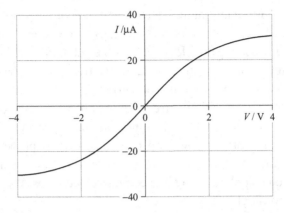

Figure 10.3 A double-probe characteristic for a plasma in argon at 5 Pa.

collected to maintain current continuity, effectively taking I_i back into the plasma. Therefore, at the more positive probe,

$$-I_i = eA\left[-\frac{n_0 \bar{v}_e}{4}\exp\left(\frac{e(\Delta\phi - \phi_p)}{kT_e}\right) + n_s u_B\right],$$

where $\Delta\phi$ is the required small positive shift in floating potential and $I_i = Aen_s u_B$. Using Eq. (10.4) this now simplifies to

$$\exp\left(\frac{e\Delta\phi}{kT_e}\right) - 1 = 1,$$

that is $\Delta\phi = \ln 2 \times kT_e/e$. This is the maximum excursion towards plasma potential by the more positive probe – all other applied voltage must be developed by the more negative electrode shifting further away from plasma potential.

Q Calculate ϕ_p, the potential difference between the floating probe and the plasma, for an argon plasma and compare it with $\Delta\phi$ if $n_0/n_s = 0.5$.
A Equation (10.4) can be simplified to obtain $\phi_p = [\ln(n_0/n_s) + 0.5\ln(M/2\pi m)] kT_e/e$. In argon at low pressure ($h_l = 0.5$), this equates to $5.4kT_e/e$ (cf. Eq. (3.32)). This is almost 8 times $\Delta\phi$.

Exercise 10.1: Double probe analysis Figure 10.3 shows a double-probe characteristic for an argon plasma at 5 Pa; the probe is formed by two single-sided disks of diameter 5 mm. Determine the electron temperature and the charged particle density of this plasma just outside the sheath. [*Hint*: The slope of $\tanh(ax)$ at $x = 0$ is a.]

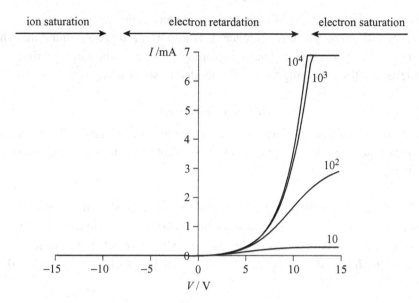

Figure 10.4 Current–voltage (I–V) characteristics (Eq. (10.2)) for asymmetrical planar probes having area ratios 10, 10^2, 10^3 and 10^4 in the same plasma as that for Figure 10.3.

10.1.3 Asymmetrical (double) probe

Now consider the case when the two collecting areas are not identical. The asymmetrical arrangement means that when the larger surface, say A_1, draws its ion saturation current ($n_s u_B e A_1$), then the smaller surface, in drawing an equivalent net current over a smaller area (A_2), must collect a larger flux of electrons. To do this the surface potential must shift closer to plasma potential than the maximum $\ln 2 \times kT_e/e$ found for the symmetrical case. The larger the ratio A_1/A_2, the larger the shift in ϕ_2 until at some stage it reaches ϕ_p – the electron flux to the surface is no longer retarded by the Boltzmann factor and ions are no longer attracted to it. The model developed so far does not apply when $\phi_2 > \phi_p$, and in strict planar geometry it must be assumed that the current collected by the smaller area thereafter remains fixed in this condition at $eA_2 n_0 \bar{v}_e/4$.

Figure 10.4 shows what happens to the entire $I(V)$ curve as the area of surface 1 is increased 10 000 times, mapping the current collected by the fixed-area smaller surface 2, as a function of the voltage between it and the larger surface. Notice that all curves pass through zero current when the applied voltage is zero and also that for any given positive voltage the current collected increases as the area ratio is increased.

Since the smaller surface is more localized, it can now be called the 'probe' and the larger surface can be regarded as a 'reference electrode'. Notice that electron collection in this system constitutes positive current. The applied voltage V is the probe bias with respect to the reference. It is important to keep in mind the whole system – probe *and* reference *and* intervening plasma – when interpreting probe characteristics; the following tasks are intended to show why.

Reference electrode

If one chooses a sufficiently large area ratio then the smaller probe always controls the shape of the current–voltage characteristic. That raises the question of just what is 'sufficient'.

Q Based on Figure 10.4, suggest how large an area should be used for the reference electrode (A_1), relative to the probe area (A_2) in the given plasma if the probe current is to pass discontinuously into electron saturation.
A From the figure it is clear that a factor of 1000 is sufficient, whereas 100 is not.

The biggest current that a planar probe can draw is the constant electron saturation current, $I = eA_2 n_0 \bar{v}_e / 4$. Provided that at all voltages $\exp(eV/kT_e) \ll A_1/A_2$, the probe current will reach electron saturation, unimpeded by the collection of current at the reference electrode. Then Eq. (10.2) shows that the current will follow the Boltzmann exponential up to the maximum current:

$$I = n_s u_B e A_2 \left[\exp\left(\frac{eV}{kT_e}\right) - 1\right] \qquad I \leq eA_2 \left[\frac{n_0 \bar{v}_e}{4}\right].$$

The useful limit of the Boltzmann exponential factor is reached when the probe potential reaches the local plasma potential, $V = \phi_p$ (see Figure 10.4). Equation (10.4) shows that in argon, at low pressure, $\phi_p = 5.4 kT_e/e$ so the area ratio condition for an argon plasma becomes

$$\exp(e\phi_p/kT_e) \ll A_1/A_2 \quad \equiv \quad A_1 \gg \sim 200 A_2. \tag{10.5}$$

When the reference electrode is chosen to satisfy this criterion, the smaller electrode is often called 'a single Langmuir probe'. The current–voltage characteristic for a single probe can then also be written

$$I = eA_2 \left[\frac{n_0 \bar{v}_e}{4} \exp\left(\frac{e(V - \phi_p)}{kT_e}\right) - n_s u_B\right] \qquad I \leq eA_2 \left[\frac{n_0 \bar{v}_e}{4}\right]. \tag{10.6}$$

When the positive bias takes the single probe above plasma potential it is said to be in electron saturation – only electrons are collected from an electropositive plasma, though as will be shown in the next section, the electron saturation current in non-planar geometry is not expected to be constant. A probe biased well below the floating potential is in ion saturation. Between these two extremes is the electron retardation region; later it will be shown that this region captures information on the electron energy distribution function.

Exercise 10.2: 'Single probe' design Rework the area ratio criterion for a low-pressure hydrogen plasma, assuming all positive ions are H^+.

The reference electrode is required to ensure that the current drawn from the plasma is returned to it with negligible changes in potential between the reference and the surrounding plasma. A large-area surface is certainly one way to achieve this. The rôle can also be fulfilled by a thermionically emitting surface that is able freely to release electrons into the surrounding plasma to supply the return current, again without significant changes in potential. A so-called emissive probe is formed from a loop of incandescent wire, conveniently heated with a floating power unit [207] or a focused laser beam [208]. Depending on the material, sufficient emission can be obtained from wires heated to between red and white heat 1000–2000 K. The hot surface will be surrounded by a reservoir of charge in an electron-rich sheath and as a result the area of a hot reference can then be much smaller than a cold one. As its temperature increases, an emissive probe floats closer and closer to the potential of surrounding plasma – if it were to be at a lower potential it would lose negative charge from its thermionic sheath into the plasma, thereby rising in potential – if it were to rise above the local DC plasma potential it would receive a significant flux of negative charge from the plasma, thereby falling in potential. Thus, a self-regulation for its natural DC floating (zero net current) condition sets it at plasma potential and it then forms an excellent reference electrode. Emissive probes have also been used widely to map out the spatial variation of plasma potential in various RF plasmas [208–210].

Separation of probe and reference electrodes

If a probe and its reference electrode were too widely separated then some of the applied voltage would be dropped across the plasma. Probe currents are generally small and plasmas are good conductors, but it is still prudent to keep the current path in the plasma 'local'. Widely separated electrodes may be exposed to 'different' plasmas. For instance, as was revealed in Chapter 3, the plasma is not an exact equipotential even in idealized 1-D models – in real systems therefore one cannot expect that the potential at which a surface floats with respect to the plasma will be the same throughout. The consequence is that where ϕ_p has been included in the modelling, one would be wiser to use ϕ_{p1} near the reference electrode and ϕ_{p2}

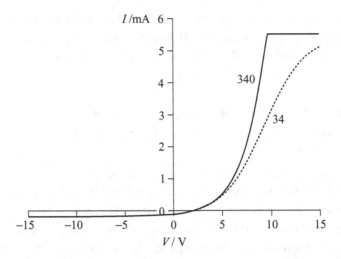

Figure 10.5 Current–voltage (I–V) characteristics (Eq. (10.2)) obtained with two asymmetrical planar probe arrangements (area ratios 34 and 340) in a low-pressure hydrogen plasma – see Exercise 10.3.

nearer the probe. This would then be apparent in the I–V characteristic having the zero current at $V = \phi_{p2} - \phi_{p1} = V_f$ rather than at $V = 0$.

Practical analysis

Given the above background one should now be able to deduce various plasma parameters from data obtained with asymmetrical planar probes. In fact, for reasons to be discussed in the next section, planar probes are difficult to realize so the following examples use 'artificial' data. Nevertheless, it is worth practising on these simple cases.

Exercise 10.3: 'Single probe analysis' 1 Figure 10.5 shows I–V characteristics for a hydrogen plasma, using area ratios of 34 and 340, with a probe electrode area of 1.9×10^{-5} m^2 in both cases. Comment on the nature of the two curves and then determine (i) the plasma potential, (ii) the floating potential, (iii) the plasma density at the sheath/plasma boundary based on the ion saturation current and (iv) the electron density in the undisturbed plasma.

The current that enters at the probe electrode ($V < \phi_p$) comprises the voltage-dependent arrival of electrons partially compensated by the steady flow of positive ions. To examine the electron component alone one can remove the ion component: $I_e = I + I_0$. What remains according to the planar model developed here, provided the reference area is large enough, should be a purely exponential electron current:

$$I_e = \frac{n_s \bar{v}_e}{4} \exp\left(\frac{e(V - \phi_p)}{kT_e}\right) eA_2;$$

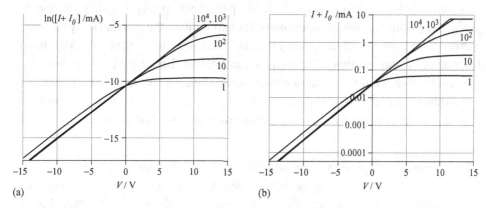

Figure 10.6 (a) The semi-log plot of electron current (in mA) at the probe electrode for reference electrodes having area ratios 1, 10, 10^2, 10^3 and 10^4, in the same plasma as that for Figure 10.3; (b) the same data presented as a plot of current on a $\log_{10}$ scale.

so a plot of $\ln(I + I_0)$ against V should be linear throughout the range up to the plasma potential, with slope e/kT_e. Exercise 10.4 examines the 'semi-log' plot further.

Exercise 10.4: 'Single probe analysis' 2 Figure 10.6(a) shows the semi-log plot of the electron current for the data in Figure 10.4. Comment on the nature of the different curves and then determine (i) the plasma potential and (ii) the electron temperature.

Instead of plotting $\ln(I + I_0)$ one can simply plot $I + I_0$ on a log scale as in Figure 10.6(b). To determine the electron temperature from this plot one then needs to account for the change of base for the logarithms, thus $kT_e/e = \Delta V/2.3\Delta(\log_{10}[I + I_0])$. In other words, for a Maxwellian distribution one expects that the electron temperature in eV is the voltage range for one decade change of current, divided by 2.3. So for instance over 4 decades of current the voltage changes by about 19.5 V, which would give $kT_e/e = 5.0/2.3 = 2.2$ V.

Area of probe electrode

The lower limit on size of a planar disk probe has been discussed – the diameter must be much larger than the thickness of the sheath. If the probe were much smaller than this, then it would appear from the neighbouring plasma to be almost indistinguishable from a hemispherical collector. Non-planar geometries are considered in the next section. The last point to consider in this section therefore is how large a planar probe electrode can realistically be without causing a serious perturbation to the environment that it is designed to sense. One problem is a shadowing effect, which means that the distribution of plasma particles is locally depleted because the probe itself both absorbs charged particles and inhibits the arrival of replacement particles. A related problem arises if the current flowing to the probe is a significant fraction of the current that sustains the entire plasma. Either the probe is diverting the current that would otherwise contribute to the regeneration of the plasma, or it is adding current that will enhance those processes – either way, the probe is making a serious perturbation to the equilibria that define the plasma.

> **Q** Suppose that a plasma of density 10^{16} m^{-3}, and electron temperature 2 eV, is sustained by a power supply that couples 50 W into it. Estimate the maximum diameter for a single planar disk probe if it is to divert no more than 1% of this power when in electron saturation, given that the mean energy invested in the production of an electron–ion pair is $e\varepsilon_T \equiv 50$ eV.
>
> **A** The limit is set by the rate at which energy is drained away by removing electrons from the bulk, given that on average $e\varepsilon_T$ is invested in the production of an electron–ion pair – Eq. (2.46). Thus the power drained by a single-sided disk of diameter d in electron saturation must be such that
>
> $$\frac{n_0 \bar{v}_e \pi d^2}{16} e\varepsilon_T < 0.5 \text{ W}.$$
>
> This can be evaluated to deduce that the limit on the diameter is $d < 6$ mm.
> *Comment: The maximum diameter is only marginally greater than the minimum size specified on the basis of maintaining planar geometry.*

The operational constraints on the diameter of planar (disk) probes is quite small: too small and they look hemispherical rather than planar (see next section); too large and if they were allowed to approach electron saturation they would divert a significant fraction of the power that sustains the plasma; large probes may also create a significant shadow. That's a pity because the method and the analysis are both relatively simple. They are nevertheless useable in the double-probe configuration where currents never exceed the ion saturation level. The next section looks at the added complexity that comes with trying alternative geometries as a means of getting accurate data on the charged particle populations without serious perturbation to the plasma in question.

10.1.4 Cylindrical and spherical geometry

Spheres and cylinders with radii that are much larger than the thickness of a sheath will behave rather like planar collectors since the rejection (or acceleration) of

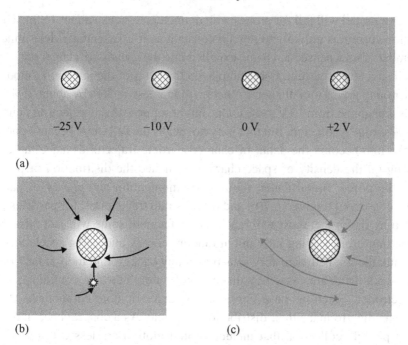

Figure 10.7 Schematics of a cylindrical (or spherical) probe. (a) Sheath development around a biased probe – the potential indicated is referenced to the plasma (ϕ_p); (b) example ion trajectories ($\phi < \phi_p$); (c) example electron trajectories in electron saturation ($\phi > \phi_p$). The lighter regions represent net space charge: positive if $\phi_p < 0$ and negative if $\phi > \phi_p$.

charges will then happen over a narrow range of radius, so that the effects of surface curvature will be negligible. For plasmas at 10^{16} m^{-3} and 2 eV that means that a planar probe model might be applicable for spheres and cylinders with diameters much larger than 1 mm.

Q Taking the area of a 5 mm diameter disk (about 2×10^{-5} m^2) as the basis for comparison, equivalent collection areas are obtained with a sphere of 1.25 mm radius and a 10 mm long cylinder of 0.3 mm radius. Will the planar model of the previous section be applicable to a probe made from a 1.25 mm radius spherical bead or from 10 mm of 0.3 mm straight wire?

A Take the earlier estimate for the thickness of a 25 V sheath in a 2 eV plasma ($5\lambda_{De} \sim 0.5$ mm). The wire is certainly too fine to ignore the differential curvature of the sheath and the surface of the probe – they differ in area by a factor of about 2.5 – see Figure 10.7. The bead, though less so, is also too small for a planar model, with sheath and probe areas differing by almost a factor of 2.

The planar model will fail for probes with diameters less than about 1 mm. The ion saturation current is unlikely to remain constant as it is essentially determined by the number of ions per second being expelled into the sheath, so a larger area sheath will collect a larger current. For example, on the basis of the ion matrix model the sheath around a 1.25 mm diameter sphere will increase in area by about 25% when the bias is changed from 10 V to 25 V. One might suppose that the electron saturation region is similarly affected. But that is by no means the full story, because one must also take into account the detail of charged particle trajectories since these have a bearing on the density of space charge and hence the distribution of potential around the probe. In particular, the angular momentum and energy of particles must be conserved. Figure 10.7(b) and (c) illustrate trajectories, for particles being attracted by the probe: most will reach the surface but some will just miss, being on a passing orbit. Having little initial random energy and being accelerated into the sheath, the ion paths are very sensitive to any background motion far from the probe – such drifts are caused by the ambipolar field between confining surfaces and electrodes. Electrons have considerably larger random components to their motion, tending to make their distribution isotropic. As a consequence, models of charged particle collection that include orbital motion are less controversial for electron saturation currents than they are for ion currents.

Electron saturation current limited by orbital motion

Many electrostatic probes use fine wires as the charge-collecting surface, so this geometry will be considered here. It is supposed that the radius of the wire is less than a few Debye lengths, so that simply neglecting the curvature of the probe is not an option. When the potential of the wire is positive with respect to the plasma, electrons will be attracted to it by the electric field that develops around it, but unless the initial path is exactly radial the trajectory will be an orbit along which energy and angular momentum are simultaneously conserved. In collision theory, with a target placed on the axis, the perpendicular distance of the initial path from the axis, before any forces have deflected the incident particle, is called the 'impact parameter' (see Figure 10.8) – this concept is used here to describe the encounter between charged particles and a wire. At large impact parameter the electrons are merely deflected as they follow passing orbits. Electrons with a low impact parameter collide with the probe. The aim here is to obtain an expression for the current collected by a wire, taking account of this orbital motion; the assumptions of isotropic distributions and the absence of collisions with the background gas are particularly important. The probe is a wire of radius r_c and length $l \gg r_c$; the potential on the wire is $V_c > \phi_p$ – the subscript 'c' identifies the collecting surface, i.e., the probe (see Figure 10.8). Consider the trajectory of an electron (charge $-e$) on a path that just grazes the probe surface with an impact parameter h_{graze}. Energy

10.1 Electrostatic probes

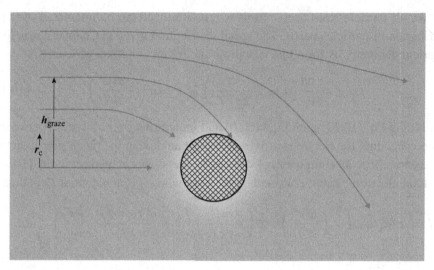

Figure 10.8 Trajectories of particles with the same energy, starting with different impact parameters. Those with $h \leq h_\text{graze}$ are collected by the probe; those with $h > h_\text{graze}$ miss the probe.

will be conserved, so that for an electron that is accelerated to just graze the probe at the surface,

$$\frac{1}{2}mv^2 = \frac{1}{2}mv_c^2 - e(V_c - \phi_p). \tag{10.7}$$

At the same time angular momentum must be conserved, so that for a particle that is going to graze the probe,

$$mvh_\text{graze} = mv_c r_c. \tag{10.8}$$

These two expressions can be combined to give the impact parameter for grazing incidence as a function of the initial speed:

$$h_\text{graze} = r_c \left(1 + \frac{2e(V_c - \phi_p)}{mv^2}\right)^{1/2}. \tag{10.9}$$

Note that for a large initial speed the impact parameter equals r_c, so only those high-speed particles that start off travelling directly towards the probe are collected, whereas very slow particles are likely to be collected, even when their trajectory far from a very small probe is not directed straight at it. The current that is contributed by particles at any particular speed v is $dI_e = evA\,dn$, with $A = 2l\,h_\text{graze}$, such that

$$dI_e = 2el \times r_c \left(1 + \frac{2e(V_c - \phi_p)}{mv^2}\right)^{1/2} v\,dn. \tag{10.10}$$

To take all particles into account this expression needs to be integrated over the distribution of particle speeds. The speed distribution given in Chapter 2 was for a spherical geometry. A form that is appropriate for a cylindrical geometry is

$$f_{\text{s-cyl}} = \frac{dn}{dv} = n_0 \left(\frac{m}{2\pi k T_e}\right) 2\pi v \exp\left(-\frac{mv^2}{2kT_e}\right). \quad (10.11)$$

Combining Eq. (10.11) with Eq. (10.10) and integrating over all speeds between 0 and ∞ accounts for electrons arriving from all angles, in the plane perpendicular to the wire; the axial component of velocity is already accounted for in Eq. (10.10). The total electron current collected by the wire when it is at potential V_c is then

$$I_e = e\, 4 r_c l n_0 \int_0^\infty \left(\frac{mv^2}{2kT_e}\right) \exp\left(-\frac{mv^2}{2kT_e}\right)\left(1 + \frac{e(V_c - \phi_p)}{kT_e}\frac{2kT_e}{mv^2}\right)^{1/2} dv. \quad (10.12)$$

Q Show that the current collected by a cylindrical probe in electron saturation at potential V_c is

$$I_e = e\, 2\pi r_c l \frac{n_0 \bar{v}}{4}\left(2\sqrt{\frac{\eta}{\pi}} + \exp\eta\,\text{erfc}\sqrt{\eta}\right), \quad (10.13)$$

where $\eta = e(V_c - \phi_p)/kT_e$.

A First recast Eq. (10.12) using

$$\eta = e(V_c - \phi_p)/kT_e \quad \text{and} \quad u^2 = mv^2/2kT_e + \eta$$

so that $v\,dv = (2kT_e/m)\,u\,du$. Then the current integral becomes

$$I_e = e\, 4 r_c l n_0 \sqrt{2kT_e/m}\, \exp\eta \int_{\sqrt{\eta}}^\infty (u^2 \exp(-u^2))\,du.$$

The integration can be completed by parts:

$$\int_{\sqrt{\eta}}^\infty (u \exp(-u^2))\, u\, du = \left[\frac{\exp(-u^2)}{-2} u\right]_{\sqrt{\eta}}^\infty + \int_{\sqrt{\eta}}^\infty \frac{\exp(-u^2)}{2}\, du.$$

The remaining integral can be written in terms of the complementary error function as $(\sqrt{\pi}/4)\text{erfc}\sqrt{\eta}$. Rearranging then recovers the given result.

The expression for the saturation current, Eq. (10.13), can be approximated to within 1% for $\eta > 2$ by the following:

$$I_e = e\, 2\pi r_c l \frac{n_0 \bar{v}}{4} 2\sqrt{\frac{1 + e(V_c - \phi_p)/kT_e}{\pi}}; \quad (10.14)$$

in fact this approximation is no more than 13% adrift for $e(V_c - \phi_p)/kT_e < 2$.

10.1 Electrostatic probes

This result for the orbital-motion-limited current was first obtained for cylindrical probes by Mott-Smith and Langmuir [211]. They divided the space into plasma and sheath as in the planar geometry case above but they did not then solve for the potential in the sheath – in fact, that is why the approach here did not need to appeal to the existence of the sheath [211,212]. It is possible to solve the complete problem of cylindrical and spherical collectors in plasmas, but the analysis relies upon numerical integrations of Poisson's equation – there is not much more insight gained by doing so, though it does show that the above analysis is valid in most of the parameter space [212].

Plasma potential

A probe collects zero net current when it is at its floating potential and as has been seen earlier in electropositive, cold ion ($T_i \ll T_e$) plasmas, this corresponds with strong Boltzmann factor retardation of electrons. At more positive potentials the electron current rapidly increases until there is no retardation of electrons when it is at the local space potential in the plasma – i.e., at the DC plasma potential. At higher potentials on cylindrical and spherical probes the exponential retardation factor is replaced by the collection of electrons under orbital motion, which changes less rapidly with potential. The transition between these two contrasting regimes of electron collection can be identified as a 'knee' in the probe characteristic – it is helpful to highlight the changeover by taking the derivative of the characteristic (dI/dV) which peaks just close to the plasma potential (see the curves on the left of Figure 10.9). In practice, the peak in the first derivative tends to place the plasma potential too high and it is perhaps more 'reasonable' to take a second derivative and then to identify the plasma potential as being between the peak and the zero-crossing of d^2I/dV^2.

Exercise 10.5: Applying OML analysis Figure 10.9 shows the electron current characteristic from a fine wire, 'single' Langmuir probe for the same plasma as Figures 10.3, 10.4 and 10.6, replotted on axes of I_e^2 against V. The fine wire has an exposed surface area of 2.0×10^{-5} m^2. Use the electron saturation region ($V > \phi_p$) to deduce the density and temperature of electrons in the surrounding plasma.

Values of electron density and electron temperature can thus be deduced from the electron saturation portion of the I–V characteristic of a cylindrical probe. However, operating a probe in electron saturation risks perturbing the local plasma by drawing too much current from it. For this reason, analysis of the ion current is worthwhile.

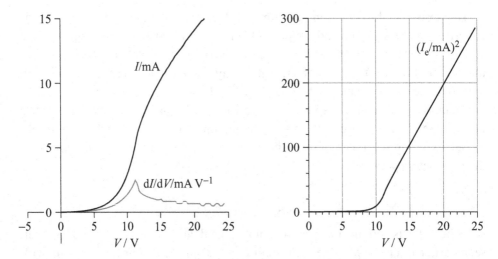

Figure 10.9 Left: The electron current to a cylindrical probe of area 2.0×10^{-5} m^2 in the same plasma as that for Figure 10.3 and its derivative, plotted against probe voltage. Right: The same probe current plotted as the square of the electron component of the current against voltage.

Ion saturation current to cylindrical probes

The ion current to a cylindrical probe does not simply saturate at a constant level (the sheath area grows as more bias is applied), so for the ion portion of the characteristic ($V < \phi_p$) it is tempting simply to replicate the analysis of electron collection, changing the charge, mass and temperature accordingly. Indeed, many practitioners do just this but for a number of reasons it may not be an appropriate course of action. For one thing it turns out that for low-energy particles (remember normally in the plasma $T_i \ll T_e$), the potential around the probe may in fact capture ions that graze a radius somewhat larger than that of the actual probe (and not just those that graze the actual probe). This arises from the orbital mechanics and the shape of the potential around the probe. To be certain of the effective radius of the probe under these conditions one must solve Poisson's equation to find the potential structure [213] and then fit probe data to I–V curves calculated with different values of ion density. In the previous section the form of the potential was not required because the thermal energy of electrons is large enough to avoid this issue.

Another consequence of the low energy of ions in the plasma is that they have very little angular momentum and they collide most often with neutral gas atoms that are of comparable mass. This means that collisions may effectively bring an accelerating ion to rest; then the radial field around the probe accelerates it so that it approaches the cylindrical (or spherical) probe with purely radial motion. An alternative to the orbital motion limited model is the radial flow model known as 'ABR' after Allen, Boyd and Reynolds [214]. This also requires Poisson's equation to be solved and then ion currents can be determined as a function of probe potential, for which Chen [215] has reported a comprehensive set of computations. The ion density is then deduced by fitting the ion portion of the characteristic ($V < \phi_p$) to calculated curves. There is considerable evidence that the radial motion model is more appropriate for confined plasmas like those described earlier in this book [216].

The effects of collisions

The orbital motion model is more appropriate for electron collection than it is for ion currents because of the relatively low mass and high thermal energy of electrons that tends to maintain their motion around the probe, even when there are collisions between charged particles and the background gas. One effect of collisions is to cause some particles to become trapped in orbits that do not intercept the probe – trapped populations will build up until collisional de-trapping establishes an equilibrium. The space charge arising from trapped orbits would then affect the local potential, but the OML analysis does not directly refer to the local potential. It seems that OML models for electron currents are satisfactory until collisions are so frequent that all semblance of orbital motion is destroyed ($\lambda_e < r_c$). For the collection of ions the situation is somewhat different, as the orbital motion is much more fragile for heavy particles – collisions with the gas instantly disrupt orbital motion and knock passing ions into paths that intercept the probe. There are extensions to OML that model the collisional regime analytically for ion collection [217], treating the dominant process as charge exchange. The principal effect is as anticipated – collisions tend to increase ion current at any particular potential. The model appears to work satisfactorily only for very fine wire probes with $r_c/ \ll \lambda_{DE}$ – in view of this, the model of Allen et al. [214] appears to be more robust.

Intermediate review of electrostatic probes

So, the initial appeal of the simplicity of electrostatic probes is reduced by the complexity and contentiousness of the analysis. Truly planar probes are difficult to achieve for localized measurements. The electron saturation of fine wire probes is readily modelled, but it risks perturbation of the plasma through the high level of current that is drawn. On the other hand, the ion saturation region is much less invasive but rather more difficult to analyse in non-planar geometry. The intermediate ('electron-retardation') region is perhaps the most appealing in terms of both implementation and analysis: for planar, cylindrical and spherical geometry it yields an electron temperature through the semi-log plot of the I–V characteristic. In fact this region is richer still and will reveal the whole electron energy distribution, as will be shown in the next section.

10.1.5 The electron energy distribution

The electron current collected by retarding probes (i.e., $V < \phi_p$) is worth a closer look. It turns out that provided the distribution of electrons is isotropic and the collector is convex, then the equations developed for an element of surface in Section 10.1.4 can be integrated over a retarding surface without directly appealing to Maxwellian distributions. This leads to a means of determining the energy distribution function from the probe characteristic. That is important because this distribution interacts with the gas to sustain the plasma. This section steps through the analysis to show the connection between $f(\varepsilon)$ and $d^2 I/dV^2$.

Consider a small surface ΔA immersed in an isotropic plasma. It is supposed that the probe is sufficiently small that it does not seriously deplete the local distribution of particles. The electron current that reaches the surface when it is biased below the potential of the plasma can be found by integrating over all angles and all speeds that contribute to the flux at the surface; a minimum speed is set by the retarding potential of the surface and, depending on the angle at which an electron moves relative to the element of area, there is for any speed a maximum angle of incidence beyond which the component towards the probe is insufficient to overcome the potential barrier:

$$\Delta I_e = e \Delta A \int_{v_{min}}^{\infty} \int_{0}^{\theta_{max}(v)} \int_{0}^{2\pi} f(v)\, v \cos\theta\, v d\varphi\, v \sin\theta d\theta\, dv, \quad (10.15)$$

where $v \cos\theta_{max} = v_{min}$ and $v_{min} = (2e(\phi_p - V)/m)^{1/2}$, and the speed distribution, $f(v)$, is that of electrons in the undisturbed plasma around the probe. The integrations over the angular ranges leads directly to

$$\Delta I_e = \pi e \Delta A \int_{v_{min}}^{\infty} v^3 f(v) \left(1 - \frac{v_{min}^2}{v^2}\right) dv. \quad (10.16)$$

Since the distribution is isotropic and the surface of the probe is small and convex, the total current will scale with the collection area, so ΔI_e and ΔA can be replaced by I_e and A. At this stage it is convenient to recast the problem in kinetic energy $\varepsilon = mv^2/2$ rather than speed, since the retardation is brought about through the surface potential. The distribution in electron speed $f(v)$ can be expressed in terms of the distribution in electron energy $f_\varepsilon(\varepsilon)$ using the defining expression

$$4\pi v^2 f(v) dv = f_\varepsilon(\varepsilon) d\varepsilon.$$

Therefore,

$$I_e = \frac{1}{4}eA \int_{\varepsilon_{\min}}^{\infty} \left(\frac{2\varepsilon}{m}\right)^{1/2} f_\varepsilon(\varepsilon)\left(1 - \frac{\varepsilon_{\min}}{\varepsilon}\right) d\varepsilon, \qquad (10.17)$$

where the threshold energy above which particles are collected is $\varepsilon_{\min} = e(\phi_p - V)$ – therefore, $\varepsilon_{\min}$ changes as the voltage on the probe is scanned. In a sense, the relationship between the energy distribution function, the probe voltage and the probe current is now complete, but ideally one wants it in the explicit form, i.e., '$f_\varepsilon(v) =$'. The next step in the analysis is not obvious but it eventually achieves what is required. Note that the integration involves one limit that is a function of one of the controlling parameters, V, so it is useful to make use of Leibniz's rule for differentiating under an integral:

$$\frac{d}{dy}\int_{b(y)}^{a(y)} F(x, y)dx = F(a, y)\frac{\partial a}{\partial y} - F(b, y)\frac{\partial b}{\partial y} + \int_{b(y)}^{a(y)} \frac{\partial F}{\partial y}dx. \qquad (10.18)$$

Applying this formula, the first term is zero because the upper limit is independent of V, the second term is zero because the integrand evaluated at the lower limit vanishes, leaving only the third term. Using $d\varepsilon_{\min}/dV = -e$, this simplifies the expression:

$$\frac{dI_e}{dV} = \frac{1}{4}eA\left(\frac{2}{m}\right)^{1/2} \int_{\varepsilon_{\min}}^{\infty} \varepsilon^{1/2} f_\varepsilon(\varepsilon)\left(\frac{-1}{\varepsilon}\right)(-e)d\varepsilon \qquad (10.19)$$

$$= \frac{1}{4}e^2 A\left(\frac{2}{m}\right)^{1/2} \int_{\varepsilon_{\min}}^{\infty} \varepsilon^{-1/2} f_\varepsilon(\varepsilon)d\varepsilon. \qquad (10.20)$$

This still leaves a limit that depends on V, but on applying the Leibnitz formula again only the second term is non-zero:

$$\frac{d^2 I_e}{dV^2} = -\frac{1}{4}e^2 A\left(\frac{2}{m}\right)^{1/2}(-e)\varepsilon_{\min}^{-1/2} f_\varepsilon(\varepsilon_{\min})$$

$$= \frac{1}{4}e^3 A\left(\frac{2}{m}\right)^{1/2}\left[\frac{f_\varepsilon(\varepsilon_{\min})}{\varepsilon_{\min}^{1/2}}\right], \qquad (10.21)$$

with $\varepsilon_{\min} = e(\phi_p - V)$. The quantity in square brackets in Eq. (10.21) is often called the 'electron energy probability function' (EEPF).

This procedure thus arrives at the very useful result that the electron energy distribution function (EEDF) in the undisturbed plasma can be extracted from the second derivative of the electron current at a particular *retarding* potential V, which in turn defines the energy axis ($e(\phi_p - V)$). The implementation of this is often called the Druyvesteyn [218] method, but in fact Mott-Smith and Langmuir [211] were the first to show that the distribution function could be extracted from derivatives of the current. As noted earlier, the second derivative is also useful in identifying the plasma potential, which is in effect the zero of the energy scale for the EEDF.

Q Using probe data how could the Maxwellian character of an EEDF be tested?
A For a Maxwellian EEDF (cf. Eq. (2.9)),

$$f_\varepsilon(\varepsilon) \propto \varepsilon^{1/2} \exp(-\varepsilon/kT_e). \qquad (10.22)$$

The EEPF for a Maxwellian is thus a simple exponential. Therefore, a plot of $\ln(d^2 I_e/dV^2)$ against V would be linear if the EEDF is Maxwellian.

Exercise 10.6: Second derivative method Figure 5.16 shows the EEPFs extracted from data taken with a single (cylindrical) Langmuir probe. Comment on the extent to which the plasma can ever be said to contain electrons that have a Maxwellian distribution of energies.

10.2 Electrostatic probes for RF plasmas

In practice, the potential of a probe is often swept rapidly to acquire data quickly – the considerations of Chapter 4 suggest that the timescale of such changes should be kept much less than ω_{pi}^{-1} (a succession of quasi-DC states) or much more than ω_{pi}^{-1} (frozen ions) if the ion-rich sheath around an electron-retarding probe is to be considered 'steady'. Furthermore, the potential between a probe and the surrounding plasma may also change when the probe is held at a steady potential while the plasma potential changes. This section particularly considers the consequences of there being RF components of potential between the probe and the surrounding plasma.

10.2.1 Conventional probes in RF environments

The earlier analysis of RF plasmas showed that capacitive coupling across space charge sheaths introduces RF fluctuations of the plasma potential with respect to a laboratory 'ground potential', so that in general $\phi_p \sim \phi_{p0} \cos \omega t$, possibly accompanied by higher harmonics. The first thing to consider is therefore what effect this will have on a Langmuir probe that is operated with potentials that are set relative to ground. Taking just the fundamental component of plasma potential in the electron retardation region of a current–voltage characteristic for a single

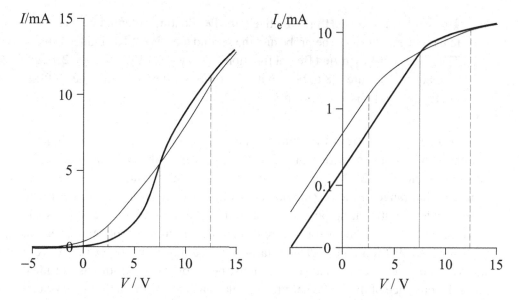

Figure 10.10 The effect of an RF fluctuation of plasma potential on the current–voltage characteristic and $\log(I_e)$–V of a Langmuir probe, area 2.0×10^{-5} m^2, in a low-pressure hydrogen plasma $n = 3.6 \times 10^{15}$ m^{-3} and $kT/e = 2.1$ V. The grey curves are the average current when $V_p = V_1 \cos \omega t$ with $V_1 = 5$ V and $\omega_{pi} < \omega < \omega_{pe}$; the black curves indicate the 'true' current that would be recorded with probe bias referenced to plasma potential. The vertical grey lines indicate the range of plasma potential during the RF cycle.

probe, Eq. (10.6) reveals the difficulty:

$$I = eA\left[-\frac{n_0 \bar{v}_e}{4} \exp\left(\frac{e(V - \phi_{p0} \cos \omega t)}{kT_e}\right) + n_s u_B\right]. \qquad (10.23)$$

This shows that there is now a non-linear RF component of current to the probe. It is not easy to measure this, as RF current will readily follow capacitive paths to ground in the probe structure before one can intercept it. Therefore, a more realistic option is to average the current over the RF period and change V more slowly and see what can be recovered from such averaged data.

Q Figure 10.10 shows the time-averaged probe current when a 5 V amplitude RF fluctuation is present. Account for the fact that the floating potential is shifted by about -2.3 V (if necessary, refer to Section 4.3.1 for inspiration).

A The floating potential can be found by averaging Eq. (10.23). Setting $\bar{I} = 0$, rearranging and taking the natural logarithm:

$$\frac{eV_f}{kT_e} = \ln\left(\frac{4n_s u_B}{n_0 \bar{v}_e}\right) - \ln I_0\left(\frac{eV_1}{kT_e}\right). \qquad (10.24)$$

> The first term on the RHS is effectively the DC floating potential with respect to the plasma far from the probe and the second term is the RF-induced shift. The value of the modified Bessel function $I_0(x)$ at $x = eV_1/kT_e = 5/2.1$ can be read from Figure 4.8 to be about 3.0; it follows therefore that the shift in floating potential is $-2.1 \times (\ln 3)$ V $= -2.3$ V.

It is also apparent from the semi-log plot in Figure 10.10 that the whole of the electron retardation section of the curve is shifted by the same amount as the floating potential (-2.3 V) up to the point where the instantaneous probe bias comes within range of the plasma potential variation. The averaging of the probe current, whether at floating potential or at some other bias, gives rise to the same shift, provided the instantaneous current remains within the exponential region. This suggests that ion and electron retardation data might still be extracted from an averaged characteristic. However, it must be recognized that this is not ideal since the presence of the probe perturbs the plasma not only through the inevitable local drain on charged particles but also by presenting a low RF impedance path to ground. Furthermore, one cannot meaningfully conduct an EEDF analysis since the full range of retardation is not available and the plasma potential is not easily identified – it is only recognizable in Figure 10.10 because the range of the plasma potential fluctuation is marked, but this is not known *a priori* in practice. Therefore, a better strategy is to take steps to prevent RF voltage appearing across the probe sheath (and prevent RF current from passing into the probe). This requires the probe tip to follow the instantaneous plasma potential, offset only by a slowly swept probe bias.

Passive compensation

The aim here is to use passive components right behind the probe tip to make the RF impedance from the tip to ground as large as possible (Figure 10.11). A common approach is to introduce components that give parallel resonances ($Z \to \infty$) at the fundamental and first few harmonics. There is not much room behind the probe tip, so a neat way to achieve this is to select wire-wound inductors that are self-resonant, owing to capacitance between the turns, at exactly ω, 2ω, 3ω, etc. This provides a notch filter at each chosen frequency. A non-resonant low-pass filter formed by an inductance and capacitance in parallel with that of the coaxial feed can also be used. Both approaches in practice can be made more effective by designing additional capacitive coupling to the plasma across the probe's sleeve (C_sleeve in Figure 10.11) – RF current coupled this way importantly does not cross the sheath at the exposed tip where the probe bias controls the particle current.

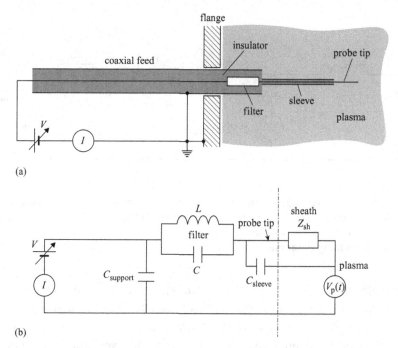

Figure 10.11 (a) Schematic diagram of a passively compensated probe for an RF plasma environment with a filter circuit close up behind the probe tip. Here the reference electrode is the flange and any conducting surface connected to it that is also exposed to the plasma. (b) An equivalent circuit of the resonant passive compensation filter; the RF fluctuations of plasma potential are included in $V_p(t)$.

If carefully matched in frequency, passive compensation will restore the probe characteristic to its 'DC' shape, as shown dotted in Figure 10.10. The filter characteristics are awkward to tune *in situ* and in practice it is difficult to achieve complete compensation, especially when there is a large fluctuation of plasma potential (e.g., several tens of volts), such as occurs in symmetric CCPs. In that case an active compensation scheme may be more effective.

Active compensation

A probe can be actively compensated by driving it from an external RF source with a signal that synthesizes the RF fluctuation of the plasma potential (Figure 10.12). In the simplest implementation the probe is driven at the fundamental frequency of the main plasma excitation source [219]. The amplitude and phase of the active RF bias must be set to match the plasma potential fluctuation. The tuning condition is based on maximizing the floating potential, which indicates that the RF potential between

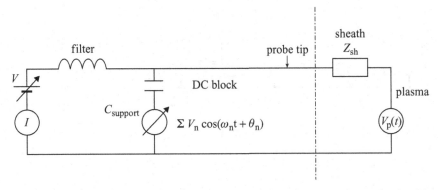

Figure 10.12 The equivalent circuit of an actively compensated probe circuit with RF bias waveform synthesized from a number of harmonics each with independently adjustable phase and amplitude. The tuning condition is based on maximizing the floating potential.

probe and plasma is minimized – cf. Eq. (10.24). In practice, the harmonic content of the plasma potential leaves residual RF between the plasma and the tip. A three-harmonic active compensation has been shown to be more effective [220]. A seven-harmonic version has been devised with the tuning of seven amplitudes and seven phases being accomplished through goal-seeking algorithms in the controlling software [221]. Although one could simply measure the plasma potential variation and feed that direct to the probe tip, there would be cable-length phase-shift effects in the higher harmonics – the *in situ* tuning method side-steps this issue.

Further review of electrostatic probes

In DC plasmas, single Langmuir probes are powerful tools for making local measurements of the EEDF. To use the same double-derivative method in RF plasmas one needs to compensate for the effect of RF fluctuations between the plasma potential and the probe tip potential. Passive methods have been devised to offset the effects of at least the first three harmonics – in highly symmetrical discharges, where the plasma potential variation can be very large, the passive compensation may not achieve the goal of attenuating the RF across the probe sheath to much less than kT_e/e. Active methods allow an optimization of the compensation signal enabling higher fidelity, but at the expense of considerable complexity in the apparatus, compromising the simplicity of the Langmuir probe method. There are further factors that frustrate the routine use of Langmuir probes in RF plasmas that are used for processing – these are the issues of the compatibility of probe materials and probe shadowing effects in an etching or a deposition environment. The next section describes a method that attempts to deal with these limitations.

10.2.2 Electrostatic probes for real processing environments

> **Q** Real plasma processing environments are primarily intended for the deposition or etching of conductors or insulators. (i) Identify the issues that should be addressed in devising electrostatic probes for these environments; (ii) comment on the effect the presence of the probe may have on a plasma process.
>
> **A** (i) The materials and design of the probe must be compatible with the plasma chemistry and the consequences of material deposition need to be mitigated. (ii) The presence of the probe will locally perturb the plasma as the surface of the probe and its support will act as an additional site for recombination within the plasma volume, thereby affecting the various global equilibria.

One strategy to avoid introducing new materials into a processing chamber when sampling the charged particles is to build a probe out of the existing structure. So for instance in a CCP the electrodes themselves might be used directly; in an ICP a substrate holder could be used. For example, measurements of RF currents and voltages have been used in combination to infer ion fluxes and ion energy within an ICP, employing sensors that were wholly outside the chamber, with models that carefully account for the inductance and the capacitance of the electrical feed between the sensing point and the plasma [222, 223]. These methods require accurate calibration but are the basis of assessments of the global RF models described in earlier chapters.

It is also feasible to use the self-bias effect introduced in Chapter 4 to obtain a bias potential for smaller, more localized probes that are mounted in existing surfaces [224]. This uses RF signals that are not related to the excitation of the plasma under investigation, which might be a CCP, an ICP, a helicon plasma or any other form of low-pressure plasma. Figure 10.13 illustrates the principle of operation of a self-biased surface probe that is 'charged' to an RF floating potential by the application of a burst of RF signal, amounting to many RF cycles, coupled through an external capacitance. The capacitance blocks DC current and allows a self-bias potential to be established on the surface during the early stages of the burst. The self-bias sets the surface potential so that the steady arrival of ions over one cycle is exactly compensated by the arrival of a pulse of electrons during the brief moments when the instantaneous surface potential is close to plasma potential (Section 4.3.1). After the end of the RF burst the surface potential is initially the self-bias voltage, now without superimposed RF, so electrons no longer reach it and the charge on the capacitor is changed initially by the arrival of positive ions until it approaches the RF-free floating potential, when electrons once more reach the surface.

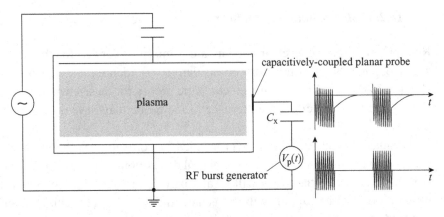

Figure 10.13 Schematic diagram of a self-biased planar probe, embedded in a grounded surface. The probe bias is derived from a burst of RF applied to the probe, during which a self-bias voltage is established across the external capacitance C_x. At the end of the burst the RF is switched to zero and the capacitor discharges at a rate initially determined by the flux of positive ions from the plasma onto the probe.

Analytic expressions can be obtained that describe both the charging and the discharging transients in electropositive plasmas in terms of a few parameters related to the properties of the plasma and the external environment. The charging of the probe with different RF amplitudes effectively maps out the modified Bessel function of Figure 4.9, and so can be used to deduce an effective electron temperature. The decay of the bias after the end of the RF burst is similarly related to the ion flux and just the tail of the electron energy distribution function – the decay of the bias effectively sweeps out the ion characteristic of the planar probe from ion saturation to floating potential. The decay phase is easier to analyse and to interpret because the applied RF is zero during this period. The current in the external capacitor (C_x) is equal to the net particle current arriving at the surface, provided $|dV/dt| \ll \omega_{pi} kT_e/e$, so that the displacement current is negligible (cf. Section 4.2). The starting point is the differential equation that describes this slowly changing potential:

$$C_x \frac{dV}{dt} = eA \left[n_s u_B - \frac{n_0 \bar{v}_e}{4} \exp\left(\frac{-e(V - \phi_p)}{kT_e} \right) \right]. \tag{10.25}$$

For the decay phase the initial condition is that at $t = t_1$, when the applied RF amplitude is switched from V_{RF} to zero, the potential across the capacitor is the RF self-bias established by the burst when $t < 0$, that is

$$V(t_1) = -kT_e/e \times \ln(I_0(eV_{RF}/kT_e)). \tag{10.26}$$

10.2 Electrostatic probes for RF plasmas

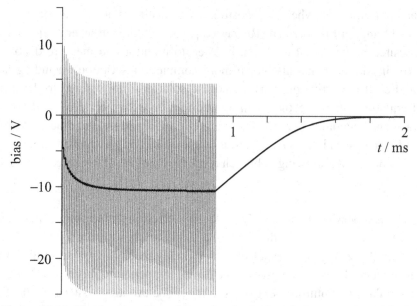

Figure 10.14 The instantaneous (unresolved, grey) and slow transient (black) potential of a self-biased planar probe of area $1.8 \times 10^{-4}\,\text{m}^2$ embedded in the grounded wall of a plasma source. Outside the chamber the probe is driven with a 0.9 ms burst of RF at 12 MHz and 15 V amplitude, applied through a 22 nF capacitor.

The solution of Eq. (10.25) can then be shown to be

$$\frac{eV}{kT_e} = \left(\frac{t-t_1}{\tau}\right) - \ln\left\{\exp\left(\frac{t-t_1}{\tau}\right) - 1 + I_0(eV_{RF}/kT_e)\right\}, \quad (10.27)$$

where $\tau = C_x kT_e / n_s u_B e^2 A$ is a characteristic time for the decay process.

For the period when the RF burst is active, the slow transient potential can be followed as the self-bias builds up by including the average of the instantaneous electron current and starting with the initial condition $V = 0$. Figure 10.14 shows the complete solution for surface potential of a planar probe, starting with a 0.9 ms period during which the RF burst is active.

> **Q** Show that the initial decay of the transient potential is linear, remembering that in Eq. (10.27), V_{RF} is the RF voltage during the 'on' phase and therefore defines the initial conditions for the decay phase.
>
> **A** In Eq. (10.27), if $t \approx t_1$, then the Bessel function dominates the logarithm leaving only the linear time dependence of $(t-t_1)/\tau$.

The use of bursts of RF and transients means that quasi-DC measurements can be made through insulated surfaces. The method is particularly appealing for 'dirty' plasma environments, wherein deposition of insulating material can poison conventional Langmuir probes. The RF-biased probe can tolerate several micrometres of deposited material – it can even be used to monitor and measure deposition rates for insulating materials. For a more complete description, including large bias and electronegative plasmas, numerical solutions are required to the differential equations that, together with appropriate boundary conditions, determine the transient phenomena [225]. The method provides simple and accurate measures of ion flux and effective electron temperature and has proved to be a very sensitive means of following subtle changes in a variety of low-pressure plasma sources.

> **Q** Why is it possible to deduce the 'effective' electron temperature with this RF probe? What part of the distribution actually reaches the probe?
>
> **A** When the voltage decays, electrons gradually reach the probe, so the curve in the late decay phase gives the electron temperature, much as it does for a regular Langmuir probe. However, the probe eventually sits at the floating potential, that is at a potential substantially smaller than the plasma potential: only the tail of the electron energy distribution reaches the probe.

Exercise 10.7: RF self-biased planar probe With reference to Figure 10.14, (i) say how the electron temperature could be found from the steady -10.5 V self-bias created by 15 V of RF and (ii) deduce the ion flux onto the surface of the probe from the rate of decay of the self-bias potential at the end of the burst.

The RF-biased planar probe combines a relatively uncomplicated technique with a simple analysis; a major advantage is its tolerance of insulating films [226], and indeed it can even be used as a film thickness monitor. The initial transient collects a constant, saturated ion current, essentially rejecting electrons and attracting positive ions with a large negative bias. However, the collector is unable to discriminate among positive ions that reach the surface with different energies – the next technique manages to do this by taking the ions into a region that is electrostatically screened from the plasma.

10.3 A retarding field analyser (RFA)

The analysis of the speeds or energies of charged particles in a plasma can be done in various ways. One is by selecting particles according to particular spatial trajectories that correspond with a narrow range of speeds. Another is simply by

10.3 A retarding field analyser (RFA)

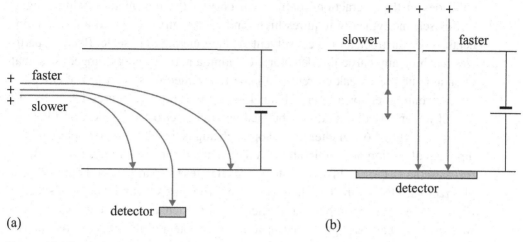

Figure 10.15 (a) A deflecting filter selects particles within a specific (narrow) range of energy; (b) a retarding filter passes particles with energy above a threshold level,

blocking those with energies below a specific level. In this section, after briefly exploring both of these types, attention is focused on the latter, which is simpler to implement for direct measurements in the physical environments of RF plasma devices. It will be shown that the electric field between narrowly spaced grids at different potentials provides a means of assessing the energy and speed distributions of ions arriving at surfaces exposed to low-pressure plasmas [210].

10.3.1 Basic principles

When a charged particle enters a region of electrostatic field it is accelerated in the direction of the field. However, the trajectory that it subsequently follows depends also upon its speed in directions perpendicular to the field (Figure 10.15). The motion of the particles along the field direction is slowed or speeded up depending on the sign of the charge and the direction of the potential gradient. The motion across the field is not changed, but the trajectory is deflected by the acceleration in the field direction and the path depends on the charge-to-mass ratio Ze/M.

So, taking a collimated sample of charged particles one could, for instance, spread their paths through space by introducing them to a region where their motion is *deflected* by an electric field. It is simplest to consider a case in which the initial motion is perpendicular to the field (Figure 10.15(a)). The resolution that can be achieved is a combination of geometry and surface potentials – higher resolutions require more space and will inevitably produce lower signals. Any collisions in the interaction volume will destroy the information being sought, so this scheme typically requires high (and ultra-high) vacuum to allow components to be constructed on a convenient vacuum-engineering scale. Because particles reaching any particular detector are within a narrow energy range, the signal often amounts to a weak current that is best amplified by some form of cascaded electron multiplier (or a micro-channel plate), which also requires UHV. Examples of ion energy distributions obtained with an electrostatic deflection filter are shown in Figure 4.6. An alternative approach aligns the electrostatic field with the direction of motion of a collimated beam of charged particles (Figure 10.15(b)). In this case the electric field accelerates or decelerates the particles without a change of direction. The electrostatic field is derived from grids that are biased at specific potentials relative to the plasma. Particles of a particular kinetic energy, ε, in the source region can pass grids that have a lower potential than that of the source ($V < \phi_{source}$), gaining kinetic energy as they go. However, in regions where the field *retards* the motion ($V > \phi_{source}$), particles can only reach where the potential energy, eV, is less than their initial kinetic energy. This allows an energy threshold to be set on (or near) a collector such that only particles with initial kinetic energy in excess of a specific value are collected. The retarding field analyser is based on this arrangement.

Figure 10.16 shows one implementation of an RFA that can be used in low-pressure plasma systems. The grid across the entrance must have a hole size that is smaller than λ_{De} so that it is not penetrated by quasi-neutral plasma, which would prevent proper operation of the analyser. The broad aperture allows a simple current detection system to be used without an electron multiplier, but this makes effective differential pumping difficult, so the device must be operated at chamber pressure. Collisions in the analyser again destroy energy information and this limits the operational pressure range.

> **Q** If an RFA is to operate with positive ions drawn from an argon plasma without differential pumping, what is the maximum operating pressure if the aperture-to-collector distance is 0.75 mm?
>
> **A** The charge exchange mean free path (λ_i) for argon ions is about 1 mm at 5 Pa (cf. Eq. (2.30)). So the operating pressure should be less than this if differential pumping is not available [46].

The current that reaches the collector of a retarding field analyser is made up of those particles with sufficient energy to overcome the discriminating potential (whether it be that of a scanned collector or that of a separate discriminator grid), attenuated also by the transparency of the grids. For a two-grid design biased to

10.3 A retarding field analyser (RFA)

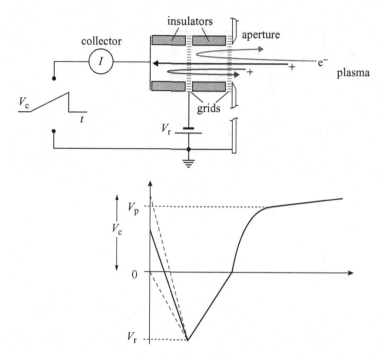

Figure 10.16 A two-grid RFA, with a plan of the potential distribution. The aperture grid screens the plasma from potentials set within the analyser; the repeller grid is set sufficiently negative to reject virtually all electrons; the collector potential is scanned, rejecting ions that have insufficient energy to reach it. In a three-grid design the scanned collector is replaced by a further grid that is scanned to reject ions below a certain energy and a collector at fixed potential.

collect ions, in terms of the velocity distribution at the surface, $f(v)$,

$$I = \beta_{RFA}\theta_a\theta_r \int_{v_{min}}^{\infty} vf(v)dv, \qquad (10.28)$$

where $v_{min} = \sqrt{2eV_c/M}$ and $\theta_{a,r}$ are the fractional open areas of the grids; the constant β_{RFA} can be determined from the fact that when $v_{min} = 0$ the integral should equal the ion saturation current into the area of the aperture, $eAn_s u_B$. The upper part of Figure 10.17 shows the current–voltage characteristic of an RFA for ions arriving at a plasma boundary with a narrow distribution of energies (Section 4.2.3) and then falling freely through a sheath potential of 9.6 V onto the front surface of the RFA. The energy resolution of the analyser is determined in part by the extent to which the grids ensure planar equipotentials within the analyser: smaller holes favour finer resolution at the expense of decreased transmission. The next section considers how the current is related to the distributions of ion energies and speeds at the entrance to the analyser.

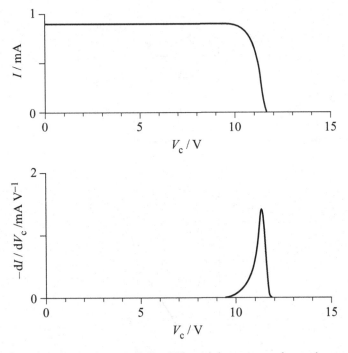

Figure 10.17 Current–voltage and the differential current–voltage characteristics for an RFA with an aperture of 10^{-4} m^2, when the ions enter the sheath with a distribution like that in Figure 4.5 and are then accelerated through a sheath potential of 9.6 V. This particular RFA has an energy resolution of 0.4 eV.

10.3.2 The ion velocity and energy distributions

The integral in the expression for the analyser current does not depend on potential, though the lower limit does. In that case the Leibnitz rule for differentiating under the integral (Section 10.1.5) gives the following result:

$$\frac{dI}{dV_c} = -\beta_{RFA}\theta_a\theta_r \sqrt{2eV_c/M}\ f\left(\sqrt{2eV_c/M}\right)\left[\frac{1}{2}\sqrt{\frac{2e}{MV_c}}\right] \quad (10.29)$$

$$= -\beta_{RFA}\theta_a\theta_r\ \frac{e}{M}\ f\left(\sqrt{2eV_c/M}\right). \quad (10.30)$$

Thus the first derivative of the current–voltage characteristic is proportional to the ion velocity distribution function, or IVDF (i.e., the distribution of speeds directed towards the analyser). Note that the IVDF is plotted in Figure 10.17 against an energy axis since eV_c is the potential energy of the collector that will collect ions having a greater initial kinetic energy. To recover the IVDF on a velocity scale, or to find the distribution in energy, the IEDF, one needs to use

$$f(v)dv = f_\varepsilon(\varepsilon)d\varepsilon \quad \text{with } \varepsilon = Mv^2/2,$$

which states that the number of particles in the speed range v to $v + dv$ must equal the number in the equivalent energy range ε to $\varepsilon + d\varepsilon$. Then it follows that

$$f_\varepsilon(\varepsilon) = \frac{1}{Mv} f(v).$$

So, in terms of the current–voltage characteristic of the collector in a two-grid RFA,

$$f_\varepsilon(eV_c) = -\frac{1}{\beta_{RFA}\theta_a\theta_r}\sqrt{\frac{M}{2eV_c}}\frac{1}{e}\frac{dI}{dV_c}. \qquad (10.31)$$

The highest velocity ions in Figure 10.17 have a kinetic energy that should closely correspond with free-falling from plasma potential, so V_p can be identified with the position of the sharply falling edge of the dI/dV plot and the width of the distribution, in the absence of collisions, can be linked to kT_e (cf. Section 4.2.3).

Exercise 10.8: RFA With reference to Figure 10.17, identify from where the potential of the plasma can be deduced and hence determine the plasma potential far from the analyser.

Q It has tacitly been assumed that the signal to an RFA comprises a single species of singly charged positive ions. What characteristics of the signal would indicate if this were not so in practice?

A (i) If there were doubly and singly charged ions of the same species leaving the plasma with the same energy, the RFA would not be able to separate them precisely because it discriminates only in terms of energy.
(ii) Likewise, a second species of a singly charged ion would gain the same increase in kinetic energy on crossing the sheath as the original species, contributing to the same peak in dI/dV as the first species; the only clue to their presence would be through processes that produce differences in the distributions of ions arising from generation and transport in the plasma.

10.3.3 The electron energy distribution

The RFA can be used to analyse the energy of negative charges that pass the aperture by reversing the repeller and collector potentials. The repeller needs to be biased above plasma potential to send the positive ions back to the plasma. The sheath at the aperture already prevents the collection of negative charge that leaves the plasma with less energy than eV_p, so the RFA will only see the high-energy tail of the distribution. If it is supposed that the negative charge reaching the collector is from a Maxwellian distribution of electron energies then, as with the simple analysis of retarding probes, a semi-log plot can be used to test this hypothesis and extract T_e.

In principle, one could also apply the full Druyvesteyn formula, Eq. (10.21), to obtain the actual shape of the EEDF tail from an RFA.

> **Q** Account for the difference between the formulas that extract distribution function data for ions and electrons from current–voltage characteristics of RFAs and probes.
>
> **A** Comparing the formulas for the IEDF, Eq. (10.31), and the EEDF, Eq. (10.21), one can see that the former is obtained from a single derivative whereas the latter is linked with the second derivative. This difference arises because the ions enter the RFA strongly directed towards it – ion motion parallel to the aperture is neglected – whereas the electron distribution is assumed to be isotropic, so account has to be taken of particles that approach from all directions outside the aperture.
>
> *Comment: In any application of the methods, these assumptions should be remembered.*

10.4 Probing with resonances and waves

The main focus of this section will be methods that are based on the interaction of a plasma with small amplitude signals close to the plasma frequency. These interactions provide further ways to probe plasma density, many of which can be implemented with less perturbation and materials that are more compatible with processing environments. Figure 10.18 shows a selection of 'microwave probes' that will be considered in this section. They are comparable in size to Langmuir probes but they do not draw net current from the plasma. As with other immersive probes, however, their presence introduces a local drain on the plasma particles that recombine on their surfaces and on that basis they are not attractive options for an active processing environment. Nevertheless, the models developed in earlier chapters and numerical simulations need devices like these to validate them. The first step is to look at the microwave properties of a small sphere in a large plasma.

10.4.1 Microwave impedance of a small spherical probe

The density of the RF-generated plasmas described in this text generally falls in the range 10^{15}–10^{19} m^{-3}, which corresponds with electron plasma frequencies that are

10.4 Probing with resonances and waves

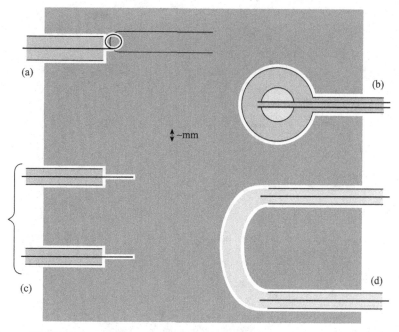

Figure 10.18 Various types of microwave probes in cross-section: (a) hairpin resonator (the hairpin is supported in a plane slightly behind that of the loop, from which it is DC-isolated); (b) multipole resonator; (c) transmission cut-off; (d) surface waveguide.

at the lower end of the microwave region of the spectrum, in the range 0.3–30 GHz. It is interesting, therefore, to examine the response of a plasma to signals in this frequency range. Straightaway the restriction 'small' can be quantified, since at 10 GHz a spherical probe must have dimensions ≪ 3 cm if it is considered to be all at the same potential at the same instant of time (cf. Chapter 6). In contrast to the modelling related to the production and confinement of RF plasma, the consideration of the response of plasmas to microwaves will presume that signals are of low amplitude and that the responses are linear.

It was established in Chapter 2 that electromagnetic signals will only propagate in an *unbounded* plasma if their frequency exceeds the plasma frequency. Later chapters have been concerned with *bounded* plasma and the introduction of microwave signals into such plasmas necessarily involves consideration of boundaries. As a simple example of a plasma with internal and external boundaries, consider a spherical chamber filled with plasma at the centre of which is a small spherical probe (see Figure 10.19). The task here is to quantify the impedance that the probe would present to a 50 Ω microwave source coupled to it by a 50 Ω coaxial cable. In Section 2.4.5 the reciprocal of the impedance (i.e., the admittance) of a

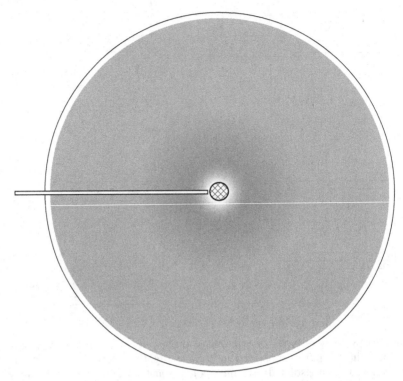

Figure 10.19 A spherical probe coupling a coaxial cable to a large volume of plasma.

plasma slab was shown to be

$$\frac{1}{Z} = i\omega C_0 \varepsilon_r = i\omega C_0 \left(1 - \frac{\omega_{pe}^2}{\omega(\omega - i\nu_m)}\right), \qquad (10.32)$$

where $C_0 = \varepsilon_0 A/d$ is the capacitance of the same slab but with vacuum between the boundaries. This was then recast into a combination of vacuum capacitance, plasma inductance and plasma resistance:

$$\frac{1}{Z_p} = i\omega C_0 + \frac{1}{i\omega L_p + R_p}. \qquad (10.33)$$

The inductance of the plasma slab, which results from the electron inertia, was

$$L_p = \frac{1}{\omega_{pe}^2 C_0}.$$

The resistance of the plasma slab, which results from the elastic electron–neutral collisions, was

$$R_p = \nu_m L_p = \frac{\nu_m}{\omega_{pe}^2 C_0}.$$

10.4 Probing with resonances and waves

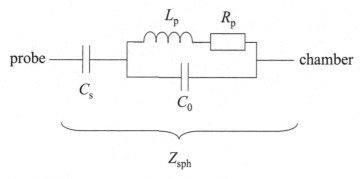

Figure 10.20 The equivalent circuit of the spherical probe in Figure 10.19.

So what is the impedance of a system in which the rectangular boundaries of the slab are replaced by the spherical surfaces of the probe and the chamber?

If the radius of the chamber is much larger than that of the spherical probe (r_0), the capacitance of the probe in vacuum (i.e., the amount of charge per volt of potential with respect to the outer boundary) is just $C_0 = 4\pi\varepsilon_0 r_0$. The admittance of the spherical plasma is then given by Eq. (10.33) with this spherical version of the 'geometric' capacitance C_0; the properties of the plasma are still summarized in L_p and R_p.

There are two further components that must be considered to complete the analysis of the impedance between the probe and the chamber – these are the sheath between the sphere and the plasma and the sheath between the plasma and the vessel. Since the area of the vessel is so much larger than that of the probe, the impedance of the latter sheath can be neglected. If $r_0 \gg \lambda_{De}$ then one can presume that the relatively thin sheath on the probe is adequately described by a planar model, so treating it as a region where $\varepsilon_r = 1$, its capacitance is $C_s = 4\pi\varepsilon_0 r_0^2/s$; for a floating sheath s is about $7\lambda_{De}$. Then the total impedance, between the sphere and the chamber Z_{sph}, as illustrated in Figure 10.20, is given by

$$Z_{sph} = \frac{1}{i\omega C_s} + \frac{1}{i\omega C_0 + \frac{1}{i\omega L_p + R_p}}. \qquad (10.34)$$

Figure 10.21 shows the magnitude and phase of Z_{sph}. There are two resonances – the upper one occurs at $\omega = \omega_{pe}$ and corresponds to the parallel resonance of the plasma through L_p and C_0 (when $Z_{sph} \to \infty$), while the lower one at $\omega < \omega_{pe}$ is a series resonance ($Z_{sph} \to 0$) between the sheath capacitance and the plasma inductance. A similar, but different, result arises in slab geometry [85] – in that work the analysis is related also to aspects of resonantly enhanced plasma generation, whereas here the emphasis is on non-perturbative diagnostic measurements.

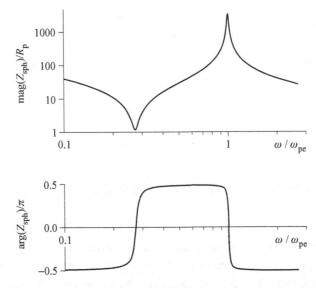

Figure 10.21 Magnitude and phase of the impedance of a floating spherical probe in a large spherical plasma-filled cavity, calculated for a uniform plasma ($n = 10^{16}$ m^{-3} and $kT_e = 2$ eV). The lower resonant frequency depends on the dimensions of the probe and its sheath: $\omega = (s/r_0)^{1/2}\omega_{pe}$.

> **Q** An electrical circuit at resonance exchanges energy between the electrostatic field in a capacitor and the magnetic field of an inductor – what two types of energy are exchanged in the series and parallel resonances of the spherical probe immersed in a plasma?
>
> **A** The resonances involve electrostatic energy, in the sheath or the plasma volume, that is being exchanged with the directed kinetic energy of the electron population in the plasma, which is what gives rise to the apparent inductance of the plasma.

> **Q** Given that the extremely low and extremely high impedances at the resonant frequencies in Figure 10.21 are likely to be identifiable by looking at the (microwave) signals reflected by the mismatch between the probe and the coaxial cable feeding it, suggest how a spherical probe could be used to determine electron density.
>
> **A** It is presumed that the probe sits within the central region of a large volume of uniform-density plasma (cf. Figure 10.19). The basic idea is to apply a swept frequency signal to the probe via the coaxial feed while monitoring the reflected power. The maximum impedance at the plasma frequency will be evident from a maximum in the signal reflected from the spherical probe (or equivalently a minimum in the transmission); electron density follows from

> $\omega_{\text{pe}}^2 = n_e e^2 / m\varepsilon_0$ – in contrast to Langmuir probe methods, one does not need to know kT_e.
>
> Also, the lower-frequency resonance corresponds to minimum impedance and therefore the greatest opportunity to excite currents in the plasma. On the basis of a thin sheath, devoid of electrons, it occurs at $\omega = (s/r_0)^{1/2}\omega_{\text{pe}}$. Linking this resonance to the electron density requires a model of the sheath on the electrode (e.g., Child–Langmuir), which then requires a value for kT_e. Alternatively, one could use the resonant frequencies to infer s and then compare that for consistency with sheath models.
>
> *Comment: With a fast oscilloscope, or a network analyser, the resonances are also easily detected from the sharp phase changes indicated in the figure [227].*

There are many closely related resonance probes which will be introduced briefly in the following sections. Note though that since resonances can be masked by heavy damping in the plasma (effectively around resonance $R_p \gg \omega L_p$), these probes are restricted to the low and intermediate pressure range.

10.4.2 Self-excited electron resonance probe

The spherical microwave resonance probe was considered in the previous section connected to an externally swept signal generator. However, it has been shown in earlier chapters that in an RF plasma there will be many harmonics of the excitation frequencies present. It is reasonable therefore to wonder under what conditions these harmonics might 'inadvertently' couple to the natural resonances of the plasma-filled chamber. Indeed, Section 5.4.3 introduced the slab-geometry counterpart of the series resonance just described, 'self-excited' by harmonics of the fundamental drive frequency. This phenomenon offers a convenient diagnostic tool. The presence of the resonance is imprinted in the current that traverses the plasma. Even when the external current is purely sinusoidal, harmonic currents can circulate through the plasma and conducting walls. A sample of this current can conveniently be taken from an isolated section of the chamber wall via a low impedance bypass and its high-frequency content analysed (Figure 10.22).

The frequency of the self-excited resonances can be linked back directly to the plasma density using a simple model of the resonance that requires an estimate of the sheath capacitance. Discussions so far have overlooked the structure of the plasma, which is not expected to have a uniform plasma density (cf. Chapter 3) and thus not one single and precisely defined plasma frequency; furthermore, the resonances are broadened by collisional damping. This leaves the plasma sensitive

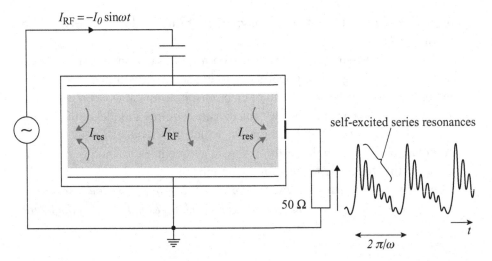

Figure 10.22 The self-excited resonance probe is effectively an isolated section of wall that takes a sample of the current flowing in the walls of the vessel, converting it to a voltage signal for analysis.

to a band of series resonances, so the drive frequency does not need to match exactly a high-quality factor resonance. The decay time of the resonance amplitude within each fundamental RF cycle is readily linked to the electron collision frequency, which in turn is a sensitive monitor of changes in gas composition. The analysis of signals from such a probe is often referred to as self-excited resonance spectroscopy (SEERS) [87, 228].

Wave phenomena must be considered when electrical signals are applied to structures that have characteristic dimensions comparable with the wavelength of electromagnetic disturbances at the same frequency. This was discussed in Section 6.2.1, where it was shown that at 200 MHz this became an issue for electrode widths (not the inter-electrode gap) of about 15 cm. Scaling up by an order of magnitude in frequency shifts that length scale down to 1.5 cm, and where microwaves are concerned objects more than a few millimetres in size will have to be considered from the perspective of electromagnetic waves. That not only sets an upper limit to the radius of a SEERS probe, but it also suggests ways to contrive other resonances that can be used to infer plasma density more directly. This is the next topic.

10.4.3 Hairpin resonator

The structure of a microwave hairpin resonator is shown in Figure 10.18(a). The resonator part is a U-shaped wire that forms a length of twin-wire transmission line that has an open circuit at one end and a half-loop short at the other. The resonator is fed from a full-loop termination of a length of coaxial cable placed adjacent to the shorted end – in practice, the components are fixed in place by a small amount of dielectric material designed to give appropriate mechanical support without undue electrical influence. The situation is analogous to an acoustic wave in an organ pipe: the resonant condition for the hairpin transmission line is when it supports a standing electromagnetic wave between the open and shorted ends; this arises when $L = \lambda/4$. So for a 2.5 cm long hairpin the resonance in vacuum occurs when

$$f_0 = \frac{c}{4L} = 3\,\text{GHz}. \tag{10.35}$$

An electromagnetic wave propagates in the space between and around the hairpin, guided by the wires, down to the open end from where it is reflected. When $L = (n+1)\lambda/4$ there is constructive interference between the forward and reflected waves that establishes a standing wave. If the hairpin were immersed in a dielectric medium then $c \to c/\sqrt{\varepsilon_r}$, which would tend to shift the quarter-wave resonance to lower frequency since, for ordinary dielectrics, $\varepsilon > 1$. In a plasma, however, neglecting collisions,

$$\varepsilon_r = 1 - \frac{f_{pe}^2}{f^2}. \tag{10.36}$$

It is convenient here to use cyclic frequency, f, rather than angular frequency, $\omega = 2\pi f$.

Q Show that in a plasma the hairpin resonance is given by

$$f_{res}^2 = f_0^2 + f_{pe}^2. \tag{10.37}$$

A The result follows from the quarter-wave resonance in a dielectric,

$$f_{res} = \frac{c}{4L\sqrt{\varepsilon_r}}. \tag{10.38}$$

Substituting for ε_r from Eq. (10.36) with $f = f_{res}$ and squaring gives

$$f_{res}^2 = \frac{(c/4L)^2}{1 - f_{pe}^2/f_{res}^2}. \tag{10.39}$$

Then with Eq. (10.35) this rearranges to Eq. (10.37).

At resonance, the standing wave is established with an antinode (maximum) of voltage and a node (zero) of current at the open end; at the closed end the current is maximum and the voltage is zero. This means that the hairpin is especially sensitive to dielectric around the open end, where the electric field is larger. At

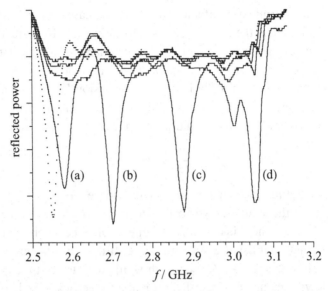

Figure 10.23 Reflected power resonance curves for a hairpin probe in vacuum (dotted) and in four plasmas with different values of electron density (a–d).

resonance the structure absorbs incident microwave power, whereas off-resonance incident microwave energy is reflected back along the coaxial feed. Thus the resonance can be detected as a minimum in the reflected power envelope. Notice that the resonant frequency always occurs above the plasma frequency. That has an important consequence for the energy absorbed by the resonator – the standing wave of current on the hairpin couples into electromagnetic waves that are able to propagate away into the surrounding plasma. The propagation of electromagnetic waves between two probes is another way to explore the plasma properties; a brief example is given in Section 10.4.5.

If the separation of the limbs is too large, then there is poor coupling between them and the energy is radiated away without building up a standing wave. At the other extreme, one does not want the inevitable sheaths around the wires of the hairpin to occupy a significant fraction of the space between them otherwise the transmission line dielectric is not comprised purely of plasma and the resonant frequency is not so easily determined. Various experimental factors (including ways to account for the sheath around the hairpin) that need to be considered in constructing and operating hairpin probes can be found in the literature [229–234].

Exercise 10.9: Hairpin probe Figure 10.23 shows data from a hairpin probe in vacuum and in a CCP, at four different RF power levels. Deduce the maximum plasma density achieved in the vicinity of the probe.

10.4 Probing with resonances and waves

The hairpin resonator is a particularly simple structure to make and to analyse. Other structures can also support standing waves and therefore can be expected to furnish other resonances. Thus it is not surprising to find a great deal of structure in the spectrum of energy reflected back from a cable terminated with a hairpin placed inside the enclosure of a plasma source. Identifying the hairpin resonance is still feasible, since this feature is first identified 'on the bench'. Then it is easily recognized after insertion and can be tracked when the plasma is formed, as it shifts to higher frequency when the plasma density is increased. By contrast, resonances of parts of the structure away from the plasma do not change with plasma density and can be 'subtracted out'. Other plasma-linked resonances such as the series resonance can still be excited, and these and higher modes of the hairpin can be avoided by restricting the microwave sweep range. However, this also highlights the possibility of designing other plasma-linked resonant structures – one such example is considered next.

10.4.4 Multipole resonator

The fundamental resonance of the hairpin is above the plasma frequency and the device couples to electromagnetic waves in the plasma. Around plasma boundaries the abrupt changes associated with sheaths and the surfaces of metals and dielectrics create a complicated microwave environment. Even a simple spherical probe has been shown to have microwave resonances. The multipole resonator shown in Figure 10.18(d) is a development of the spherical probe – it looks at first sight like additional capacitance has been added in the form of a thick dielectric coating, but in fact it is more sophisticated than this. The core is not a single metallic sphere, being in fact a pair of separate hemispheres that are connected to a balanced feed with the microwave input signal on the two hemispheres in anti-phase. The analysis of this structure is complicated [235], and it reveals that electromagnetic waves guided around the surface of the sphere can excite resonant modes. These waves interact with the plasma over a distance of a few times the outer radius of the probe. Figure 10.24 shows the results of calculations showing readily identifiable resonances that can be linked to the plasma properties, principally electron density in the vicinity of the probe. The multipole probe is able to meet the constraints on materials exposed to the plasma since the dielectric layer and the outer surface of the support shaft can be made from material chosen solely for its compatibility with the plasma environment.

One of the important points about all the hairpin and multipole probes is that resonances are formed by standing waves guided by the plasma boundary. The closing part of this chapter looks briefly at these waves themselves.

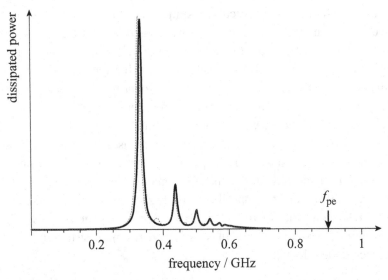

Figure 10.24 Calculated absorption spectrum of a multipole resonator.

10.4.5 Microwave transmission and cut-off methods

In a vacuum, any frequency of electromagnetic wave will propagate. In a plasma, plane-wave solutions to Maxwell's equations can be found only for $\omega > \omega_{pe}$. The presence of free electrons forced into oscillation by an incident wave effectively excludes frequencies below ω_{pe}, reflecting the incident energy – the plasma frequency marks the limit beyond which a population of electrons is unable to keep up and waves can penetrate. The sheath regions that form at plasma boundaries are largely free of electrons (so $\varepsilon = 1$) and are therefore important channels in which wave solutions to Maxwell's equations can be found. Modes are found that propagate parallel to the boundary in some of the range of $\omega < \omega_{pe}$ that is excluded by the bulk. Although the wave does not escape into the body of the plasma, the fields decay over a skin depth or so inwards from the boundary as a disturbance travels along the boundary.

Detailed analysis [236] shows that there are sharp bands of frequency in which waves will or will not propagate along the boundary, depending on the boundary structure. For instance, waves are guided down a sheath between a dielectric and a plasma for $\omega_{pe}/\sqrt{2} < \omega < \omega_{pe}$ whereas they will only pass along a sheath between metal and plasma in the complementary region $\omega < \omega_{pe}/\sqrt{2}$ and $\omega > \omega_{pe}$. A diagnostic probe based on this has been demonstrated – it is illustrated in Figure 10.18(d). A band of transmission is observed between a 'transmitter' and a 'receiver' that are linked by a rod of dielectric, which is itself shrouded by the plasma and a sheath that guides the waves. The ratio of the upper and lower limits of transmission is close to the expected $\sqrt{2}$. This 'transmission' probe is not as localized as the hairpin and multipole resonators, and the edges of the pass band are less easy to discern than resonance peaks, especially in the presence of collisional damping.

A variation on this idea places a transmission probe in one arm of an interferometer and measures the phase difference between a path guided along a plasma boundary and one outside the plasma in coaxial cable [237]. In this version the waves are guided along a sheath around a section of coaxial line from which the dielectric has been stripped back to expose the surface of the inner conductor. A frequency is chosen above the expected plasma frequency and waves are guided along the bare conductor, although some energy propagates away because $\omega > \omega_{pe}$. The phase shifts for an interferometer with a 3 cm path through the plasma correspond conveniently with electron densities at the rate of about $10°$ per $10^{16}\,\mathrm{m}^{-3}$ of electron density.

The final configuration in this range is that of Figure 10.18(c), in which one simply tries to find ω_{pe} by determining the frequency below which wave propagation from one probe to another across a region of plasma [238] is cut off. This is not as simple as it first sounds because, as has already been discussed, just as the plasma path is cut off, the dielectric waveguide modes switch on and the best one can hope for is a sharp dip in transmission at ω_{pe}, which is inevitably blurred by variations in electron density along the path. In fact, a bare wire probe is not necessarily the best way to launch and detect microwaves in a plasma and the cut-off and interferometer methods can also be done using horns to transmit and receive signals across the plasma [229, 239]. Interferometry with unguided microwaves is a well-established, non-invasive technique, especially in high-temperature plasmas, though the method necessarily gives a line-of-sight integration of plasma density.

10.5 Summary of important results

A simple refractory, conducting probe is a convenient means of quantifying the charge composition of a plasma. Langmuir tried this in the 1920s and was one of the first to develop an electrical probe method. There are now numerous variations on the theme, including planar, cylindrical and spherical geometry. There are also probes that resonate near the plasma frequency and others that launch waves. Some probes are electrostatic and others are electromagnetic; some are effectively wireless; most absorb but some emit. An important feature is that a probe should function in steady and transient plasmas, while special schemes have been devised for RF plasmas. Magnetized plasmas pose further challenges. Each configuration is accompanied by assumptions that constrain both the applicability and the

Table 10.1 *Summary of electrical measurement capabilities: solid circles imply a high degree of effectiveness and suitability, whereas open circles indicate a lesser degree*

Method	n_i	$n_i u_B$	T_e	n_e	V_f	V_p	EEDF	IVDF	RF-compatible	Comment
Planar double probe		●	○	●	●				○	Tail temperature only; needs RF compensation
Cylindrical single probe	○		●	●	●	●	●		○	Complex for ion density; needs RF compensation
Emissive probe						●			●	
Self-bias probe		●	○	●	○	●			●	V_{f-RF} and V_{f-DC}
RFA		●						●	●	Tail temperature only
SEERS				●					●	Also measures collisionality
Hairpin resonator				●					●	
Multipole resonator				●					●	
Microwave transmission				●					●	
Microwave cut-off				●					●	
External I and V	○		○	○	○	○		○	●	Global quantities only

10.5 Summary of important results

analytical methods that translate the measured currents and voltages variously into charge densities, space potentials, particle fluxes, energy distributions and measures of collisionality. This chapter looked at electrical measurement methods in general for local characterization of charged particles in the environment of the non-equilibrium plasmas used in materials processing. Table 10.1 sets out the main methods discussed in this chapter.

There is a complementary toolbox based on optical techniques that can be used in measurements on charged *and neutral* populations and for measuring electric and magnetic fields. It is in the nature of these methods that they involve specific atoms and molecules in specific environments, so they are often not universally applicable. However, their capacity to probe locally and non-invasively makes them attractive partners to electrical methods.

Appendix: Solutions to exercises

Chapter 2

Solution 2.1: Using Eq. (2.11) with $kT/e = 2\,\text{V}$ gives $\bar{v} = 9.5 \times 10^5\,\text{m s}^{-1}$. Then, using Eq. (2.20) with $\Delta\phi = 10\,\text{V}$, $kT/e = 2\,\text{V}$ and $n = 10^{16}\,\text{m}^{-3}$:

$$Q_w = \left[\frac{10^{16}\,\text{m}^{-3} \times 9.5 \times 10^5\,\text{m s}^{-1}}{4} \exp(-5)\right](4\,\text{V}) \times 1.6 \times 10^{-19}\,\text{C}$$

$$\approx 10\,\text{W m}^{-2}.$$

Solution 2.2: Use $n_g = p/kT_g$, $kT_e/e = 2\,\text{V}$, $\bar{v}_e = (8kT_e/\pi m)^{1/2}$, $kT_i/e = 0.05\,\text{V}$ and $\bar{v}_i = (8kT_i/\pi M)^{1/2}$. Then

frequencies ($K = \nu/n_g$):

$$\nu_{iz} = 2.7 \times 10^4\,\text{s}^{-1},$$
$$\nu_{exc} = 2.0 \times 10^5\,\text{s}^{-1},$$
$$\nu_m = 1.5 \times 10^8\,\text{s}^{-1},$$
$$\nu_i = 1.4 \times 10^6\,\text{s}^{-1};$$

mean free paths ($\lambda = \bar{v}/\nu$):

$$\lambda_{el} = 6.5 \times 10^{-3}\,\text{m},$$
$$\lambda_i = 4 \times 10^{-4}\,\text{m}.$$

Solution 2.3: Using the mean free path and frequencies calculated for the same conditions in the previous example,

$$\lambda_{el} = 4 \times 10^{-3}\,\text{m},$$
$$\nu_{iz} = 2 \times 10^4\,\text{s}^{-1},$$
$$\nu_{exc} = 1.55 \times 10^5\,\text{s}^{-1},$$
$$\nu_m = 2.3 \times 10^8\,\text{s}^{-1},$$

and the data given in Table 2.1, one obtains $\lambda_\varepsilon = 0.067$ m. This length is greater than the typical plate separation in capacitive discharges.

Chapter 3

Solution 3.1:

$$\lambda_{De} = \sqrt{\varepsilon_0 k T_e/(n_{e0}e^2)} = \sqrt{\varepsilon_0 (k T_e/e)/(n_{e0}e)}$$
$$= \sqrt{8.9 \times 10^{-12} \times 2.0 \times /(1.0 \times 10^{16} \times 1.6 \times 10^{-19})} \text{ m}$$
$$= 1.1 \times 10^{-4} \text{ m}.$$

Solution 3.2:

$$\frac{eV_0}{kT_e} = \frac{e^2 n_{i0}}{2\varepsilon_0 k T_e} s^2 = \frac{1}{2}\left(\frac{s}{\lambda_{De}}\right)^2, \text{ since } n_{i0} = n_{e0},$$

$$s/\lambda_{De} = \sqrt{2eV_0/kT_e}. \tag{A.3.1}$$

So the thickness of an ion matrix sheath would be $\sqrt{200} = 14$ Debye lengths.

Solution 3.3: Divide both sides by λ_{De} and rearrange to make s/λ_{De} the subject, then simplify the RHS by expanding λ_{De} and explicitly grouping V_0 with e/kT_e to obtain

$$\frac{s}{\lambda_{De}} = \left(\frac{4\sqrt{2}\, n_{e0}\, e\, \sqrt{kT_e/M}}{9 J_i}\right)^{1/2} \left(\frac{eV_0}{kT_e}\right)^{3/4}. \tag{A.3.2}$$

The Child–Langmuir sheath under the given conditions is then found to be about 25 Debye lengths thick.

Solution 3.4: Equation (3.23) gives

$$\frac{s}{\lambda_{De}} = \left(\frac{8}{9\pi}\frac{\lambda_i}{\lambda_{De}}\right)^{1/5} \left(\frac{n_{e0} e \sqrt{kT_e/M}}{J_i}\right)^{2/5} \left(\frac{5}{3}\frac{eV_0}{kT_e}\right)^{3/5}.$$

Inserting values gives

$$\frac{s}{\lambda_{De}} = \left(\frac{8}{9\pi} \times 3\right)^{1/5} \times 1 \times \left(\frac{5}{3}\frac{200}{2}\right)^{3/5} = 20.84.$$

Solution 3.5: First check that the sheath will be collisionless:

$$\lambda_{De} = \sqrt{\frac{\varepsilon_0 k T_e}{ne^2}} \sim 10^{-4} \text{ m} \ll \lambda_i, \quad \text{so the sheath is collisionless.}$$

Next calculate the ion flux using the idea that the ion density at the boundary is reduced by the (collisionless) acceleration to the Bohm speed (Eq. (3.30)):

$$\Gamma_{\text{boundary}} \equiv \Gamma_{\text{surface}} \approx 0.6 n_0 \sqrt{\frac{kT_e}{M}} \quad \text{ions m}^{-2} \text{s}^{-1}.$$

Then, evaluating the potential drop across the sheath using Eq. (3.32) and including the ion energy $Mu_B^2/2$ at the sheath edge, the ion energy is

$$w_{\text{surface}} = kT_e \frac{1}{2} \left| \ln\left(\frac{2\pi m}{M}\right) \right| + \frac{1}{2} kT_e \quad \text{joules per ion}$$

and so the net ion energy flux to the surface is

$$Q_{\text{surface}} = \frac{1}{2} kT_e \left[\left| \ln\left(\frac{2\pi m}{M}\right) \right| + 1 \right] \times 0.6 n_0 \sqrt{\frac{kT_e}{M}} \approx 22 \text{ W m}^{-2},$$

remembering that $2 \text{ eV} = 3.2 \times 10^{-19}$ J.
Comment: Compare this with the electron energy flux to a surface calculated for the same situation at the end of Section 2.1.2.

Chapter 4

Solution 4.1: Using Eqs (4.1) and (4.2), with $M = 40$ amu (argon) gives

$$\tau_i \approx 50 \text{ ns}$$

$$\tau_e \approx 0.2 \text{ ns}.$$

The period of a 13.56 MHz waveform is $\tau_{RF} = 74$ ns, so that $\tau_e \ll \tau_i \lesssim \tau_{RF}$.

Solution 4.2: Using Eq. (4.3) the results are H: $\omega_{pi} = 130 \times 10^6$ s^{-1}, H$_2$O: $\omega_{pi} = 31 \times 10^6$ s^{-1}, Ar: $\omega_{pi} = 21 \times 10^6$ s^{-1}.
Comment: These are angular frequencies so the unit is s^{-1} rather than Hz.

Solution 4.3: The total DC bias is the floating potential given by Eq. (4.15):

$$V_{f_{RF}} = 2 \left[\frac{1}{2} \ln\left(\frac{2\pi \times 9.1 \times 10^{-31}}{40 \times 1.7 \times 10^{-27}}\right) - \ln I_0(25) \right] \text{V} = -54.3 \text{ V}.$$

Inserting the magnitude of this DC voltage into Eq. (A.3.1) for a DC matrix sheath gives

$$s/\lambda_{\text{De}} = \sqrt{2 \times 54.3/2} \sim 7.$$

So one needs to compare the mean free path with $7\lambda_{De}$. For the given plasma the Debye length is 10^{-4} m, which is only 1% of the mean free path so the ions would pass through more or less without collisions.

Comment: This simple estimate of sheath size can be improved by building a proper model of the RF sheath region – see Section 4.4.

Solution 4.4: The ions will arrive at a rate given by the Bohm flux and with the mean sheath energy, so the ion energy flux is

$$Q_{\text{surface}} = h_1 n_0 \, u_B \, eV_{f_{RF}}.$$

Taking the centre-to-edge density ratio from the low-pressure side of Figure 3.11:

$$Q_{\text{surface}} = 0.5 \times 10^{16} \times \sqrt{2 \times 1.6 \times 10^{-19}/(1.67 \times 10^{-27} \times 40)}$$
$$\times 1.6 \times 10^{-19} \times 54.3 \sim 100\,\text{W m}^{-2}.$$

Solution 4.5: To control the energy of ions arriving at an insulating substrate one must use the self-bias effect of an RF modulation, as the substrate blocks DC. The pressure must be set low enough to avoid collisions in the sheath, e.g., using $\lambda_{De} n_g \sigma_{i-n} \ll 1$; a low electron temperature is also desirable to keep the distribution at the plasma/sheath boundary as narrow as possible, though this may not be easily controlled as the pressure has already been constrained. Then one must choose a frequency high enough ($\omega \gg \omega_{pi}$) or sufficiently asymmetric to promote a single-peaked IEDF.

Comment: Wang and Wendt [50] proposed a voltage waveform that combined a slow negative-going ramp and a short positive-going pulse for application to a substrate via a coupling capacitor, which achieves a steady, narrow IEDF with a steady ion flux, that is neutralized on the substrate by the electrons attracted during the periodic positive pulses.

Chapter 5

Solution 5.1: The rise in temperature and fall in density follow from the reduced pressure in much the same way as with the gap size: h_1 increases, leading to enhanced losses. The sheath size deceases with increasing pressure, as indicated by Eq. (5.52).

Chapter 8

Solution 8.1: For efficient heating, $\alpha_z \leq 0.5$ m. As shown in the tables, $k_z \leq 190$ and $k_r \leq 60$ for the reference conditions, with $\omega_{ce} = 8.9 \times 10^8\,\text{s}^{-1}$. Therefore, Eqs (8.26) and (8.27) show that $\alpha_z \leq 0.5$ m is equivalent to $\nu_{\text{eff}} \gtrsim \omega_{ce}/50 \approx 2 \times 10^7\,\text{s}^{-1}$.

Appendix

Chapter 10

Solution 10.1: Since the model characteristic is $I = I_i \tanh(eV/kT_e)$, the slope at the origin is eI_i/kT_e. The first thing to do therefore is to deduce from the graph the saturation and the slope at the origin: these quantities are estimated to be 32 µA and $30/2.0$ µA V^{-1}, respectively, noting that the curve does not quite saturate in the displayed range. Thus

$$kT_e/e = 32 \times 2.0/30 = 2.1 \text{ eV}.$$

Then $u_B = \sqrt{2.1 \times 1.6 \times 10^{-19}/(40 \times 1.67 \times 10^{-27})} = 2200 \text{ m s}^{-1}$ and $I_0 = n_s e u_B A$ gives

$$n_s = 32 \times 10^{-6}/[1.6 \times 10^{-19} \times 2200 \times \pi \times 25/4 \times 10^{-6}] = 0.46 \times 10^{16} \text{ m}^{-3}.$$

Comment: This is the density at the plasma boundary. Given the relatively low pressure one can take $h_1 \approx 0.5$, so the density beyond the immediate vicinity of the probe, n_0, will be double this value.

Solution 10.2: For $M \equiv 1$ amu the plasma potential is expected to be $3.5kT_e/e$ above the potential of a reference surface. In that case the area criterion, Eq. (10.5), becomes

$$A_1 \gg 34 A_2,$$

which is a markedly easier thing to achieve for a reference electrode that is genuinely in the vicinity of the probe electrode.

Solution 10.3: The first thing to do is to reject the data for the area ratio of 34 as this is insufficient to ensure an unimpeded electron saturation (i.e., this is not the characteristic of a single probe). One can use the curve with area ratio of 340 since this meets the criterion for a single probe in hydrogen.

(i) At the plasma potential there is a sharp discontinuity: on the '340' graph this is at 9.6 V.

(ii) At the floating potential the current is zero: in the figure this is at 1.9 V (whatever the area ratio). *Comment: In most real systems the floating potential is indeed not that of the local ground for one or more of the following reasons – the plasma is non-uniform so that $\phi_{p2} \neq \phi_{p1}$; the vessel walls are not conducting or are set at a potential different from ground; the object that is floating does not have the simple planar geometry presumed here.*

(iii) The ion density follows from $n_s = I(-15 \text{ V})/eA_2 u_B$. From the graph $I(-15 \text{ V}) = 0.15$ mA, but to evaluate the Bohm speed one needs the electron temperature. This can be estimated from the difference between the plasma and floating potential:

$$kT_e/e = (V_p - V_f)/0.5 \ln(M/2\pi m) = \frac{6.2}{2.8} \approx 2.2\,\text{V}.$$

Then one can evaluate $u_B = 1.4 \times 10^4\,\text{m s}^{-1}$ and finally

$$n_s = 0.15\,\text{mA}/[1.6 \times 10^{-19}\,\text{C} \times 2.0 \times 10^{-5}\,\text{m}^2 \times 1.4 \times 10^4\,\text{m s}^{-1}]$$
$$= 3.5 \times 10^{15}\,\text{m}^{-3}.$$

(iv) The electron density in the undisturbed plasma follows from $n_0 = 4I(9.6\,\text{V})/eA_2\bar{v}_e$; the mean thermal electron speed is $\bar{v}_e = \sqrt{8kT_e/\pi m} = 9.8 \times 10^5\,\text{m s}^{-1}$. This gives

$$n_0 = 4 \times 5.3\,\text{mA}/[1.6 \times 10^{-19}\,\text{C} \times 2.0 \times 10^{-5}\,\text{m}^2 \times 9.8 \times 10^5\,\text{m s}^{-1}]$$
$$= 7.1 \times 10^{15}\,\text{m}^{-3}.$$

Solution 10.4: Linearity on a semi-log plot implies that a function is exponential. All curves show a linear portion at very small current, but only the two that meet the criterion for 'single probe' operation are linear throughout.
(i) The discontinuity locates the plasma potential at 11.3 V.
(ii) From the slope of the semi-log plot, $kT_e/e = 20/9 \approx 2.2$ V.

Solution 10.5: According to Eq. (10.14), a few volts above plasma potential, the plot of I_e^2 against V should be linear with gradient

$$\frac{d(I_e^2)}{dV} = (e\,2\pi r_c l n_0)^2\,e/\pi^2 m$$

and intercept ($I_e = 0$) at

$$V_{\text{int}} = kT_e/e + \phi_p.$$

The plasma potential is clearly seen as a knee in both the I–V curve and the I^2–V curve (and as a peak in dI/dV) at 11.3 V. After this the transition to linear behaviour of I^2–V begins. Reading from the graph, the gradient well above the knee is $100 \times 10^{-6}/5.5\,\text{A}^2\,\text{V}^{-1}$ and the intercept is 9.3 V. It follows from these measurements that $n_0 = 7.2 \times 10^{15}\,\text{m}^{-3}$ and $kT_e = 2\,\text{eV}$.

Solution 10.6: The Maxwellian character is indicated by linear semi-log plots of the EEPF. In Figure 5.16 simple linear behaviour is only really evident around 40 Pa. Below that the distribution seems to show two different linear portions. Treating the two regions as separate Maxwellians the gentler slope at high energy corresponds to a higher-temperature distribution whereas the steeper slope at low energy is associated with a cooler component. The temperatures can be deduced by the same means as the semi-log analysis of I–V curves introduced in Section 10.1.3. Above 40 Pa, the distribution in the high-energy tail is much cooler than the bulk. [Closer scrutiny reveals that the distribution here scales with $\exp(-av^4)$ – a form that is called a Druyvesteyn distribution.]

Solution 10.7: (i) The electron temperature could be found from a graphical solution of Eq. (10.26) with $V_{RF} = 15$ V – the result is 2.5 eV.
(ii) The slope of the initial decay rate is

$$\left.\frac{dV}{dt}\right|_{\text{initial}} = \frac{eAn_s u_B}{C_x}.$$

From the graph the initial slope is 2 V/0.1 ms, so inserting values for C_x and A gives $n_s u_B = 1.5 \times 10^{19}$ ions m^{-2} s^{-1}.

Solution 10.8: The highest velocity ions in Figure 10.17 have a kinetic energy that should closely correspond with falling freely from plasma potential, so V_p can be identified with the position of the sharply falling edge of the dI/dV plot. The lower side of the IVDF is located at the 9.6 V given as the sheath potential and the distribution extends above that to 11.6 V, which corresponds with the plasma potential far from the RFA. According to the Tonks–Langmuir formulation the width of the IEDF (which is equivalent to the width of the IVDF on an energy scale) is $0.854\,kT_e$, so that in this case $kT_e \sim (11.6 - 9.6)/0.854 = 2.3$ V.
Comment: Because of the limited energy resolution in the analyser this is slightly larger than the 2.0 V used to calculate the initial distribution for Figure 4.5.

Solution 10.9: The maximum-density case corresponds to the highest resonant frequency: $f_{\text{res}} = 3.05$ GHz. The vacuum resonance f_0 is at 2.55 GHz. Using Eq. (10.37),

$$f_{\text{pe}}^2 = (3.05^2 - 2.55^2)(\text{GHz})^2.$$

Therefore

$$n_e = \frac{m\varepsilon_0}{e^2}(2\pi f_{\text{pe}})^2 = 3.5 \times 10^{16} \text{ m}^{-3}.$$

References

[1] J. W. Coburn and H. F. Winters. *J. Appl. Phys.*, 50(5):3189–96, 1979.
[2] M. A. Lieberman and A. J. Lichtenberg. *Principles of Plasma Discharges and Materials Processing*. John Wiley & Sons, 2nd edition, New York, 2005.
[3] J. P. Booth, G. Cunge, P. Chabert and N. Sadeghi. *J. Appl. Phys.*, 85:3097, 1999.
[4] G. Cunge and J. P. Booth. *J. Appl. Phys.*, 85:3952, 1999.
[5] X. Detter, R. Palla, I. Thomas-Boutherin, E. Pargon, G. Cunge, O. Joubert and L. Vallier. *J. Vac. Sci. Technol. B*, 21(5):2174–83, 2003.
[6] G. Cunge, M. Kogelschatz and N. Sadeghi. *Plasma Sources Sci. Technol.*, 13(3):522–30, 2005.
[7] S. Bouchoule, G. Patriarche, S. Guilet, L. Gatilova, L. Largeau and P. Chabert. *J. Vac. Sci. Technol. B*, 26:666, 2008.
[8] C. Y. Duluard, R. Dussart, T. Tillocher, L. E. Pichon, P. Lefaucheux, M. Puech and P. Ranson. *Plasma Sources Sci. Technol.*, 17:045008, 2008.
[9] P. Chabert, N. Proust, J. Perrin and R. W. Boswell. *Appl. Phys. Lett.*, 76:2310, 2000.
[10] P. Chabert, G. Cunge, J.-P. Booth and J. Perrin. *Appl. Phys. Lett.*, 79:916, 2001.
[11] P. Chabert. *J. Vacuum. Sci. Technol. B*, 19:1339, 2001.
[12] J. Schmitt, M. Elyaakoubi and L. Sansonnens. *Plasma Sources Sci. Technol.*, 11:A206, 2002.
[13] D. M. Goebel and I. Katz. *Fundamentals of Electric Propulsion*. John Wiley & Sons, Hoboken, NJ, 2008.
[14] V. V. Zhurin, H. R. Kaufmann and R. S. Robinson. *Plasma Sources Sci. Technol.*, 8:R1, 1999.
[15] J. P. Squire, F. R. C. Diaz, T. W. Glover, V. T. Jacobson, D. G. Chavers, E. A. Bering, R. D. Bengtson, R. W. Boswell, R. H. Goulding and M. Light. *Fusion Sci. Technol.*, 43:111–17, 2003.
[16] C. Charles, R. W. Boswell and M. A. Lieberman. *Appl. Phys. Lett.*, 89:261503, 2006.
[17] A. Aanesland, A. Meige and P. Chabert. *IOP J. Phys.: Conf. Ser.*, 162:012009, 2009.
[18] M. M. Turner and M. A. Lieberman. *Plasma Sources Sci. Technol.*, 8:313, 1999.
[19] M. A. Lieberman, A. J. Lichtenberg and A. M. Marakhtanov. *Appl. Phys. Lett.*, 75:3617, 1999.
[20] P. Chabert, A. J. Lichtenberg, M. A. Lieberman and A. M. Marakhtanov. *Plasma Sources Sci. Technol.*, 10:478, 2001.

[21] P. Chabert, J.-L. Raimbault, P. Levif, J.-M. Rax and M. A. Lieberman. *Phys. Rev. Lett.*, 95:205001, 2005.
[22] P. Chabert. *J. Phys. D: Appl. Phys.*, 40:R63–R73, 2007.
[23] F. F. Chen. *Plasma Physics and Controlled Fusion*. XXX, 1980.
[24] B. M. Smirnov. *Physics of Ionized Gases*. John Wiley & Sons, New York, 2001.
[25] G. G. Lister, Y.-M. Li and V. A. Godyak. *J. Appl. Phys.*, 79(12):8993, 1996.
[26] H. Abada, P. Chabert, J.-P. Booth, J. Robiche and G. Cartry. *J. Appl. Phys.*, 92:4223, 2002.
[27] V. A. Godyak. *IEEE Trans. Plasma Sci.*, 34(3):755, 2006.
[28] I. B. Bernstein and T. Holstein. *Phys. Rev.*, 94:1475, 1954.
[29] L. D. Tsendin. *Sov. Phys. JETP*, 39:805, 1974.
[30] V. I. Kolobov and V. A. Godyak. *IEEE Trans. Plasma Sci.*, 23:503, 1995.
[31] U. Kortshagen, C. Busch and L. D. Tsendin. *Plasma Sources Sci. Technol.*, 5:1, 1996.
[32] C. D. Child. *Phys. Rev (ser I)*, 32:492–511, 1911.
[33] I. Langmuir. *Phys. Rev.*, 2:450–86, 1913.
[34] M. S. Benilov. *Plasma Sources Sci. Technol.*, 18:014005, 2008.
[35] D. Bohm. *The Characteristics of Electrical Discharges in Magnetic Fields*. McGraw-Hill, New York, 1949.
[36] J. E. Allen. *J. Phys. D: Appl. Phys.*, 9:2331–2, 1976.
[37] L. Tonks and I. Langmuir. *Phys. Rev.*, 34:876, 1929.
[38] W. Schottky. *Phys. Z.*, 25:635, 1924.
[39] V. A. Godyak. *Soviet Radiofrequency Discharge Research*. Delphic Associates, Fall Church, VA, 1986.
[40] S. A. Self and H. N. Ewald. *Phys. Fluids*, 9:2486, 1966.
[41] V. A. Godyak and N. Sternberg. *IEEE Trans. Plasma Sci.*, 18:159, 1990.
[42] W. B. Thompson and E. R. Harrison. *Proc. R. Soc. Lond.*, 74:145–52, 1959.
[43] R. N. Franklin. *J. Phys. D: Appl. Phys.*, 36:2660–61, 2003.
[44] J.-L. Raimbault, L. Liard, J.-M. Rax, P. Chabert, A. Fruchtman and G. Makrinich. *Phys. Plasmas*, 14:013503, 2007.
[45] P. Chabert, A. J. Lichtenberg and M. A. Lieberman. *Phys. Plasmas*, 14:093502, 2007.
[46] S. G. Ingram and N. St. J. Braithwaite. *J. Phys. D: Appl. Phys.*, 21:1496–503, 1998.
[47] M. A. Sobolewski. *Phys. Rev. E*, 21:8540–53, 2000.
[48] E. Kawamura, V. Vahedi, M. A. Lieberman and C. K. Birdsall. *Plasma Sources Sci. Technol.*, 8:R45–R64, 1999.
[49] S. B. Radovanov, J. K. Olthoff, R. J. Van Brunt and S. Djurovic. *J. Appl. Phys.*, 78:746–58, 195.
[50] S. B. Wang and A. E. Wendt. *J. Appl. Phys.*, 88:643–6, 2000.
[51] D. Vender and R. W. Boswell. *J. Vac. Sci. Technol. A*, 10:1331–8, 1992.
[52] C. M. O. Mahony, R. AlWazzan and W. G. Graham. *Appl. Phys. Lett.*, 71:608–10, 1999.
[53] H. B. Vallentini. *J. Appl. Phys.*, 86:6665–72, 1999.
[54] P. R. J. Barroy, A. Goodyear and N. St. J. Braithwaite. *IEEE Trans Plasma Sci.*, 30:148–9, 2002.
[55] J. Schulze, Z. Donko, B. G. Heil, D. Luggenhoelscher, T. Mussenbrock, R. P. Brinkmann and U. Czarnetzki. *J. Phys. D: Appl. Phys.*, 41:105214, 2008.
[56] M. A. Lieberman. *IEEE Trans. Plasma Sci.*, 16:638, 1988.
[57] V. A. Godyak. *Sov. Plasma Phys.*, p. 141, 1976.
[58] M. A. Lieberman. *IEEE Trans. Plasma Sci.*, 17:338, 1989.
[59] V. A. Godyak. *IEEE Trans. Plasma Sci.*, 19:660, 1991.
[60] M. M. Turner. *Phys. Rev. Lett.*, 75:1312, 1995.

[61] G. Gozadinos, M. M. Turner and D. Vender. *Phys. Rev. Lett.*, 87:135004, 2001.
[62] G. R. Misium, A. J. Lichtenberg and M. A. Lieberman. *J. Vacuum Sci. Technol. A*, 7:1007, 1989.
[63] P. Chabert, J.-L. Raimbault, J.-M. Rax and M. A. Lieberman. *Phys. Plasmas*, 11(5):1775, 2004.
[64] M. M. Turner and P. Chabert. *Appl. Phys. Lett.*, 89:231502, 2006.
[65] O. A. Popov and V. A. Godyak. *J. Appl. Phys.*, 57:53, 1985.
[66] V. A. Godyak and R. B. Piejak. *Phys. Rev. Lett.*, 65:996, 1990.
[67] V. A. Godyak. *Sov. Phys. – Tech. Phys.*, 16:1073–6, 1972.
[68] M. A. Lieberman and V. A. Godyak. *IEEE Trans. Plasma Sci.*, 26:955, 1998.
[69] E. Kawamura, M. A. Lieberman and A. J. Lichtenberg. *Phys. Plasmas*, 13:053506, 2006.
[70] G. Gozadinos, D. Vender, M. M. Turner and M. A. Lieberman. *Plasma Sources Sci. Technol.*, 10:1, 2001.
[71] M. Surendra and D. B. Graves. *Phys. Rev. Lett.*, 66:1469, 1991.
[72] M. Surendra and M. Dalvie. *Phys. Rev. E*, 48:3914, 1991.
[73] M. Surendra and D. B. Graves. *Appl. Phys. Lett.*, 59:2091, 1991.
[74] M. Meyyappan and M. J. Colgan. *J. Vac. Sci. Technol. A*, 14:2790, 1996.
[75] A. Perret. *Effets de la fréquence d'excitation sur l'uniformité du plasma dans les réacteurs capacitifs grande surface*. PhD thesis, École Polytechnique, Palaiseau, France, June 2004.
[76] J.-P. Boeuf. *Phys. Rev. A*, 36:2782, 1987.
[77] A. Fiala, L. C. Pitchford and J.-P. Boeuf. *Phys. Rev. E*, 49:5607, 1994.
[78] J.-P. Boeuf and L. C. Pitchford. *Phys. Rev. E*, 51:1376, 1995.
[79] D. Vender and M. M. Turner. Epic simulations. In *Invited Paper, 16th ESCAMPIG and 5th ICRP Proceedings*, Vol. 2, p. 3, Grenoble, France, July 2002.
[80] D. Field, Y. Song and D. F. Klemperer. *J. Phys. D: Appl. Phys.*, 23(6):673–81, 1989.
[81] V. A. Godyak, R. B. Piejak and B. M. Alexandrovich. *Phys. Rev. Lett.*, 68:40, 1992.
[82] V. A. Godyak and A. S. Khanneh. *IEEE Trans. Plasma Sci.*, PS-14:112, 1986.
[83] N. St. J. Braithwaite, F. A. Haas and A. Godyear. *Plasma Sources Sci. Technol.*, 7(4):471–7, 1998.
[84] Ph. Belenguer and J.-P. Boeuf. *Phys. Rev. A*, 41:4447, 1990.
[85] V. P. T. Ku, B. M. Annaratone and J. E. Allen. Part I. *J. Appl. Phys.*, 84:6536–45, 1998.
[86] V. A. Godyak and O. Popov. *Sov. J. Plasma Phys.*, 5:227, 1979.
[87] M. Klick, W. Rehak and M. Kammeyer. *Jpn. J. Appl. Phys.*, 36:4625–31, 1997.
[88] T. Mussenbrock, R. P. Brinkmann, M. A. Lieberman, A. J. Lichtenberg and E. Kawamura. *Phys. Phys. Lett.*, 101:085004, 2008.
[89] J. Schulze, B. G. Heil, D. Luggenhoelscher, R. P. Brinkmann and U. Czarnetzki. *J. Phys. D: Appl. Phys.*, 41:195212, 2008.
[90] A. Perret, P. Chabert, J. Jolly and J.-P. Booth. *Appl. Phys. Lett.*, 86(1):021501, 2005.
[91] A. Perret, P. Chabert, J.-P. Booth, J. Jolly, J. Guillon and Ph. Auvray. *Appl. Phys. Lett.*, 83(2):243, 2003.
[92] H. H. Goto, H.-D. Lowe and T. Ohmi. *J. Vac. Sci. Technol. A*, 10(5):3048, 1992.
[93] T. Kitajima, Y. Takeo, Z. Lj. Petrovic and T. Makabe. *Appl. Phys. Lett.*, 77(4):489, 2000.
[94] J. Robiche, P. C. Boyle, M. M. Turner and A. R. Ellingboe. *J. Phys. D: Appl. Phys.*, 36:1810, 2003.
[95] M. M. Turner and P. Chabert. *Phys. Rev. Lett.*, 96:205001, 2006.

[96] M. M. Turner and P. Chabert. *Plasma Sources Sci. Technol.*, 16:364, 2007.
[97] T. Gans et al. *Appl. Phys. Lett.*, 89:261502, 2006.
[98] J. Schulze et al. *J. Phys. D: Appl. Phys.*, 40:7008–18, 2007.
[99] P. Levif. *Excitation multifréquence dans les décharges capacitives utilisées pour la gravure en microélectronique*. PhD thesis, École Polytechnique, Palaiseau, France, November 2007.
[100] H. C. Kim and J. K. Lee. *Phys. Rev. Lett.*, 93(8):085003–1, 2004.
[101] Y. J. Hong et al. *Comput. Phys. Commun.*, 177:122–3, 2007.
[102] B. G. Heil, U. Czarnetzki, R. P. Brinkmann and T. Mussenbrock. *J. Phys. D: Appl. Phys.*, 41:165202, 2008.
[103] J. Schulze, E. Schöngel, D. Luggenhoelscher, U. Czarnetzki and Z. Donkù. *J. Phys. Appl. Phys.*, 106:3223310, 2009.
[104] P. Chabert, J.-L. Raimbault, P. Levif, J.-M. Rax and M. A. Lieberman. *Plasma Sources Sci. Technol.*, 15:S130, 2006.
[105] J. E. Stevens, M. J. Sowa and J. L. Cecchi. *J. Vaccum Sci. Technol. A*, 14:139, 1996.
[106] G. A. Hebner, Ed. V. Barnat, P. A. Miller, A. M. Paterson and J. P. Holland. *Plasma Sources Sci. Technol.*, 15:879, 2006.
[107] L. Sansonnens, A. Pletzer, D. Magni, A. A. Howling, Ch. Hollenstein and J. P. M. Schmitt. *Plasma Sources Sci. Technol.*, 6:170, 1997.
[108] M. A. Lieberman, J.-P. Booth, P. Chabert, J.-M. Rax and M. M. Turner. *Plasma Sources Sci. Technol.*, 11:283, 2002.
[109] L. Sansonnens and J. Schmitt. *Appl. Phys. Lett.*, 82(2):182, 2003.
[110] H. Schmidt, L. Sansonnens, A. A. Howling, Ch. Hollenstein, M. Elyaakoubi and J. P. M. Schmitt. *J. Appl. Phys.*, 95(9):4559, 2004.
[111] P. Chabert, J.-L. Raimbault, J.-M. Rax and A. Perret. *Phys. Plasmas*, 11:4081, 2004.
[112] A. A. Howling, L. Sansonnens, J. Ballutaud, Ch. Hollenstein and J. P. M. Schmitt. *J. Appl. Phys.*, 96:5429, 2004.
[113] A. A. Howling, L. Derendinger, L. Sansonnens, H. Schmidt, Ch. Hollenstein, E. Sakanaka and J. P. M. Schmitt. *J. Appl. Phys.*, 97:123308, 2005.
[114] A. A. Howling, L. Sansonnens, H. Schmidt and Ch. Hollenstein. *Appl. Phys. Lett.*, 87:076101, 2005.
[115] L. Sansonnens. *J. Appl. Phys.*, 97:063304, 2005.
[116] L. Sansonnens, B. Strahm, L. Derendinger, A. A. Howling, Ch. Hollenstein, Ch. Ellert and J. P. M. Schmitt. *J. Vac. Sci. Technol. A*, 23:0734–2101, 2005.
[117] L. Sansonnens, H. Schmidt, A. A. Howling, Ch. Hollenstein, Ch. Ellert and A. Buechel. *J. Vac. Sci. Technol. A*, 24:0734–2101, 2006.
[118] L. Sansonnens, A. A. Howling and Ch. Hollenstein. *Plasma Sources Sci. Technol.*, 15:302, 2006.
[119] P. A. Miller, Ed. V. Barnat, G. A. Hebner, A. M. Paterson and J. P. Holland. *Plasma Sources Sci. Technol.*, 15:889, 2006.
[120] A. Lapucci, F. Rossetti, M. Ciofini and G. Orlando. *IEEE J. Quant. Electron.*, 31(8):1537, 1995.
[121] Y. P. Raizer and M. N. Schneider. *IEEE Trans. Plasma Sci.*, 26(3):1017, 1998.
[122] S. Ramo, J. R. Whinnery and T. Van Duzer. *Fields and Waves in Communication Electronics*. John Wiley & Sons, New York, 1965.
[123] I. Lee, D. B. Graves and M. A. Lieberman. *Plasma Sources Sci. Technol.*, 17:015018, 2008.
[124] P. Chabert, H. Abada, J.-P. Booth and M. A. Lieberman. *J. Appl. Phys.*, 94:76, 2003.

[125] P. Chabert, A. J. Lichtenberg, M. A. Lieberman and A. M. Marakhtanov. *J. Appl. Phys.*, 94:831, 2003.
[126] A. M. Marakhtanov, M. Tuszewski, M. A. Lieberman, A. J. Lichtenberg and P. Chabert. *J. Vac. Sci. Technol. A*, 21:1849, 2003.
[127] Jackson. *Classical Electrodynamics*. John Wiley & Sons, New York, 1960.
[128] R. B. Piejak, V. A. Godyak and B. M. Alexandrovich. *Plasma Sources Sci. Technol.*, 1:179, 1992.
[129] V. A. Godyak, R. B. Piejak and B. M. Alexandrovich. *J. Appl. Phys.*, 85:703, 1999.
[130] L. J. Mahoney, A. E. Wendt, E. Barrios, C. J. Richards and J. L. Shohet. *J. Appl. Phys.*, 76:2041, 1994.
[131] P. Colpo, T. Meziani and F. Rossi. *J. Vacuum Sci. Technol.*, A23:270, 2005.
[132] S. Lloyd, D. M. Shaw, M. Watanabe and G. J. Collins. *Jpn. J. Appl. Phys.*, 38:4275, 1999.
[133] V. A. Godyak. *Proceedings of the XVth International Conference on Gas Discharge and their Applications*, p. 621, 2004.
[134] E. S. Weibel. *Phys. Fluids*, 10:741, 1967.
[135] M. M. Turner. *Phys. Rev. Lett.*, 71:1844, 1993.
[136] V. A. Godyak. *Phys. Plasmas*, 12:055501, 2005.
[137] V. A. Godyak and R. B. Piejak. *J. Appl. Phys.*, 82:5944, 1997.
[138] R. Piejak, V. Godyak and B. Alexandrovich. *J. Appl. Phys.*, 81:3416, 1997.
[139] G. Cunge, B. Crowley, D. Vender and M. M. Turner. *J. Appl. Phys.*, 89:3580, 2001.
[140] G. J. M. Hagelaar. *Phys. Rev. Lett.*, 100:025001, 2008.
[141] V. A. Godyak and V. I. Kolobov. *Phys. Rev. Lett.*, 79:4589, 1997.
[142] V. A. Godyak, R. B. Piejak, B. M. Alexandrovich and V. I. Kolobov. *Phys. Rev. Lett.*, 80:3264, 1998.
[143] V. A. Godyak and V. I. Kolobov. *Phys. Rev. Lett.*, 81:369, 1998.
[144] M. Tuszewski. *Phys. Rev. Lett.*, 77:1286, 1996.
[145] A. Smolyakov, V. A. Godyak and A. Duffy. *Phys. Plasmas*, 7(11):4755, 2000.
[146] D. R. Hartree. *Proc. Cambridge Phil. Soc.*, 27:143, 1931.
[147] E. V. Appleton. *J. Inst. Elec. Engrs.*, 71:642, 1932.
[148] P. Aigrain. *Proc. Int. Conf. Semiconductor Physics, Prague, Czeckoslovakia*, p. 224, 1960.
[149] R. W. Boswell. *Phys. Lett.*, 33A:470, 1970.
[150] R. W. Boswell and F. F. Chen. *IEEE Trans. Plasma Sci.*, 25(6):1229–44, 1997.
[151] F. F. Chen and R. W. Boswell. *IEEE Trans. Plasma Sci.*, 25(6):1245–57, 1997.
[152] P. Zhu and R. W. Boswell. *Phys. Rev. Lett.*, 63(26):2805–7, 1989.
[153] F. F. Chen. *Plasma Phys. Control. Fusion*, 33(4):339–64, 1991.
[154] C. Charles, R. W. Boswell and H. Kuwahara. *Appl. Phys. Lett.*, 67:40, 1995.
[155] A. J. Perry, D. Vender and R. W. Boswell. *J. Vac. Sci. Technol. B*, 9(2):310, 1991.
[156] M. A. Lieberman and R. W. Boswell. *J. Phys. IV France*, 8:145–63, 1998.
[157] T. H. Stix. *Waves in Plasmas*. John Wiley & Sons, New York, 1992.
[158] H. A. Blevin and P. J. Christiansen. *Aust. J. Phys.*, 19:501, 1966.
[159] F. F. Chen, M. J. Hsieh and M. Light. *Plasma Sources Sci. Technol.*, 3:49, 1994.
[160] B. N. Breizman and A. V. Arefiev. *Phys. Rev. Lett.*, 84(17):3863, 2000.
[161] T. Watari et al. Rf plugging of a high-density plasma. *Phys. Fluids*, 21:2076, 1978.
[162] T. Shoji. *IPPJ Annu. Rep., Nagoya Univ.*, p. 67, 1986.
[163] Y. Celik, D. L. Crintea, D. Luggenholscher and U. Czarnetzki. *Plasma Phys. Control. Fusion*, 51:124040, 2009.

[164] A. W. Degeling, C. Jung, R. W. Boswell and A. R. Ellingboe. *Phys. Plasmas*, 3:2788–96, 1996.
[165] A. Fruchtman, G. Makrinich, P. Chabert and J.-M. Rax. *Phys. Rev. Lett.*, 95:115002, 2005.
[166] B. Clarenbach, B. Lorenz, M. Krämer and N. Sadeghi. *Plasma Sources Sci. Technol.*, 12:345, 2003.
[167] D. B. Hash, D. Bose, M. V. V. S. Rao, B. A. Cruden, M. Meyyapan and S. P. Sharma. *J. Appl. Phys.*, 90:2148, 2001.
[168] M. Shimada, G. R. Tynan and R. Cattolica. *Plasma Sources Sci. Technol.*, 16:193, 2007.
[169] D. Bose, D. Hash, T. R. Govindan and M. Meyyappan. *J. Phys. D: Appl. Phys.*, 34:2742, 2001.
[170] C.-C. Hsu, M. A. Nierode, J. W. Coburn and D. B. Graves. *J. Phys. D: Appl. Phys.*, 39:3272, 2006.
[171] L. Liard, J.-L. Raimbault, , J.-M. Rax and P. Chabert. *J. Phys. D*, 40:5192–5195, 2007.
[172] J.-L. Raimbault and P. Chabert. *Plasma Sources Sci. Technol.*, 18:014017, 2009.
[173] L. Liard, J.-L. Raimbault and P. Chabert. *Phys. Plasmas*, 16:053507, 2009.
[174] N. St. J. Braithwaite and J. E. Allen. *J. Phys. D: Appl. Phys.*, 21:1733, 1988.
[175] T. E. Sheridan, P. Chabert and R. W. Boswell. *Plasma Sources Sci. Technol.*, 8:457, 1999.
[176] P. Chabert and T. E. Sheridan. *J. Phys. D*, 33:1854, 2000.
[177] A. Kono. *J. Phys. D: Appl. Phys.*, 32:1357, 1999.
[178] I. G. Kouznetsov, A. J. Lichtenberg and M. A. Lieberman. *J. Appl. Phys.*, 86:4142, 1999.
[179] R. N. Franklin. *J. Phys. D: Appl. Phys.*, 32:L71, 1999.
[180] R. N. Franklin. *Plasma Sources Sci. Technol.*, 9:191, 2000.
[181] R. N. Franklin. *J. Phys. D: Appl. Phys.*, 36:R309, 2003.
[182] D. D. Monahan and M. M. Turner. *Plasma Sources Sci. Technol.*, 17:045003, 2008.
[183] S. Kim, M. A. Lieberman, A. J. Lichtenberg and J. T. Gudmundsson. *J. Vac. Sci. Technol. A*, 24:2025–40, 2006.
[184] I. G. Kouznetsov, A. J. Lichtenberg and M. A. Lieberman. *Plasma Sources Sci. Technol.*, 5:662, 1996.
[185] A. J. Lichtenberg, I. G. Kouznetsov, M. A. Lieberman and T. H. Chung. *Plasma Sources Sci. Technol.*, 9:45, 2000.
[186] R. N. Franklin. *Plasma Sources Sci. Technol.*, 11:A31, 2002.
[187] R. N. Franklin. *J. Phys. D: Appl. Phys.*, 34:1243, 2001.
[188] R. N. Franklin. *J. Phys. D: Appl. Phys.*, 34:1834, 2001.
[189] R. Kawai and T. Mieno. *Jpn. J. Appl. Phys.*, 36:L1123, 1997.
[190] P. Chabert, T. E. Sheridan, J. Perrin and R. W. Boswell. *Plasma Sources Sci. Technol.*, 8:561, 1999.
[191] R. N. Franklin and J. Snell. *J. Phys. D: Appl. Phys.*, 32:1031, 1999.
[192] G. Leray, P. Chabert, A. J. Lichtenberg and M. A. Lieberman. *J. Phys. D: Appl. Phys.*, 42:194020, 2009.
[193] A. J. Lichtenberg, P. Chabert, M. A. Lieberman and A. M. Markhtanov. *Bifurcation Phenomena in Plasmas*, p. 3, 2001.
[194] P. G. Steen, C. S. Corr and W. G. Graham. *Plasma Sources Sci. Technol.*, 12:265, 2003.
[195] P. G. Steen, C. S. Corr and W. G. Graham. *Appl. Phys. Lett.*, 86:141503, 2005.

[196] J.-P. Booth, H. Abada, P. Chabert and D. B. Graves. *Plasma Sources Sci. Technol.*, 14:273, 2005.
[197] C. Charles and R. W. Boswell. *Appl. Phys. Lett.*, 82:1356, 2003.
[198] C. Charles and R. W. Boswell. *Phys. Plasmas*, 11:1706, 2004.
[199] C. Charles. *Plasma Sources Sci. Technol.*, 16:R1–R25, 2007.
[200] F. F. Chen. *Phys. Plasmas*, 13:034502, 2006.
[201] M. A. Lieberman, C. Charles and R. W. Boswell. *J. Phys. D: Appl. Phys.*, 39:3294, 2006.
[202] A. Fruchtman. *Phys. Rev. Lett.*, 96:065002, 2006.
[203] N. Plihon, C. S. Corr and P. Chabert. *Appl. Phys. Lett.*, 86:091501, 2005.
[204] N. Plihon, C. S. Corr, P. Chabert and J.-L. Raimbault. *J. Appl. Phys.*, 98:023306, 2005.
[205] M. Tuszewski. *J. Appl. Phys.*, 79:8967, 1996.
[206] M. Tuszewski and S. P. Gary. *Phys. Plasmas*, 10:539, 2003.
[207] N. Hershkowitz and M. H. Cho. *J. Vac. Sci. Technol. A*, 6:2054–9, 1988.
[208] K. Teii, M. Mizumura, S. Matsumura and S. Teii. *J. Appl. Phys.*, 93:5888–92, 2003.
[209] T. Lho, N. Hershkowitz, G. H. Kim, W. Steer and J. Miller. *Plasma Sources Sci. Technol.*, 9:5–11, 2000.
[210] T. Lafleur, C. Charles and R. W. Boswell. *Phys. Plasmas*, 16:044510, 2009.
[211] H. M. Mott-Smith and I. Langmuir. *Phys. Rev.*, 28:727–63, 1926.
[212] J. E. Allen. *Phys. Scr.*, 45:497–503, 1991.
[213] J. G. Laframboise. Numerical computations for ion probe characteristics in a collisionless plasma. *U.T.I.A.S. Report No. 100, University of Toronto*, 1966.
[214] J. E. Allen, R. L. F. Boyd and P. Reynolds. *Proc. Phys. Soc. Lond. B*, 70:297–304, 1957.
[215] F. F. Chen. *J. Nucl. Energy C*, 7:41–67, 1965.
[216] B. M. Annaratone, M. W. Allen and J. E. Allen. *J. Phys. D: Appl. Phys.*, 25:417–24, 1992.
[217] Z. Sternovsky, S. Robertson and M. Lampe. *J. Appl. Phys.*, 94:1374–81, 2003.
[218] M. J. Druyvesteyn. *Z. Phys.*, 64:781, 1930.
[219] N. St. J. Braithwaite, N. M. P. Benjamin and J. E. Allen. *Phys. Plasmas*, 20:1046–9, 1987.
[220] A. Dyson, P. M. Bryant and J. E. Allen. *Measurement Sci. Technol.*, 11:554–9, 2000.
[221] L. Nolle, A. Goodyear, A. A. Hopgood, P. D. Picton and N. St. J. Braithwaite. *Knowledge-Based Syst.*, 15:349–54, 2002.
[222] M. A. Sobolewski. *J. Appl. Phys.*, 90:2660–71, 2001.
[223] M. A. Sobolewski. *J. Vac. Sci. Technol. A*, 5:677–84, 2006.
[224] N. St. J. Braithwaite, J. P. Booth and G. Cunge. *J. Plasma Sources Sci. Technol.*, 36:2837–44, 1996.
[225] N. St. J. Braithwaite, T. E. Sheridan and R. W. Boswell. *J. Phys. D: Appl. Phys.*, 36:2837–44, 2003.
[226] J. P. Booth. *Plasma Sources Sci. Technol.*, 8:249–57, 1999.
[227] D. D. Blackwell, D. N. Walker and W. E. Amatucci. *Rev. Sci. Instrum.*, 76:023503, 2005.
[228] J. Schulze, T. Kampschulte, D. Luggenholscher and U. Czarnetzki. *J. Phys. Conf. Series*, 86:012010, 2007.
[229] R. L. Stenzel. *Rev. Sci. Instrum.*, 47:603–7, 1976.
[230] G. A. Hebner and I. C. Abraham. *J. Appl. Phys.*, p. 4929–37, 2001.

[231] R. B. Piejak, V. A. Godyak, R. Garner, B. M. Alexandrovich and N. Sternberg. *J. Appl. Phys.*, 95:3785–91, 2004.
[232] R. B. Piejak, J. Al-Kuzee and N. St. J. Braithwaite. *Plasma Sources Sci. Technol.*, 14:734–43, 2005.
[233] N. S. Siefert, B. N. Ganguly, B. L. Sands and G. A. Hebner. *J. Appl. Phys.*, 100:043303, 2006.
[234] S. K. Karakari, A. R. Ellingboe and C. Gaman. *Appl. Phys. Lett*, 93:071501, 2008.
[235] C. Scharwitz, M. Boeke, J. Winter, M. Lapke, T. Mussenbrock and R. P. Brinkmann. *Appl. Phys. Lett.*, 94:011502, 2009.
[236] S. Dine, J. P. Booth, G. A. Curley, C. S. Corr, J. Jolly and J. Guillon. *Plasma Sources Sci. Technol.*, 14:777–86, 2005.
[237] C. H. Chang, C. H. Hsieh, H. T. Wang, J. Y. Jeng, K. C. Leou and C. Lin. *Plasma Sources Sci. Technol.*, 16:67–71, 2007.
[238] H. S. Jun, Y. S. Lee, B. K. Na and H. Y. Chang. *Phys. Plasmas*, 15:124504, 2008.
[239] R. H. Huddlestone and S. L. Leonard (eds). *Plasma Diagnostic Techniques*. Academic Press, New York, 1994.
[240] D. Bohm and E. P. Gross. *Phys. Rev.*, 75:1851, 1949.
[241] J. J. Thomson. *Phil. Mag.*, 4:1128, 1927.
[242] T. E. Sheridan, N. St. J. Braithwaite and R. W. Boswell. *Phys. Plasma*, 6:4375, 1999.

Index

adiabatic approximation, 39
ambipolar diffusion, 84
Ampère's theorem, 225
anisotropy, 6

Bessel function
  complex arguments, 221
  first kind, 190
  modified, 112, 221
Bohm
  speed, 74
Bohm criterion
  electronegative plasma, 300
  original, 72, 74
Boltzmann factor (or equilibrium), 64

CCP, 15, 131, 176
CCP asymmetrical, 166
CCP symmetrical, 133
charge exchange, 36
Child-Langmuir law
  collisional, 70
  collisionless, 67
  fully collisional, 69
  RF, 125, 133, 179
CMOS, 4
collision
  binary, 28
  elastic and inelastic, 27
collision frequency, 29
conductivity
  complex plasma, 47
Coulomb collisions, 280
cross-section, 29
current-driven sheath, 119
cycloidal motion, 295
cyclotron frequency
  electron, 264
  ion, 264

damping
  Landau, 282
  wave–particle, 282

DBD, *see* dielectric barrier discharge
Debye length (or distance), 64
deposition
  magnetron, 9
  PECVD, 7
  plasma, 7
  sputter, 8
dielectric barrier discharge, 2
dielectric lens, 199
dissociation, 34
dissociative ionization, 34
distribution function, *see also* EEDF, IEDF
  ion energy, 101, 105
  Maxwell–Boltzmann distribution, 21
  Maxwellian, 21
  Maxwellian energy, 25
  Maxwellian speed, 24
  Maxwellian velocity, 22
  velocity, 19
double layer, 13, 304, 313
double probe
  asymmetrical, 325
  symmetrical, 322
Druyvesteyn
  disitribution, 374
  method, 340
dual frequency
  sheath, 179

E–H transition, 202, 209
E-mode, 16, 202, 209
edge effects, 216
EEDF, 339, 346
EEPF, 169, 339
electromagnetic regime (CCP), 177
electron energy distribution, *see* EEDF
electron retardation region, 327
electron saturation
  cylindrical, 332
  planar, 326
electron temperature, 92

electronegative plasmas, 298
electrostatic regime (CCP), 177
energy relaxation length, 40
equivalent circuit
   inductive, 227
   inductive with capacitive coupling, 244
   matching network, 162
   plasma, 54
   RF sheath, 116
   symmetrical CCP, 140, 142
etching
   deep, 7
   plasma, 5
   reactive ion, 6

Faraday shield, 254
Faraday's law, 229
ferromagnetic cores, 255
floating double probe, 323
floating potential
   DC, 61, 75, 94, 324, 327, 345
   RF, 112, 341, 345
flux
   ambipolar diffusion, 85
   electron energy, 27
   ion, 75, 90
   ion energy, 76
   random thermal, 26
   wall, 26, 75

gamma mode, 170
gas heating, 291
global model
   DC, 39, 41, 75
   dual-frequency CCP, 182
   inductive, 246
   single-frequency CCP, 161
   VHF CCP, 202
group velocity, 269

H-mode, 16, 202, 209
hairpin resonator, 360
heating
   dual-frequency enhancement, 181
   dual-frequency sheath, 180
   hard wall, 150
   kinetic-fluid, 151
   ohmic (collisional), 52, 141
   stochastic, 222
   stochastic (collisionless), 141, 149, 180
   wave absorption, 277
helicon, 12, 16
helicon mode number
   azimuthal, 272
   longitudinal, 275
helicon reactor, 262

homogeneous model
   CCP, 133
   sheath, 116, 117
hysteresis, 157

ICP, 15, 219
IEDF, 105, 186
impact parameter, 332
impedance
   plasma, 54
inductance
   coil, 231
   electron inertia, 142, 231, 236
   magnetic storage, 231, 236
   mutual, 237
   plasma, 54, 142, 178
inhomogeneous model
   CCP, 146
   sheath, 125
instabilities
   E–H transitions, 309
instability, 157
ion matrix, 65
ion saturation (cylindrical), 336
ion saturation (planar), 323, 327
ion transit time, 103
ion–ion plasmas, 307
ionization
   frequency, 32
   multi-step, 28
   rate (coefficient), 32
   single-step, 28
   threshold, 31
isothermal approximation, 39
IVDF, 352

kinetic theory, 19

Larmor radius, 293

magnetized plasmas, 293
match-box, 162, 252
mean free path, 29
MERIE, 260
microwave interferometry, 365
microwave probe, 354
mode transition, 157, 170
MOSFET, 3
multi-step ionization, 28
multipole resonator, 363

negative power absorption, 258
neutral depletion, 284, 288
non-local electron kinetics, 41

OML, 332, 335, 337
orbital motion, 332

parallel resonance, 357
permittivity
   complex plasma, 47

photo-emission, 320
plasma
  ion–ion, 14
plasma frequency
  electron, 98
  ion, 98
plasma impedance, 54
plasma permittivity tensor, 264
plasma potential
  DC, 62, 216, 313, 320, 327, 335
  RF, 96, 168, 217, 253, 343
plasma transport, 78
plasma transport models
  Godyak, 87
  Schottky, 83
  Tonks–Langmuir, 82
plasma turbulence, 298
ponderomotive force, 259
power dissipation
  by electrons, 52, 141, 142
  by ions, 141
power transfer efficiency, 241, 250
Poynting theorem, 227
probe
  active compensation, 343
  analysis (planar), 328
  assumptions (planar), 320
  cylindrical, 330
  emissive, 327
  floating double, 323
  hairpin, 360
  microwave, 354
  multipole resonator, 363
  passive compensation, 342
  planar, 176
  processing environment, 345
  RF-biased planar, 348
  RF-compatible, 340
  single Langmuir, 326
  spherical, 330
  symmetrical double, 322

quasi-neutrality, 59, 76

rate coefficient, 29
reference electrode, 325
refractive index, 265
resistance
  plasma, 54
retarding field analyser, 176, 314, 348
RFA, *see* retarding field analyser

scaling laws, 158
secondary electrons, 170
secondary emission coefficient, 170
secondary processes, 320
self-bias, 113

self-excited resonance spectroscopy, 360
semi-log analysis, 329, 338, 342, 354, 374
series resonance, 171, 357
sheath
  DC, 2, 59, 61, 221
  dual-frequency, 179
  reversal, 128
  RF, 60, 96, 131, 176
sheath model
  Child–Langmuir, 67, 69, 70, 125
  current-driven, 120, 128
  homogeneous, 116, 117
  inhomogeneous, 125
  ion matrix, 65, 117
  RF, 116, 117, 125
  voltage-driven, 119, 127
sheath thickness, 94, 158
single-step ionization, 28
skin depth
  anomalous, 256
  collisional, 50
  collisionless, 57
  inertial, 57
  ordinary, 48, 220
speed
  ion acoustic (also Bohm), 52, 74
sputtering, 8
standing wave, 188, 361
stochastic heating, 141, 149, 180, 222
substrate holder, 219

TCP, 15
transformer
  coupled plasma, 221
  model, 221, 236
transition
  E–H, 202
  E–H–W, 261, 283
  global E–H, 208
  plasma/sheath, 72
  spatial E–H, 214
transmission line model, 189, 206

variable mobility, 70, 88
vibrational excitation, 35
voltage-driven sheath, 119

W-mode, 16
wavelength shortening, 196
waves
  Alfvèn, 267
  electron cyclotron, 267
  electron plasma, 51
  electrostatic, 50
  helicon, 268
  ion acoustic, 52
whistlers, 260

Printed in the United States
By Bookmasters

Printed in the United States
By Bookmasters